ARTIFICIAL INTELLIGENCE IN LOGIC DESIGN

by

Svetlana N. Yanushkevich

Department of Electrical & Computer Engineering,
University of Calgary, Canada

Partly reprinted from *Artificial Intelligence Review*,
Volume 20, Nos. 3–4 (2003)

T0135176

KLUWER ACADEMIC PUBLISHERS
DORDRECHT / BOSTON / LONDON

A C.I.P. catalogue record for this book is available from the Library of Congress.

ISBN 978-90-481-6583-4   e-ISBN 978-1-4020-2075-9

Published by Kluwer Academic Publishers,
P.O. Box 17, 3300 AA Dordrecht, The Netherlands.

Sold and distributed in the North, Central and South America
by Kluwer Academic Publishers,
101 Philip Drive, Norwell, MA 02061, U.S.A.

In all other countries, sold and distributed
by Kluwer Academic Publishers,
P.O. Box 322, 3300 AH Dordrecht, The Netherlands.

Printed on acid-free paper

Contents

About the Authors

Arturo Hernández Aguirre is Titular Researcher at the Center for Research in Mathematics, Department of Computer Science, in Guanajuato, México. He was a Visiting Associate Professor of Computer Science at the Department of Computer Science and Electrical Engineering of Tulane University, in New Orleans, Louisiana, USA. He completed his Master's degree and his PhD degree (both in Computer Science) at Tulane University in 1998 and 2000, respectively. Previously, in 1990, he completed a Master's Degree in Information Systems, and was principal investigator and Software Engineering consultant at the Arturo Rosenblueth Foundation, in Mexico City. His primary research interests are in Evolutionary Computation applied to Optimization, in Evolvable Hardware, and in Support Vector Machines for pattern classification.

Takafumi Aoki received the B. E., M. E., and D. E. degrees in electronic engineering from Tohoku University, Sendai, Japan, in 1988, 1990, and 1992, respectively. He is currently a Professor of the Graduate School of Information Sciences at Tohoku University. For 1997–1999, he also joined the PRESTO project, Japan Science and Technology Corporation (JST). His research interests include theoretical aspects of computation, VLSI computing structures for signal and image processing, multiple-valued logic, and biomolecu-lar computing. Dr. Aoki received the Outstanding Paper Award at the 1990, 2000 and 2001 IEEE International Symposiums on Multiple-Valued Logic, the Outstanding Transactions Paper Award from the Institute of Electronics, Information and Communication Engineers (IEICE) of Japan in 1989 and 1997, the IEE Ambrose Fleming Premium Award in 1994, the IEICE Inose Award in 1997, the IEE Mountbatten Premium Award in 1999, and the Best Paper Award at the 1999 IEEE International Symposium on Intelligent Signal Processing and Communication Systems.

Carlos Artemio Coello Coello received a BSc in Civil Engineering from the Universidad Autonoma de Chiapas in Mexico in 1991. Then, he was awarded a scholarship from the Mexican government to pursue graduate studies in Computer Science at Tulane University. He received a MSc and a PhD in Computer Science in 1993 and 1996, respectively. His PhD thesis was one of the first in the field now called evolutionary multiobjective optimization.

Dr. Coello has been a Senior Research Fellow in the Plymouth Engineering Design Centre (in England) and a Visiting Professor at DePauw University (in the USA). He is currently associate professor at CINVESTAV-IPN in Mexico City, Mexico.

He has published over 100 papers in international peer-reviewed journals and conferences and one book on evolutionary multiobjective optimization which is part of the Genetic Algorithms and Evolutionary Computation Series edited by David E. Goldberg. He has also served as a technical reviewer for a number of journals and international conferences and actually serves as associate editor of the IEEE Transactions on Evolutionary Computation.

He is a member of the Mexican Academy of Sciences and of the Mexican System of Researchers (as a Level 2 Researcher).

His current research interests are: evolutionary multiobjective optimization, constraint-handling techniques for evolutionary algorithms and evolvable hardware.

Kyusik Chung was born in Korea (Taejon) in 1979. He is a PhD candidate at the Korean Advanced Institute of Science and Technology in Taejon in Multimedia VLSI (MVLSI) laboratory. His professional interests are in 3D Graphics and modeling of noise and signals. Also he is active in the research in modern technologies such as quantum or reversible Logic.

Hilton Tamanaha Goi was born in São Paulo, Brazil, in 1977. He received the diploma of electronics technician from Federal Technical School (ETFSP), São Paulo, Brazil, in 1995; and the diploma of electronics and information technologist from Yatsushiro National College of Technology, Kumamoto, Japan, in 1999, as a Japan's Ministry of Education Scholar. He received the B.S.E. and M.S.E. degrees in Media and Telecommunications Engineering from Ibaraki University, Hitachi, Japan, in 2001 and 2003, respectively, as a Rotary Club

Yoneyama Fellow and Japan International Cooperation Agency Scientist. He is currently a PhD course student at the Department of Electrical Engineering and Computer Science, Korea Advanced Institute of Science and Technology (KAIST), Taejon, Korea. His research interests include smart antennas for personal wireless communications, ultra wide-band (UWB) technology, ubiquitous computing, and robotic ad-hoc networks. He is a student member of the Institute of Electronics, Information and Communications Engineers.

Tatsuo Higuchi received the B. E., M. E., and D. E. degrees in electronic engineering from Tohoku University, Sendai, Japan, in 1962, 1964, and 1969, respectively. He is currently a Professor at Tohoku Institute of Technology. From 1980 to 1993, he was a Professor in the Department of Electronic Engineering at Tohoku University. He was a Professor from 1994 to 2003, and was Dean from 1994 to 1998 in the Graduate School of Information Sciences at Tohoku University. His general research interests include the design of 1-D and multi-D digital filters, linear time-varying system theory, fractals and chaos in digital signal processing, VLSI computing structures for signal and image processing, multiple-valued ICs, multiwave opto-electronic ICs, and biomolecular computing. Dr. Higuchi received the Outstanding Paper Awards at the 1985, 1986, 1988, 1990, 2000 and 2001 IEEE International Symposiums on Multiple-Valued Logic, the Outstanding Transactions Paper Award from the Society of Instrument and Control Engineers (SICE) of Japan in 1984, the Technically Excellent Award from SICE in 1986, and the Outstanding

Book Award from SICE in 1996, the Outstanding Transactions Paper Award from the Institute of Electronics, Information and Communication Engineers (IEICE) of Japan in 1990 and 1997, the Inose Award from IEICE in 1997, the Technically Excellent Award from the Robotics Society of Japan in 1990, the IEE Ambrose Fleming Premium Award in 1994, the Outstanding Book Award from the Japanese Society for Engineering Education in 1997, the Award for Persons of scientific and technological merits (Commendation by the minister of state for Science and Technology), the IEE Mountbatten Premium Award in 1999 and the Best Paper Award at the 1999 IEEE International Symposium on Intelligent Signal Processing and Communication Systems. He also received the IEEE Third Millennium Medal in 2000. He received the fellow grade from IEEE, IEICE, and SICE.

Naofumi Homma received the B.E. degree in information engineering, and the M.S. and Ph.D. degrees in information sciences from Tohoku University, Sendai, Japan, in 1997, 1999 and 2001, respectively. He is currently a Research Associate of the Graduate School of Information Sciences at Tohoku University. For 1999–2001, he was a Research Fellow of the Japan Society for the Promotion of Science. From 2002, he also joined Japan Science and Technology Agency (JST) as a researcher for the PRESTO project. His research interests include computer arithmetic and VLSI system-level design.

Hyunkoo Jee was born in 1979. He completed his undergraduate degree in Electrical Engineering at KAIST in Taejon, Korea. He has interests in computer and electrical engineering, rock music and playing quitar. He is a member of Neural Networks and Machine Intelligence Laboratory at KAIST and a master degree student, interested in efficient program development.

Byung-Guk Kim was born in Korea (Taejon) in 1980. He is a master candidate at the Korean Advanced Institute of Science and Technology in Taejon in Multimedia VLSI (MVLSI) laboratory. His professional interests are in Digital Circuits design and testing.

Yong-Duk Kim was born in Korea (Taejon) in 1977. He is actually a PhD candidate at the Korean Advanced Institute of Science and Technology in Taejon. His professional interests are in Artificial Intelligence, Entertainment and Humanoid Robots, Robotic hardware and Technology. Also as part of his PhD education he belongs to the Robot Intelligence Technology Lab at Kaist where he is an active member of Decision Making for Robots Development group.

Christian Lang is 35 from Chemnitz, Germany. He is a senior engineer for the design of microelectronic circuits at the Institute for Microelectronic and Mechatronic Systems in Erfurt, Germany. Previously he was a research associate at the Institute of Computer Science of the Freiberg University of Mining and Technology in Freiberg (Saxony), Germany. From this University he received his PhD (Dr.-Ing.) in Computer Science in 2003. His Diploma (Dipl.-Ing.) in Electrical Engineering was obtained from the Chemnitz University of Technology in 1995. He completed a Master's Degree (M.S.E.)

in Electrical Engineering at the Western Michigan University, Kalamazoo, USA in 1994. His research interests are in design and verification of analog and mixed-signal circuits with emphasis on formal and symbolic methods.

Martin Lukac was born in Bratislava, Slovakia in 1974. He is a PhD candidate in Electrical and Computer Engineering Department at Portland State University (Portland, Oregon, USA). He obtained his Master Degree in Cognitive Science from Polytechnic School (Paris, France) in 1999. From 1999 till 2001 he worked as a research assistant in the Department of Informatics at the Comenius University in Bratislava. His primary research interests are Artificial Life, Artificial Intelligence, and Adaptive Systems. Since 2001 (beginning of his doctoral studies) his interests are applications of biologically-inspired and quantum-inspired algorithms and tools in robotics and Logic Synthesis.

Denis V. Popel is an Assistant Professor in the Computer Science Department at Baker University, Baldwin City, Kansas, U.S.A. Previously, he held teaching and research positions at the University of Wollongong, Australia, Dubai Campus, and at Technical University, Szczecin, Poland. He completed a Master's degree with Honors in Computer Engineering (biometrics and contextual support n document understanding systems) at the State University of Informatics and Radioelectronics, Minsk, Belarus, and a PhD with Honors in Computer Science (information theory, logic design and decision making) at Technical University, Szczecin, Poland.

His current job responsibilities include research advising, teaching courses related to data structures, database design, image processing and bioinformatics, and administration in the Computer Science Department, which often involves strategic planning and curricula changes. His main research interests are in Data Mining (knowledge representation, optimization and data retrieval), Biometrics (image processing, recognition and simulation) and Logic Design (uncertainty evaluation and timing analysis).

Marek Perkowski is a professor of Electrical and Computer Engineering at Portland State University. He got his PhD. from Warsaw University of Technology, Poland and was on the faculties of Warsaw University of Technology, and University of Minnesota in Minneapolis. He has Professor appointment at Korea Advanced Institute of Science and Technology, Taejon, Korea, for years 2002/2002 and 2003/2004. He has been on sabbatical appointments at LIRM, CNRS, University of Montpelier, France, Kyushu Institute of Technology, Japan, and Technical University of Eindhoven, The Netherlands. He worked for several software and hardware companies and was a summer professor for GTE, Waltham, MA, Intel Scientific Computers, Sharp Microelectronics, Cypress Semiconductor, Wright Laboratories of Air Force, Dayton, Ohio, and others. His main research interest is designing efficient algorithms. Since 1970 he has been working on logic synthesis and CAD. His algorithms are parts of some industrial EDA systems and scientific software. His recent interests are in Intelligent Humanoid Robots and Quantum Computing.

Vlad P. Shmerko is 54 and he is a Full Professor at Technical University of Szczecin, Poland, and an Adjunct Professor in the Department of Electrical and Computer Engineering, University of Calgary, Canada. He is a Senior member of the IEEE. He has obtained his PhD in Computer Science in 1985 from the State University of Informatics and Radioelectronics, Belarus, and his Habilitation (Doctor of Eng. Sciences) degree from Latvian Academy of Sciences, USSR, in 1990. His research interests are in advanced digital design techniques, including spectral techniques, artificial intelligence, neural networks. He is also interested in image processing and synthesis, pattern recognition and application to biometric technologies.

Bernd Steinbach is 51 and from Chemnitz, GERMANY on the East of the EU. He is Chair in Software Engineering and Programming Technique at the Institute of Computer Science, Faculty of Mathematics and Computer Science, University "TU Bergakademie Freiberg" in Freiberg (Saxony), Germany. Previously he was Associated Professor in Design Automation at the University of Technology in Chemnitz. He get extensive practical experiences in testing of large digital Systems through his work as research engineer in the German computer company Robotron.

He completed his qualification for lectureship (habil.) in 1985 at the University of Technology in Chemnitz. Four years before, he completed his PhD in Information Technique (Dr.-Ing.) at the same University. His Diploma (Dipl.-Ing.) was obtained from the University of Technology in Chemnitz in 1977. He has been awarded with the Barkhausen-Award of the Technical University of Dresden for his research achievements in 1984.

His main research interests are focused to Boolean Problems and include the analysis, the synthesis and testing of digital hardware systems as well as efficient data structures and algorithms to solve such and further Boolean Problems by Software. He is coauthor of the book "Logic design using XBOOLE" and General Chair of a series of "International Workshops on Boolean Problems", which sixth meeting will take place in September 2004 in Freiberg. Strong applications of his research cover software quality, testing of software, productivity of software design, UML-based co-design for runtime reconfigurable architectures and design of digital circuits using decomposition methods.

Svetlana Yanushkevich is 36 and she is currently an Assistant Professor in the Department of Electrical and Computer Engineering, University of Calgary, Canada. She is also a Professor at Technical University of Szczecin, Poland. Her MSc in Electrical Engineering and her PhD in Computer Science were obtained from the State University of Informatics and Radioelectronics, Belarus, in 1989 and 1992 respectively. Her Habilitation degree (Computer Science) was obtained from Warsaw University of Technology Poland, in 1999. Her primary research interests are in Electronic Design Automation, including artificial intelligence, decision making, and machine learning aspects. Her recent interests are in the design of nano-electronic devices. She also heads the Biometric Technologies Laboratory at University of Calgary.

Chung-Hyo Yu was born in Korea (Taejon) in 1979. He is a PhD candidate at the Korean Advanced Institute of Science and Technology in Taejon in Multimedia VLSI (MVLSI) laboratory. His professional interests are in Digital circuits and MPEG 4, Signal processing and regular structures.

Editorial – Artificial Intelligence in Logic Design

This issue represents a collection of selected "pieces" created by "masters" in the area of logic design. Although relatively few people in the world participate in the research on artificial intelligence aspects of digital design, this is no impairment to the truly original research. We believe that artificial intelligence approach is an important aspect of digital systems design, because it includes advanced theory (such as multiple-valued logic, fuzzy logic and evolutionary algorithms) and is flexible to advanced and new technologies that could have far-reaching effects on future computers and information systems.

These papers combine four main topics towards the advanced VLSI and ULSI circuit design and nanotechnology era: (i) Multiple-valued logic circuit synthesis and analysis under the concept of learning, (ii) Evolutionary circuit design, (iii) Information measures in circuit design, and (vi) logic network representation in nanometric space.

The fundamentals of logic function manipulation based on artificial intelligence paradigm are the subject of papers

(i) *Multiple-Valued Logic and Artificial Intelligence. Fundamentals of Fuzzy Control Revisited* by C. Moraga, E. Trillas, and S. Guadarrama

(ii) *Bi-Decomposition of Function Sets in Multiple-Valued Logic for Circuit Design and Data Mining* by C. Lang, and B. Steinbach, and

(iii) *Exploiting the Functional Properties of Boolean Function for Optimal Multi-level Design by Bi-Decomposition* by B. Steinbach and C. Lang.

These papers are focused on multiple-valued logic that proves itself to be a platform for an evolution of traditional binary (Boolean) domain of computation towards MVL.

Evolutionary circuit design approach is introduced in the papers

(i) *Evolutionary Synthesis of Arithmetic Circuit Structures* by T. Aoki, N. Homma, T. Higuchi

(ii) *Evolutionary Approach to Quantum and Reversible Circuits Synthesis* by M. Lukac, M. Perkowski, H. Goi, C. H. Yu, K. Chung, H. Jee, B.-G. Kim, Y.-D. Kim, and M. Pivtoraiko,

7

(iii) *Evolutionary Synthesis of Logic Circuits Using Information Theory* by A. H. Aguirre, and C. A. Coello Coello.

The evolutionary algorithms as a nature inspired framework for search technique, hold their honorary leadership among "intelligent" optimization paradigm in many areas, including the logic circuit design.

Information measures in circuit design are another area of multidisciplinary research:

(i) *Advanced AI Search Techniques in Modern Digital Circuit Synthesis* by L. Jozwiak, and

(ii) *Conquering Uncertainty in Multiple-valued Logic Design* by D. Popel.

Artificial intelligence paradigm would not be completed without **neural computation**. The paper

Three-Dimensional Feedforward Neural Networks and Realization by Nano-Devices, by V. Shmerko and S. Yanushkevich,

contributes in the extension of this domain to the next generation of logic devices created by nanotechnology.

The authors offer their creations in the hope that they will spawn innovative ideas that enrich our ongoing investigation in artificial intelligence and its application to logic design. We are confident that through the author's contribution new ideas will emerge from this issue, thus advancing research in both artificial intelligence and logic design.

The editors gratefully acknowledges the important support given by our reviewers. They spent their valuable time and put incredible efforts to review the papers.

We also highly appreciate support of Professor Paul McKevitt, Editor-in-Chief of the *Artificial Intelligent Review* journal, through the entire process of assembling the issue. And, finally, we thank Clarisse da Silva Pereira, Senior Desk Editor, for her efforts in preparing this publication.

Guest Editor,

Svetlana N. Yanushkevich
University of Calgary
Canada

Multiple-Valued Logic and Artificial Intelligence Fundamentals of Fuzzy Control Revisited

CLAUDIO MORAGA, ENRIC TRILLAS and SERGIO GUADARRAMA
Departamento de Inteligencia Artificial Universidad Politécnica de Madrid, Campus de Montegancedo, 28660 Madrid, Spain

Abstract. This paper reviews one particular area of Artificial Intelligence, which roots may be traced back to Multiple-valued Logic: the area of fuzzy control. After an introduction based on an experimental scenario, basic cases of fuzzy control are presented and formally analyzed. Their capabilities are discussed and their constraints are explained. Finally it is shown that a parameterization of either the fuzzy sets or the connectives used to express the rules governing a fuzzy controller allows the use of new optimization methods to improve the overall performance.

Keywords: approximate reasoning, fuzzy control, fuzzy if-then rules

1. Introduction

Starting in the 70s of the past century, an area of research has grown under the name "Multiple-valued Logic"; which comprises much more than just Logics other than the Aristotelian Logic (see e.g., Rescher 1969; Dunn and Epstein 1977), interval-valued logics up to infinite-valued logics; including from discrete algebras to appropriate hardware (see e.g., *Multiple-valued Logic* (2002) and predecessors, as well as Rine (1977)). The main scope of this special issue of the Journal of Artificial Intelligence is the application of methods of Artificial Intelligence to the problem of Logic Design of Multiple-valued Digital Circuits. The Editors wanted however a balanced issue and gave us the opportunity and the challenge to show some relevant aspects of the other side of the medal, that is, the contributions of multiple-valued logics to Artificial Intelligence.

Among the main concerns of Artificial Intelligence, Knowledge Representation, Management of Uncertainty and Approximate Reasoning constitute relevant, close related research areas where logic plays a very important role. The very early work of Lukasiewicz (1923) on a Logic with three truth values *{true, false, unknown}* and an appropriate concept of negation $(\neg true = false; \neg false = true; \neg unknown = unknown)$ may be considered to be possibly the first step towards a consistent generalization of classical logic allowing a more realistic modelling of reasoning by including

9

uncertainty. One of the most appropriated formalisms to work with uncertainty in the present days is fuzzy logic (Zadeh 1965). It is interesting to recall that in this seminal paper Lotfi A. Zadeh wrote: "If the values of $\mu_A(x)$ are interpreted as truth values, the later case corresponds to a multiple-valued logic with a continuum of truth values in the interval [0, 1]" (The meaning of the notation $\mu_A(x)$ is explained below).

The basic rules of reasoning used in classical logic is the modus ponens for forward reasoning and modus tollens for backwards reasoning, whose symbolic expressions are

$$\frac{\begin{array}{c} A \to B \\ A \end{array}}{B} \qquad\qquad \frac{\begin{array}{c} A \to B \\ \neg B \end{array}}{\neg A}$$

meaning that if the rule "from A follows B" is given – (in the language of logic this kind of rule is expressed as "if A then B") – and the event A is observed, then the event B should also be observed. (Similarly, given the rule "if A then B" and ¬B is observed then ¬A should be expected.) These processes are referred to in the literature as inference. In the case of fuzzy logic, a generalization of modus ponens is used, based on fuzzy sets. Given a universal set X, a fuzzy set A on X is defined by its characteristic function $\mu_A : X \to [0, 1]$ and for all $x \in X$, $\mu_A(x)$ gives the degree of membership of x to A or the degree with which x fulfills the concept represented by A. In this paper fuzzy sets have a semantic role: they represent the way in which a statement *"x is A"* is used in a given context.

The generalized modus ponens used in fuzzy logic may be given in its simplest expression as

$$\frac{\begin{array}{c} A \to B \\ A^* \end{array}}{B^*}$$

where A, A^*, B and B^* are fuzzy sets, A and A^* are defined on a same universe, but they are not necessarily equal. Similarly for B and B^*. The meaning in this case is the following: given a rule "if A then B" and observing an event A^* which is *similar* to A, an event B^* is expected, which should also be *similar* to B. Similarity of fuzzy sets has been studied from two different perspectives: given a proper metric, the closer two fuzzy sets, the more similar they are (Cross and Sudkamp 2002); on the other hand, the less distinguishable two fuzzy sets, the more similar they are (Trillas and Valverde 1985a). To allow more specified situations, in $A \to B$, A may stand for a set of conditions that have to be fulfilled at the same time. The formal representation is a conjunction of fuzzy premises. For the computation of conjunctions, operations belonging to the class of triangular norms are used.

Triangular norms, or simply t-norms, may be traced back to work of Karl Menger (1942) and were formally defined by Schweizer and Sklar (1983); however their use in fuzzy reasoning was introduced among others by Alsina, Trillas and Valverde (1983). If $T : [0, 1]^2 \rightarrow [0, 1]$ is a t-norm, then it is non-decreasing, associative, commutative and has 1 as identity. In the rest of this paper we will only consider continuous t-norms, since some useful characterizations are available for them (see e.g., Klement et al. 2000). The other needed operation is the "then-arrow" in "if A then B". In the case of fuzzy logic this operation is not only an extension of the classical implication and its characterization has been thoroughly studied (see e.g., Trillas and Valverde 1985b; Pradera et al. 2000; Trillas and Guadarrama 2001).

It is beyond saying that fuzzy reasoning represents a very important area of research in Artificial Intelligence. A treatment in depth is however beyond the scope of this paper. For further reading, the following references may be suggested: Baldwin (1985), Skala (1978), Skala et al. (1984), Smet et al. (1988), Whalen and Schott (1985), Teodorescu et al. (1991).

1.1. *Automatic control*

Probably one of the most successful developments of fuzzy reasoning, from the industrial point of view, is the design of fuzzy control systems, also called *linguistic* control systems, or simpler, the applications of fuzzy controllers. Controllers are special purpose dedicated pieces of hardware of digital, analog or hybrid realization (see e.g., Watanabe et al. 1990; Yamakawa 1988; Huertas et al. 1994), with a functionality whose specification will be discussed in details below.

The first published work suggesting the application of fuzzy logic for automatic control purposes is due to L. Zadeh (1972) however the first works on the design of fuzzy control systems are to be thanked to E. Mamdani and his colleagues (1974, 1974, 1976) in England. Probably the first known application of fuzzy control at industrial level, is the control of a furnace in the process of cement production in Denmark (Holmblad and Østergaard 1982). Remarkable is the fact that both the initial theoretical developments and the first industrial application took place in Europe. The real break through came however a decade later, when the Japanese industry started to use fuzzy control both in home appliances and in complex systems; the fuzzy controlled train of Sendai being one of the best known examples of this epoch (Yasunobu et al. 1985), followed by the fuzzy control of a water sewage plant (Yagishita et al. 1999) and the fuzzy control of a flying helicopter (Sugeno 1999; Sugeno and Murofushi 1991; Sugeno et al. 1993).

A fuzzy control system is based on a set of fuzzy *"if-then"* rules of behavior (as the ones introduced in the former section) that consider the kind of stimuli from the environment, that the system will receive, meanwhile at a given time, the values of these stimuli represent the facts, that the rules have to consider to offer proper actions.

A fuzzy controller has the general structure shown in Figure 1 (More elaborated models may be found in e.g., Lee (1990), Pedrycz (1992), Driankov et al. (1996)). Three main blocks may be distinguished: a data base, a rule base and a processing unit.

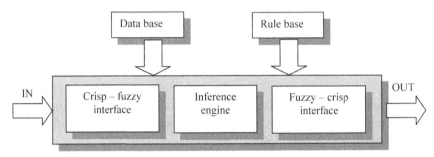

Figure 1. Model of a simple fuzzy controller.

The data base contains information related to the "high level" parameters of the controller, such as the availability of possibly different kinds of crisp-fuzzy and fuzzy-crisp conversion (including the case of no conversion), and availability of operations for the numerical calculation of the conjunction or disjunction of premises. Similarly with respect to the availability of operations for the implication, that realizes the "then" connective of the rules, as well as for the operation that computes the aggregation of the conclusions of several rules that might be simultaneously activated by the prevailing conditions of the environment. Furthermore the data base also has information related to the number, shape and distribution of the fuzzy sets specifying the meaning of the linguistic terms of each linguistic variable associated to a physical variable stimulating the system to be controlled. In summary, the data base contains all information needed to specify a particular configuration of the fuzzy controller.

The rule base contains the set of rules that will govern the behavior of the controller. The definition of the rule base is one of the main tasks of the designer of a fuzzy controller. There are several strategies to achieve this goal, e.g.:

— Choose the rules as to represent the knowledge of a control engineer.
— Select the rules to give a fuzzy model of the plant.
— Design the rules to model the knowledge of *an experienced operator of the plant.*
— *Learn* the rules from examples of behavior.

— If the plant may be partitioned into functionally well defined blocks (i.e., if the "divide and conquer" strategy is applicable, which is often the case in technical problems), different subsets of rules may be designed, one for each block. Moreover every subset of rules may be designed following a different strategy thus leading to a final hierarchical hybrid rule base.

It should be noticed, that the emphasis is given by the words "knowledge" and "model". ("Learning" is understood as the process of acquiring knowledge.) The goal is then the use of "intelligent control" based knowledge. Fuzzy sets provides for an adequate formalism to represent and process this knowledge.

The processing unit cares for the compatibility of data – (input and output interfaces) – and for the execution of the rules – (inference engine) –, which is mostly done through pointwise numerical calculations of implications, as will be formally discussed in details below.

1.2. *Simple illustrative example*

The following example is presented at a "phenomenological" level to allow an intuitive understanding of the main issues. A detailed formal treatment will be the topic of the next sections.

Assume that a heating system has to be controlled. The system is based on warm water circulating at constant speed and passing through well distributed heating panels. Moreover assume that the water temperature should be proportional to the heating demand. The following rules reflect the knowledge of an experiences operator of such a heating system:

```
R1:   if the external temperature is freezing,
      then the heating demand is large
R2:   if the external temperature is cold,
      then the heating demand is average
R3:   if the external temperature is medium,
      then the heating demand is low
```

It becomes apparent that the *meaning* of the predicates *freezing*, *cold* and *medium* is different if the system is intended for Ottawa, Madrid or Dakar, since they would not be *used* in exactly the same way. Similarly, the *meaning* of a *large*, *average* or *low* heating demand (in a KW scale) is different for a system meant to heat an office room, a conference hall for 200 people or a 25 stories office building. The relative ordering of the terms *large*, *average* and *low* will certainly be the same, and their shapes will probably be the same, in all three mentioned cases. Consider one instance of the problem as illustrated in Figure 2 and assume the situation that the external temperature is 6 °C.

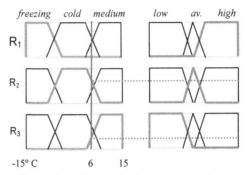

Figure 2. Rule base for the heating problem.

It is fairly obvious that if the external temperature is 6 °C, the first rule does not apply, since 6 ° is not considered to be *freezing*, but "more *cold* than *medium*" (see Figure 2). The degree of satisfaction of the premises or conditions stated in the "if-part" of rules 2 and 3 will affect the strength of the corresponding conclusions. This will be specified by the "then"-operation. As will be discussed in the next sections, the product is one such "then"-operation frequently used in fuzzy control. The effect is shown in Figure 3, where it may be seen that rules 2 and 3 are activated *to a certain degree* and give proportional suggestions for action. These have to be combined into one by means of an aggregation operation. From the many aggregation operations that may be used for this purpose (see e.g., (Dubois and Prade 1985)) the pointwise maximum is usually the first choice. The aggregated result is also shown in Figure 3.

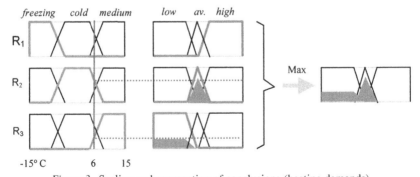

Figure 3. Scaling and aggregation of conclusions (heating demands).

Since it was assumed that finally the water temperature should be proportional to the heating demand, the fuzzy set representing the aggregation of activated heating demands has to be converted into a real value. The established slang speaks of *defuzzification*. It becomes apparent that this cannot be done without information loss, since it is analogous to representing a signal with only one coefficient of its Fourier power spectrum. However, experimental results have shown that "approximating" a fuzzy set by the abscise of its gravity center or the abcise of its center of area leads to an adequate control performance (see e.g., Driankov et al. 1996). Furthermore, rules may be given a more specified structure, providing directly a numerical conclusion. In that case the (fuzzy – crisp) output interface simply realizes the identity function.

2. The Meaning of Conditionals and Implications: The Case of Fuzzy Logic

As it is well know (see Klir and Yuan 1995), the success of fuzzy logic is mainly due to the representation of elementary statements *"x is P"* ($x \in X$ and P a precise or imprecise predicate or linguistic label on X) by a function $\mu_P : X \rightarrow [0, 1]$, in the hypothesis that $\mu_P(x)$ is the degree up to which x is P, or x verifies the property named P. In the same view, a rule *"If x is P, then y is Q"* ($x \in X$, $y \in Y$) is represented by a function of the variables μ_P and μ_Q, by

$$R(\mu_P, \mu_Q)(x, y),$$

a number in [0,1] once P, Q, x and y are fixed.

Functions R can be or can be not functionally expressible, that is, it can exist or it cannot exist a numerical function $J : [0, 1] \times [0, 1] \rightarrow [0, 1]$ such that

$$R(\mu_P, \mu_Q)(x, y) = J(\mu_P(x), \mu_Q(y))$$

for all $x \in X$, $y \in Y$. **Fuzzy logic works mainly within the positive supposition**, and several families of such functions J have been proposed (see Tanaka 1996) and are actually used. Such numerical functions are called implication functions, and their diverse types are derived from the linguistic meaning of the conditional frase (the rule) *"If x is P, then y is Q"*, that is, from its use in the universe $X \times Y$ at each particular problem.

2.1. *The problem of inference from imprecise rules*

Let us suppose a dynamic (in the sense of non-rigid) systems with input variables $(x_1, \ldots, x_n)$ and output variable y, whose behavior is known by m imprecise rules R_i:

R_1:   If x_1 is P_{11} and x_2 is $P_{12} \cdots$ and x_n is P_{1n}, then y is Q_1
R_2:   If x_1 is P_{21} and x_2 is $P_{22} \cdots$ and x_n is P_{2n}, then y is Q_2

$\vdots$

R_m: If x_1 is P_{m1} and x_2 is $P_{m2} \cdots$ and x_n is P_{mn}, then y is Q_m

The question is the following: If we observe that the input variables are in the "states" "x_1 is P_1", "x_2 is P_2", ..., "x_n is P_n", respectively, what can be inferred for variable y? That is, supposing that it lies in the "state" "y is Q^*", what is Q^*?

Without loss of generality let us consider only the case $n = m = 2$:

R_1: If x_1 is P_{11} and x_2 is P_{12}, then y is Q_1
R_2: If x_1 is P_{21} and x_2 is P_{22}, then y is Q_2

---

x_1 is P_1^* and x_2 is P_2^*

---

y is Q^*,   what is Q^*?

where $x_1 \in X_1$, $x_2 \in X_2$, $y \in Y$, and let us represent it in fuzzy logic by means of the functions R_1^* and R_2^*. This leads to:

$$R_1^*(\mu_{P_{11}} \wedge \mu_{P_{12}}, \mu_{Q_1})((x_1, x_2), y) = J_1(T_1(\mu_{P_{11}}(x_1), \mu_{P_{12}}(x_2)), \mu_{Q_1}(y))$$
$$R_2^*(\mu_{P_{21}} \wedge \mu_{P_{22}}, \mu_{Q_2})((x_1, x_2), y) = J_2(T_2(\mu_{P_{21}}(x_1), \mu_{P_{22}}(x_2)), \mu_{Q_2}(y))$$

$\mu_{P_1^*}(x_1)$ and $\mu_{P_2^*}(x_2)$

$$\mu_{Q^*}(y) = ?$$

for convenient continuous t-norms T_1, T_2 and convenient implication functions J_1, J_2. Convenient, in the sense of adequate to the use of the conditional phrases R_1 and R_2 relative to the problem under consideration within the given system S.

To find a solution to this problem, the problem of linguistic control, that is a part of what can be called intelligent systems' control, we need to pass throughout several steps.

2.1.1. First step: Functions J

Implication functions $J : [0, 1] \times [0, 1] \rightarrow [0, 1]$ are (to be) obtained through the interpretation and representation of the rule's use, that is, from the actual meaning of these conditional phrases: For example, if in case the rule 'If α, then β' ($\alpha \rightarrow \beta$) is used as equivalent to 'not α or β', then the function J is to be chosen among those in the family $J(a, b) = S(N(a), b)$, for all a, b in [0,1], some strong negation function N and some continuous t-conorm S.

These implication functions are called S-implications (S is for *strong*) (see Klir and Yuan 1995).

Provided $\alpha \to \beta$ is used as the lowest statement γ such that '*If γ and α, then β*', J belongs to the family $J_T(a, b) = Sup\{z \in [0, 1]; T(z, a) \le b\}$, for all a, b in [0,1] and some continuous t-norm T. These implication functions are called R-implications (R is for *residuated*) (see Klir and Yuan 1995).

If $\alpha \to \beta$ is used as '*not α or (α and β)*', J belongs to the family of functions $J(a, b) = S(N(a), T(a, b))$, called Q-implication (Q is for *quantum*, since these implications were used in the so called quantum logic (Trillas et al. 2000a)).

If $\alpha \to \beta$ is used as '*α and β*' (because, for example, it is never the case that *not α*), J is in the family of functions $J(a, b) = T(a, b)$, called ML-implications (ML is for Mamdani-Larsen, the names of the researchers who introduced this kind of implications, respectively)(see Campo and Trillas 2000). As in fuzzy control mostly ML-implications are considered, let us analyze the different types of these functions.

Since T is a continuous t-norm, it is $T = Min$, or $T = Prod_\varphi = \varphi^{-1} \circ Prod \circ (\varphi \times \varphi)$, or $T = W_\varphi = \varphi^{-1} \circ W \circ (\varphi \times \varphi)$, with φ an order-automorphism of the unit interval ($[0, 1], \le$), $Prod(x, y) = x.y$ and $W(x, y) = Max(0, x + y - 1)$. Of course, T can be also an ordinal-sum (see Klement et al. 2000), but these t-norms have never been considered in fuzzy logic. Hence, a ML-implication belongs to the types: $J_M(a, b) = Min(a, b)$, $J_L(a, b) = Prod_\varphi(a, b)$ and $J_W(a, b) = W_\varphi(a, b)$. Only in the third type we can have $J(a, b) = 0$ with $a \ne 0$ and $b \ne 0$, since it is $W_\varphi(a, b) = 0$ whenever $\varphi(a) + \varphi(b) \le 1$, and as in fuzzy control it is desirable not only that $a = 0$ implies $J(a, b) = 0$ but also that $a \ne 0$ and $b \ne 0$ imply $J(a, b) \ne 0$, the third type is rarely used and only $J_M(a, b) = Min(a, b)$ (Mamdani implication) and $J_L(a, b) = a.b$ (Larsen implication) are almost always considered. Notice that for all S, R, and Q implication functions it is: $J(0, b) = 1$.

2.1.2. *Second step: Modus Ponens*

Rules are used in our problem to infer μ_{Q^*}, and this inference requires that when the states of the input variables $x_1, \ldots, x_n$ are exactly those appearing in the antecedent part of one of the m rules, say rule number i, then μ_{Q^*} should be the consequent μ_{Q_i} of this rule. That is, each rule should satisfy the meta-rule of *Modus Ponens*:

$$\frac{\text{If } x \text{ is } P, \text{ then } y \text{ is } Q \quad | \quad R(\mu_P, \mu_Q)(x, y)}{\begin{array}{c|c} x \text{ is } P & \mu_P(x) \\ \hline y \text{ is } Q & \mu_Q(y) \end{array}}$$

This meta-rule is satisfied when there is a continuous t-norm T_1 such that

$$T_1(\mu_P(x), R(\mu_P, \mu_Q)(x, y)) \leq \mu_Q(y),$$

for all $x \in X$, $y \in Y$. When $R(\mu_P, \mu_Q)(x, y) = J(\mu_P(x), \mu_Q(y))$, the last inequation is

$$T_1(\mu_P(x), J(\mu_P(x), \mu_Q(y))) \leq \mu_Q(y),$$

for all $x \in X$, $y \in Y$.

Hence, for each type of implication function J we need to know which T_1 allows the verification of the *Modus Ponens* inequality:

$$T_1(a, J(a, b)) \leq b, \text{ for all } a, b \text{ in } [0, 1].$$

For example, with an S-implication, since $T_1(a, S(N(a), b)) \leq b$ implies (with $b = 0$) $T_1(a, N(a)) = 0$, it should be $T_1 = W_\varphi$ for some automorphism φ of $([0, 1], \leq)$.

With R-implications J_T, since

$$T(a, J_T(a, b)) = Min(a, b) \leq b,$$

the same T in J_T allows to have that inequality.

With Q-implications, since

$$T_1(a, S(N(a), T(a, b))) \leq b$$

also implies $T_1(a, N(a)) = 0$ for all $a \in [0, 1]$, it should be also $T_1 = W_\varphi$.

Concerning ML-implications, since

$$T_1(a, T(a, b)) \leq Min(a, Min(a, b)) = Min(a, b) \leq b,$$

because both $T_1 \leq Min$ and $T \leq Min$, the *Modus Ponens* inequality is verified for all t-norms T_1 and, hence, for $T_1 = Min$ (the biggest t-norm) (see Trillas and Alsina 2001). This is a privileged situation.

If T_1 verifies $T_1(a, J(a, b)) \leq b$, because of the well-known result (see Gottwald 1993) that for left-continuous t-norms T_1, $T_1(a, t) \leq b$ is equivalent to $t \leq J_{T_1}(a, b)$, it results that the inequality is equivalent to $J(a, b) \leq J_{T_1}(a, b)$. Hence, among the functions J verifying the *Modus Ponens* inequality with a continuous t-norm T_1, the R-implication J_{T_1} is the biggest one and, consequently, $T_1(a, J_{T_1}(a, b))$ is closer to b than $T_1(a, J(a, b))$. In particular, it is

$$J_M(a, b) = Min(a, b) \leq J_{Min}(a, b) = \begin{cases} 1, & \text{if } a \leq b \\ b, & \text{if } a > b \end{cases},$$

and

$$J_L(a, b) = a.b \le J_{Min}(a, b) \le J_{Prod}(a, b) = \begin{cases} 1, & \text{if } a \le b \\ \frac{b}{a}, & \text{if } a > b \end{cases}.$$

(since $a.b \le b$).

2.1.3. *Third step: Zadeh's compositional rule of inference*

Once the rule '*If x is P, then y is Q*' is represented by $J(\mu_P(x), \mu_Q(y))$, and a continuous t-norm T_1 such that $J \le J_{T_1}$ is known, the inference:

$$\text{If } x \text{ is } P, \text{ then } y \text{ is } Q$$
$$x \text{ is } P^*$$

$$\overline{}$$

$$y \text{ is } Q^*$$

is obtained (see (Trillas et al. 2000b) and (Klir and Yuan 1995)) by Zadeh's Compositional Rule of Inference (CRI):

$$\mu_{Q^*}(y) = \underset{x \in X}{Sup} \, T_1(\mu_{P^*}(x), J(\mu_P(x), \mu_Q(y))), \text{ for all } y \in Y.$$

It should be pointed out that Zadeh's CRI is not a "result" but a meta-rule. It is a "directive" allowing to reach a solution to our problem, and it should be noticed that when $P^* = P$ it is not in general $Q^* = Q$. For example, in the case of ML-implications it is:

$$\mu_{Q^*}(y) = \underset{x \in X}{Sup} \, Min(\mu_P(x), T(\mu_P(x), \mu_Q(y))) \le$$
$$\underset{x \in X}{Sup} \, T(\mu_P(x), \mu_Q(y)) = T(\underset{x \in X}{Sup} \, \mu_P(x), \mu_Q(y)) = \mu_Q(y),$$

provided that $Sup \; \mu_P = 1$, and because of $T(\mu_P(x), \mu_Q(y)) \le \mu_Q(y)$ and T is continuous. But, for example, if $T = Min$, $Sup \; \mu_P = 0.9$ and $Sup \; \mu_Q = 1$, then $\mu_{Q^*}(y) = Min(0.9, \mu_Q(y)) \ne \mu_Q(y)$. Notice that for all the cases in which μ_P is normalized ($\mu_P(x_0) = 1$ for some $x_0 \in X$), Mamdani-Larsen implications do verify $\mu_{Q^*} = \mu_Q$ whenever $\mu_{P^*} = \mu_P$.

2.1.4. *Fourth step: Numerical input*

This is the case in which μ_{P^*} is exactly $x = x_0$ or $x \in \{x_0\}$. That is "*x is P^**" is the statement '*x is x_0*' and hence

$$\mu_{P^*}(x) = \begin{cases} 1, & \text{if } x = x_0 \\ 0, & \text{if } x \ne x_0. \end{cases}$$

In that case,

$$\mu_{Q^*}(y) = \underset{x \in X}{Sup} \, T_1(\mu_{P_j}(x), J(\mu_P(x), \mu_Q(y))) = J(\mu_P(x_0), \mu_Q(y)),$$
$$\text{for all } y \in Y.$$

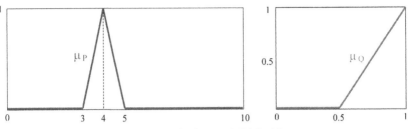

Figure 4. Left: close to 4. Right: big.

For example let J be a ML-implication, $\mu_{Q^*}(y) = T(\mu_P(x_0), \mu_Q(y))$. If $X = [0, 10]$, $Y = [0, 1]$, $P = close\ to\ 4$, $Q = big$, with uses as shown in Figure 4 and moreover $x_0 = 3.5$, with $J(a, b) = Min(a, b)$, then $\mu_{Q^*}(y) = Min(\mu_P(3.5), \mu_Q(y)) = Min(0.5, \mu_Q(y))$, as $\mu_P(x) = x - 3$ between 3 and 4. Hence, the graphic of the output μ_{Q^*} is the one shown in Figure 5.

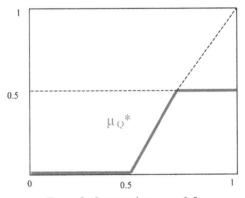

Figure 5. Output when $x_0 = 3.5$.

2.1.5. *Fifth step: Numerical consequent*

This is the case in which μ_Q is $y = y_0$ or $y \in \{y_0\}$. That is 'y is Q' corresponds to 'y is y_0', and

$$\mu_Q(y) = \mu_{y_0}(y) = \begin{cases} 1, & \text{if } y = y_0 \\ 0, & \text{if } y \neq y_0. \end{cases}$$

In this case:

$$J(\mu_P(x), \mu_{y_0}(y)) = \begin{cases} J(\mu_P(x), 1), & \text{if } y = y_0 \\ J(\mu_P(x), 0), & \text{if } y \neq y_0, \end{cases}$$

and the output μ_{Q^*} depends on the values of J, namely, on $J(a, 1)$ and $J(a, 0)$. Notice that if:

- J is an S-implication, $J(a, 1) = 1$; $J(a, 0) = N(a)$.
- J is an Q-implication, $J(a, 1) = S(N(a), a)$; $J(a, 0) = N(a)$.
- J is an R-implication, $J(a, 1) = 1$; $J(a, 0) = Sup \ \{z \in [0, 1]$; $T(z, a) = 0\}$.
- J is an ML-implication, $J(a, 1) = a$; $J(a, 0) = 0$.

Then, for example, if J is an ML-implication:

$$\mu_{Q^*}(y) = \underset{x \in X}{Sup} \ Min(\mu_{P^*}(x), J(\mu_P(x), \mu_{y_0}(y))) =$$

$$\mu_{Q^*}(y) = \begin{cases} \underset{x \in X}{Sup} \ Min(\mu_{P^*}(x), \mu_P(x)), & \text{if } y = y_0 \\ 0, & \text{if } y \neq y_0 \end{cases}$$

Notice that if the input is also numerical, i.e., $x = x_0$ then the output is:

$$\mu_{Q^*}(y) = \begin{cases} \underset{x \in X}{Sup} \ \mu_{P^*}(x), & \text{if } y = y_0 \text{ and } x = x_0 \\ 0, & \text{if } y \neq y_0 \text{ or } x \neq x_0. \end{cases}$$

2.1.6. *Sixth step: Several rules with numerical input*
The problem is now

> R1: If x is P_1, then y is Q_1
> R2: If x is P_2, then y is Q_2
> _____
> x is P^*
> _____
> y is Q^*?

By using the CRI, R1 with the input x is P^* gives an output y is Q_1^*, with $\mu_{Q_1^*}$ its membership function and R2 with the input x is P^* gives y is Q_2^* with $\mu_{Q_2^*}$ its membership function. The total output, y is Q^*, corresponds to the idea $(y$ is $Q_1^*)$ *or* $(y$ is $Q_2^*)$, and translating this *or* by means of the lowest t-conorm its value can be obtained by $\mu_{Q^*}(y) = Max(\mu_{Q_1^*}(y), \mu_{Q_2^*}(y))$.

For example, the Mandani's Method consist in taking $J(a, b) = Min(a, b)$ and the Larsen's Method, in taking $J(a, b) = a.b$. Let us consider in $X = [0, 10]$, $Y = [0, 1]$ the problem:

> R1: If x is close-to 4, then y is big
> R2: If x is small,   then y is small
> _____
> $x = 3.5$

and let us find μ_{Q^*} using both methods by supposing "close-to 4" as in the former example, $\mu_{big}(y) = y$, $\mu_{small}(x) = 1 - \frac{x}{10}$, and $\mu_{small}(y) = 1 - y$. Graphically:

The outputs Q_1^*, Q_2^* are:

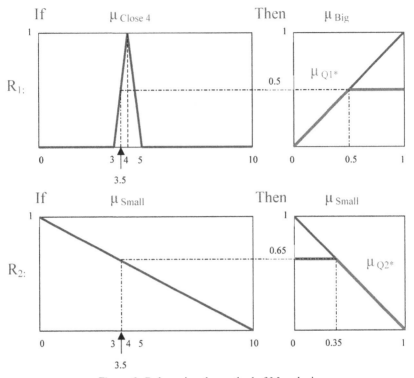

Figure 6. Rules using the method of Mamdani.

— With Mamdani's Method: $\mu_{Q_1^*}(y) = Min(\mu_{close-to4}(3.5), \mu_{big}(y)) = Min(0.5, y)$ (see Figure 6, upper right corner)

$\mu_{Q_2^*}(y) = Min(\mu_{small}(3.5), \mu_{small}(y)) = Min(0.65, 1 - y)$ (see Figure 6, lower right corner)

Hence, $\mu_{Q^*}(y) = Max(\mu_{Q_1^*}(y), \mu_{Q_2^*}(y)) = Max(Min(0.5, y), Min(0.65, 1 - y))$ (see Figure 7).

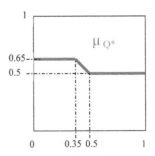

Figure 7. Ouput using the method of Mamdani.

- With Larsen's Method: $\mu_{Q_1^*}(y) = 0.5y$ (see Figure 8, left)
  $\mu_{Q_2^*}(y) = 0.65(1-y)$ (see Figure 8, middle)
  Hence, $\mu_{Q^*}(y) = Max(0.5y, 0.65(1-y))$ (see Figure 8, right)

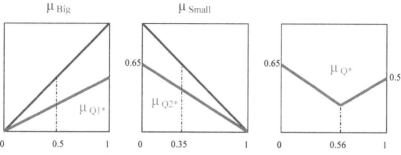

Figure 8. Rules and output using the method of Larsen.

2.1.7. *Seventh step: A more complex example with numerical inputs and consequents*

Let us consider the case:

$$
\begin{array}{llll}
\text{R1:} & \text{If} & x_1 \text{ is } P_{11} \text{ and } x_2 \text{ is } P_{12}, & \text{then} \quad y = y_1 \\
\text{R2:} & \text{If} & x_2 \text{ ix } P_{21} \text{ and } x_2 \text{ is } P_{22}, & \text{then} \quad y = y_2
\end{array}
$$

$$
\overline{x_1 = x_1^*, x_2 = x_2^*}
$$

$$
\mu_{Q^*} = ?
$$

With Larsen's method and translating the *and* in the antecedents also by $T = Prod$.

- Rule R1 is represented by $J_L(T(\mu_{P_{11}}(x_1), \mu_{P_{12}}(x_2)), \mu_{y_1}(y)) =$

$$
\mu_{P_{11}}(x_1).\mu_{P_{12}}(x_2).\mu_{y_1}(y) = \begin{cases} \mu_{P_{11}}(x_1).\mu_{P_{12}}(x_2) & , y = y_1 \\ 0 & , y \neq y_1. \end{cases}
$$

- Rule R2 is represented by $J_L(T(\mu_{P_{21}}(x_1), \mu_{P_{22}}(x_2)), \mu_{y_2}(y)) =$

$$
\mu_{P_{21}}(x_1).\mu_{P_{22}}(x_2).\mu_{y_2}(y) = \begin{cases} \mu_{P_{21}}(x_1).\mu_{P_{22}}(x_2) & , y = y_2 \\ 0 & , y \neq y_2. \end{cases}
$$

Hence, the corresponding outputs under the CRI are:

$$
\mu_{Q_1^*}(y) = \begin{cases} 0 & , y \neq y_1 \\ \mu_{P_{11}}(x_1^*).\mu_{P_{12}}(x_2^*) & , y = y_1 \end{cases}
$$

$$\mu_{Q_2^*}(y) = \begin{cases} 0 & , y \neq y_2 \\ \mu_{P_{21}}(x_1^*).\mu_{P_{22}}(x_2^*) & , y = y_2. \end{cases}$$

Consequently:

$$\mu_{Q^*} = Max(\mu_{Q_1^*}(y), \mu_{Q_2^*}(y)) = \begin{cases} \mu_{P_{11}}(x_1^*).\mu_{P_{12}}(x_2^*) & , y = y_1 \\ \mu_{P_{21}}(x_1^*).\mu_{P_{22}}(x_2^*) & , y = y_2 \\ 0 & , otherwise. \end{cases}$$

The graphical representation of the result is shown in Figure 9.

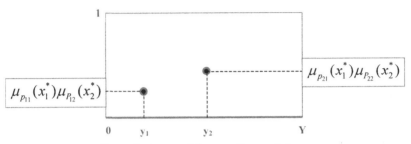

Figure 9. Output of the case discussed above.

2.1.8. *Eight step: Defuzzification*

Zadeh's CRI gives an output function μ_{Q^*}, but what it is frequently needed, mainly in control, is an output number as an "order" to be executed by the system. Hence, this step consists in compacting in the best possible way, in a single real number, the information on the system's behavior contained in μ_{Q^*}. That is, the goal is to *defuzzify* μ_{Q^*}.

In the applications the most interesting cases are those in which, respectively, either μ_{Q^*} is a non-null continuous function, or it is a non-null function at only a finite number of points in Y. For example, see Figures 7 and 9, respectively.

In the first case and among the diverse methods that have been suggested in the literature (see Nguyen and Walker 1999), that known as "center of gravity" of the area below μ_{Q^*} as well as the one known as "center of area" are perhaps the most used. In the second case, if for example,

$$\mu_{Q^*}(y) = \begin{cases} \alpha_1, & if \ y = y_1 \\ \alpha_2, & if \ y = y_2 \\ \vdots & \\ \alpha_n, & if \ y = y_n \\ 0, & otherwise \end{cases}$$

the most popular method of defuzzification is that consisting in taking the weighted mean

$$\mu_{Q*} = \frac{\alpha_1.y_1 + \alpha_2.y_2 + \cdots + \alpha_n.y_n}{\alpha_1 + \alpha_2 + \cdots + \alpha_n}$$

The case of m rules with numerical consequents and numerical inputs, with Larsen's implication function as in Section 2.1.7 and with defuzzification by the weighted mean, is the basis of the so-called Takagi-Sugeno (vid Tanaka 1996) methods of fuzzy inference of orders 1, 2, 3, ..., etc.

EXAMPLE 1. *In the case of Figure 7, the area below μ_{Q*} is easily computed by $0.35 \times 0.15 + \frac{0.15 \times 0.15}{2} + 0.5 \times 1 = 0.564$. Hence, the center of area is a point $y_0 \in (0, 1)$ such that, the areas to the left and to the right of y_0 are equal, i.e.:*

$$\frac{0.564}{2} = 0.282 = \int_0^{y_0} \mu_{Q*}(y)dy =$$

$$\int_0^{0.35} 0.65dy + \int_{0.35}^{y_0} (1-y)dy =$$

$$0.228 + y_0 - 0.35 - \int_{0.35}^{y_0} ydy$$

as the line joining the points (0.35,0.65) and (0.5,0.5) is $z = 1 - y$. Hence:

$$y_0 - \int_{0.35}^{y_0} ydy = 0.282 - 0.228 + 0.35 = 0.404$$

$$y_0 - \left[\frac{y^2}{2}\right]_{0.35}^{y_0} = y_0 - \left(\frac{y_0^2}{2} - \frac{0.35^2}{2}\right) = 0.404,$$

gives: $y_0^2 - 2y_0 + 0.686 = 0$, with positive root $y_0 = 0.43919$.

EXAMPLE 2. *In the case of Figure 9, the value y_0 is the weighted mean:*

$$y_0 = \frac{\mu_{P_{11}}(x_1^*)\mu_{P_{12}}(x_2^*)y_1 + \mu_{P_{21}}(x_1^*)\mu_{P_{22}}(x_2^*)y_2}{\mu_{P_{11}}(x_1^*)\mu_{P_{12}}(x_2^*) + \mu_{P_{21}}(x_1^*)\mu_{P_{22}}(x_2^*)}$$

Provided that $X_1 = X_2 = [0, 1]$, $Y = [0, 10]$, $\mu_{P_{11}}(x_1) = x_1$, $\mu_{P_{12}}(x_2) = 1 - x_2$, $y_1 = 6$, $\mu_{P_{21}}(x_1) = 1 - x_1$, $\mu_{P_{22}}(x_2) = x_2$, $y_2 = 4$, $x_1^* = 0.3$ and $x_2^* = 0.7$, the calculation will be:

$$y_0 = \frac{0.3 \times (1 - 0.7) \times 6 + (1 - 0.3) \times 0.7 \times 4}{0.3 \times (1 - 0.7) + (1 - 0.3) \times 0.7} = \frac{0.54 + 1.96}{0.09 + 0.49} =$$

$$\frac{2.5}{0.58} = 4.31$$

Graphically, this is illustrated in Figure 10.

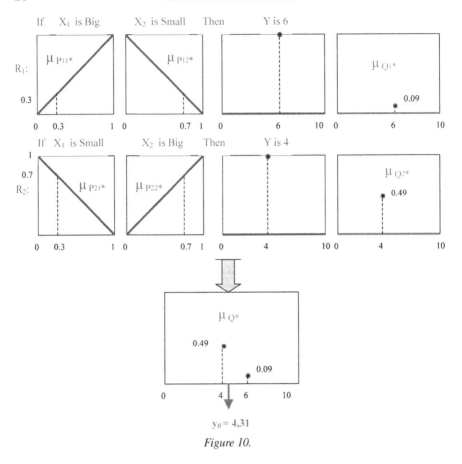

Figure 10.

2.2. *The method of Takagi-Sugeno of order 1*

The method of Takagi-Sugeno of order n (Tanaka 1996) is an immediate generalization of the last example with numerical input, numerical
consequents, ML-implication and defuzzification by a weighted mean. It is
the case in which in place of consequents $y = y_i$, y is taken as a polynomial
of degree n in the n variables $x_1, x_2, \ldots, x_n$ and with the given coefficients,
appearing in the rules' antecedents. For the sake of brevity, and without any
loss of generality, let us consider the case of $m = 2$ rules, with $n = 2$
variables:

R1:   If x_1 is P_{11} and x_2 is P_{12},   then $y = \alpha_1 x_1 + \alpha_2 x_2 + \alpha_3$
R2:   If x_2 ix P_{21} and x_2 is P_{22},   then $y = \beta_1 x_1 + \beta_2 x_2 + \beta_3$

$$x_1 = x_1^*, x_2 = x_2^*$$

$$\mu_{Q^*} = ? \quad ; y_0 = ?$$

Obviously, the case before considered is obtained with $\alpha_1 = \alpha_2 = \beta_1 = \beta_2 = 0$.

Let us shorten $y = \alpha_1 x_1 + \alpha_2 x_2 + \alpha_3$ by q_1, and $y = \beta_1 x_1 + \beta_2 x_2 + \beta_3$ by q_2. with $q_1^* = \alpha_1 x_1^* + \alpha_2 x_2^* + \alpha_3$ and $q_2^* = \beta_1 x_1^* + \beta_2 x_2^* + \beta_3$. It follows:

$$\mu_{x_1^* x_2^*}(x_1, x_2) = \begin{cases} 1, & \text{if } x_1 = x_1^* \text{ and } x_2 = x_2^* \\ 0, & otherwise, \end{cases}$$

$$\mu_{q_1}(y) = \begin{cases} 1, & \text{if } y = \alpha_1 x_1 + \alpha_2 x_2 + \alpha_3 \\ 0, & otherwise, \end{cases}$$

$$\mu_{q_2}(y) = \begin{cases} 1, & \text{if } y = \beta_1 x_1 + \beta_2 x_2 + \beta_3 \\ 0, & otherwise. \end{cases}$$

As rule R1 is represented by $J_1 = \mu_{P_{11}}(x_1).\mu_{P_{12}}(x_2).\mu_{q_1}(y)$ and rule R2 by $J_2 = \mu_{P_{21}}(x_1).\mu_{P_{22}}(x_2).\mu_{q_2}(y)$, it follows:

$$\mu_{Q_1^*}(y) = \underset{x \in X, y \in Y}{Sup} Min(\mu_{x_1^* x_2^*}(x_1, x_2), \mu_{P_{11}}(x_1).\mu_{P_{12}}(x_2).\mu_{q_1}(y)) =$$

$$\mu_{P_{11}}(x_1^*).\mu_{P_{12}}(x_2^*).\mu_{q_1}(y) = \begin{cases} \mu_{P_{11}}(x_1^*).\mu_{P_{12}}(x_2^*), & y = \alpha_1 x_1^* + \alpha_2 x_2^* + \alpha_3 \\ 0, & otherwise \end{cases}$$

$$\mu_{Q_2^*}(y) = \mu_{P_{21}}(x_1^*).\mu_{P_{22}}(x_2^*).\mu_{q_2}(y) =$$
$$\begin{cases} \mu_{P_{21}}(x_1^*).\mu_{P_{22}}(x_2^*), & y = \beta_1 x_1^* + \beta_2 x_2^* + \beta_3 \\ 0, & otherwise \end{cases}$$

Hence,

$$\mu_{Q^*}(y) = Max(\mu_{Q_1^*}(y), \mu_{Q_2^*}(y)) =$$
$$\begin{cases} \mu_{P_{11}}(x_1^*).\mu_{P_{12}}(x_2^*), & y = \alpha_1 x_1^* + \alpha_2 x_2^* + \alpha_3 \\ \mu_{P_{21}}(x_1^*).\mu_{P_{22}}(x_2^*), & y = \beta_1 x_1^* + \beta_2 x_2^* + \beta_3 \\ 0, & otherwise \end{cases}$$

And finally:

$$y_0 = \frac{\mu_{P_{11}}(x_1^*).\mu_{P_{12}}(x_2^*)(\alpha_1 x_1^* + \alpha_2 x_2^* + \alpha_3) + \mu_{P_{21}}(x_1^*).\mu_{P_{22}}(x_2^*)(\beta_1 x_1^* + \beta_2 x_2^* + \beta_3)}{\mu_{P_{11}}(x_1^*).\mu_{P_{12}}(x_2^*) + \mu_{P_{21}}(x_1^*).\mu_{P_{22}}(x_2^*)}$$

2.3. Parameterization of implication functions

The fact that fuzzy logic mainly deals with imprecise concepts an that it does not exist a unique type of this logic, any kind of flexibility in the election of the function representing the logical connectives is useful. For example, either parametric families of strong-negation functions (see Klir and Yuan 1995; Trillas and Pradera 2002; Trillas 1998; Moraga et al. 2003) like

$$N_{p,q} = \left[\frac{1 - a^q}{1 + p.a^q} \right]^{1/q}, q > 0, p > -1,$$

parametric families of t-norms, like $T_q(x, y) = [Max(0, x^q + y^q - 1)]^{1/q}, q > 0$, or parametric families of t-conorms, like $S_q(x, y) = [Min(1, x^q + y^q)]^{1/q}, q > 0$, allow to fix the parameter values by case-examples (Moraga et al. 2003).

It is also important to have parametric families of implication functions (see Moraga et al. 2003). In the case of S−implications $S(N(a), b)$ with, for example, $N = N_{p,1}$ and $S = S_q$, it is:

$$J_q(x, y) = \left[Min \left(1, \left(\frac{1 - x}{1 + px} \right)^q + y^q \right) \right]^{1/q}, q > 0$$

In the case of R−implications as, for example, $J_{W_\varphi} = \varphi^{-1} \circ J_W \circ (\varphi \times \varphi)$, with the family of order-automorphisms $\varphi(a) = a^q, q > 0$, we have the parametric family of R−implications

$$J_q(a, b) = \left[Min(1, 1 - a^q + b^q) \right]^{1/q}, q > 0$$

What about ML−implications? Given functions, $\varphi, \psi : [0, 1] \to [0, 1]$ such that $\varphi(0) = 0$, $\varphi(1) = 1$ and $\psi(a) \leq a$ for all a in $[0, 1]$ (hence, $\psi(0) = 0$), it is:

$$T_1(a, T(\varphi(a), \psi(b))) \leq Min(a, Min(\varphi(a), \psi(b))) \leq \psi(b) \leq b,$$

for all a, b in $[0, 1]$. Hence, all function $J_{\varphi\psi}(a, b) = T(\varphi(a), \psi(b))$, with a continuous t-norm T, do verify the **Modus Ponens** inequality. In particular, with $\varphi = \psi = id_{[0,1]}$, ML−implications are captured.

With either $T = Min$ or $T = Prod_\varphi$, is $J_{\varphi\psi}(a, b) = 0$ if and only if $\varphi(a) = 0$ or $\psi(b) = 0$, and, of course, $J_{\varphi\psi}(0, b) = 0$. Hence, to avoid the inconvenience of having $J_{\varphi\psi}(a, b) = 0$ with $a \neq 0$ and $b \neq 0$ it is needed to constrain '$\varphi(a) = 0$ iff $a = 0$' and '$\psi(b) = 0$ iff $b = 0$'. And to preserve $J_{\varphi\psi}(1, 1) = 1$ it is needed to require '$\varphi(a) = 1$ iff $a = 1$' and '$\psi(b) = 1$ iff $b = 1$'. Finally, to preserve monotonicity $[a_1 \leq a_2, b_1 \leq b_2$ imply $J_{\varphi\psi}(a_1, b_1) \leq J_{\varphi\psi}(a_2, b_2)]$ and avoid jumps, we will suppose that both φ and ψ are non-decreasing and continuous.

Of course, with φ and ψ we can obtain families of parametrized ML−implications. For example, with $\varphi(a) = a^{1/q} (p > 0)$ and $\psi(b) = b^q (q > 0)$, we have:

$$J_{p,q}(a, b) = T(a^{1/p}, b^q), p > 0, q > 0,$$

that originates the useful two-parametric families $J_{p,q}(a, b) = Min(a^{1/p}, b^q)$ and $J_{p,q}(a, b) = a^{1/p}.b^q$, by limiting T to the cases Min and $Prod$ as it is usual.

3. Parameterization of fuzzy sets

Provided that we can get a better representation of the rules, then we can get, e.g., a better Takagi-Sugeno model and therefore a better approximation, to a target function. The parameterization of the fuzzy sets representing the meaning of the premise, is still another way of introducing additional degrees of freedom, that properly tuned support the goal of a better approximation.

This claim is supported by the following fact. It has been shown (Castro 1995) that fuzzy controllers have the so called property of *universal approximation*. This means that any bounded continuous function may be approximated by a fuzzy controller with as much accuracy as desired.

3.1. *Using a better representation of the rules*

Fuzzy control begins with the linguistic rules describing the system's behavior, and in that kind of control it is usual to explicitly or implicitly represent the rules by means of either the Min-Implication function $J(a, b) = Min(a, b)$ – Mamdani –, or Prod-Implication function $J(a, b) = Prod(a, b)$ – Larsen –.

By using a new representation of the rules by means of operators $J(a, b) = T(a^r, b)$ then we can adjust each premise properly to obtain a better representation of the given rule. Moreover we can modify each exponent independently until the output cannot be made closer to the expected output (see Guadarrama et al. 2002).

3.2. *Case examples*

To illustrate how these changes in the representation of the rules can improve the result of a Takagi-Sugeno model in two ways, first, reducing the mean square error (MSE) between the target function and the approximating function, and second, obtaining a smoother function, we present the following two examples.

3.2.1. *A simple example*

We chose a non symmetrical function to show how we can reduce the MSE of the approximation and how we can increase the smoothness of the approximating function. Let

$$y = \frac{sin(x)}{x} \quad 0 \le x \le 8.5 \text{ see Figure 11}$$

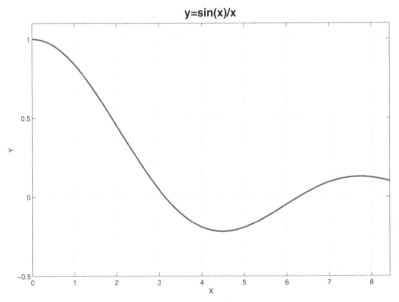

Figure 11. Target function.

Suppose that the predicate Close-to is represented by the fuzzy sets shown in Figure 12:

"*If x is Close-to 0 then y = 1*"

"*If x is Close-to 4/3 then y = $\frac{3}{4}sin(4/3)$*"

"*If x is Close-to 8/3 then y = $\frac{3}{8}sin(8/3)$*"

"*If x is Close-to 4 then y = $\frac{1}{4}sin(4)$*"

"*If x is Close-to 16/3 then y = $\frac{3}{16}sin(16/3)$*"

"*If x is Close-to 20/3 then y = $\frac{3}{20}sin(20/3)$*"

"*If x is Close-to 8 then y = $\frac{1}{8}sin(8)$*"

The Takagi-Sugeno model using the following seven rules can approximate well enough the function as shown in Figure 13 with mean square error $MSE = 0.00070459$.

If we introduce and adjust an exponent for each rule as follows:

Input Labels

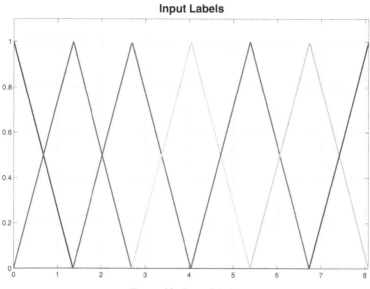

Figure 12. Input labels.

Result of Takagi–Sugeno approximation and Error

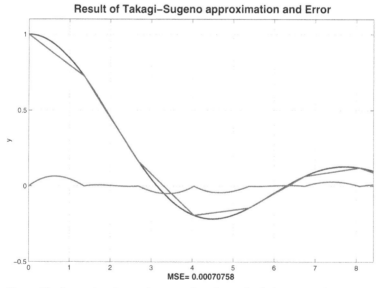

MSE= 0.00070758

Figure 13. Approximation to the target function and relative error when $\alpha = 1$.

"If x is (Close-to 0)$^{(0.9)}$ then $y = 1$"

"If x is (Close-to 4/3)$^{(0.915)}$ then $y = \frac{3}{4}sin(4/3)$"

"If x is (Close-to 8/3)$^{(0.915)}$ then $y = \frac{3}{8}sin(8/3)$"

"If x is (Close-to 4)$^{(0.6)}$ then $y = \frac{1}{4}sin(4)$"

"If x is (Close-to 16/3)$^{(0.8)}$ then $y = \frac{3}{16}sin(16/3)$"
"If x is (Close-to 20/3)$^{(0.45)}$ then $y = \frac{3}{20}sin(20/3)$"
"If x is (Close-to 8)$^{(0.95)}$ then $y = \frac{1}{8}sin(8)$"

we obtain a better approximating function with a considerable reduction of the mean square error $MSE = 0.00014209$, see Figure 14

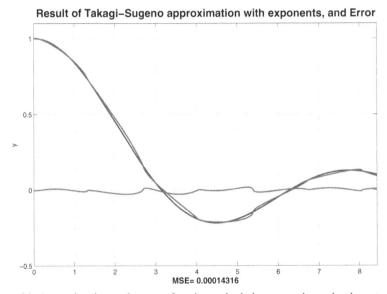

Figure 14. Approximation to the target function and relative error when α has been tuned.

3.2.2. A complex example

Let us now consider a more complex problem, that is, an approximation of the surface:

$$z = \frac{sin(x^2)e^{-x} + sin(y^2)e^{-y} + 0.2338}{0.8567},$$

for $x \in [0, 3]$ and $y \in [0, 3]$ (see Figure 15).

The T-S model for this problem is composed of 49 rules, because seven input labels for each variable are used:

"If x is (Close-to)$^\alpha$ x_1 and y is (Close-to)$^\alpha$ y_1 then $z = q_1$"

$\vdots$ $\qquad$ $\vdots$

"If x is (Close-to)$^\alpha$ x_7 and y is (Close-to)$^\alpha$ y_7 then $z = q_{49}$"

Input $(Close-to)^\alpha-$ labels X and Y are represented by the seven fuzzy sets shown in Figure 16 (with $\alpha = 1$).

$$Z = (\sin(x^2)e^{(-x)} + \sin(y^2)e^{(-y)} + 0.2338) / 0.8567$$

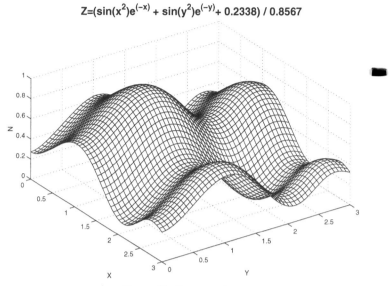

Figure 15. Target surface.

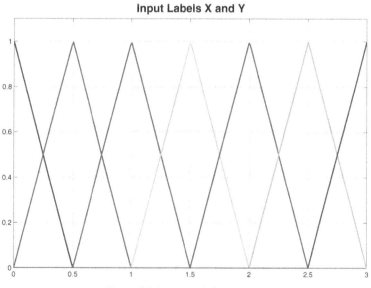

Figure 16. Inputs Labels X and Y.

In the approximation obtained using these rules, the $MSE = 0.20451039$ (see Figure 17).

In the approximation obtained using a Takagi-Sugeno model with exponents in the linguistic terms and using the same exponent for all rules $\alpha =$

Result of T–S approximation

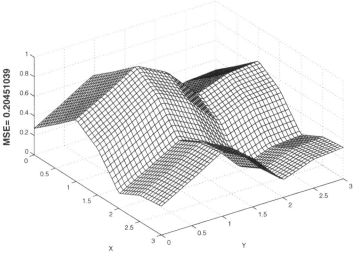

Figure 17. Approximation using $\alpha = 1$

Result of T–S approximation with Exponents

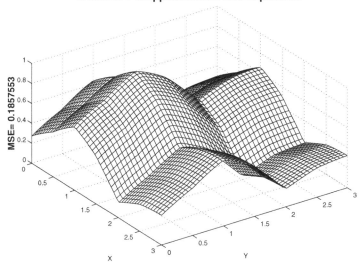

Figure 18. Approximation using $\alpha = 0.92$

0.92, the error was reduced by 10% giving a $MSE = 0.18357553$ moreover the approximated surface is smoother (see Figure 18). An individual fine tuning of the exponents could add further improvements.

Since obviously $max((Close - to)^\alpha) = 1$ for any α, a smoother approximation would require an additional parameter "β" $(0 < \beta \leq 1)$ leading to basis terms of type $\beta(Close - to)^{\alpha}$; where both α and β could be adjusted by using an evolutionary algorithm.

4. Closing Remarks

Fuzzy control is without doubts an important and successful area of Artificial Intelligence. Its very roots go however deep into fuzzy logic, which in part is a special case of multiple-valued Logic, although the inference aspects have been developed more properly in the scope of Approximate Reasoning. The thorough review of its basics shows that there is no "magic" in fuzzy control, but it is mathematically sound. New research areas of Artificial Intelligence, as neural networks and evolutionary algorithms contribute to further developments in the area of fuzzy control, by supporting, e.g., data driven modeling based on neural networks to extract fuzzy rules from numerical examples of behavior, as well as evolutionary tuning of parameters of a fuzzy controller to obtain optimal performance.

Acknowledgements

This work has been partially funded by CICYT (Spain) under project TIC2000-1420. The work of C. Moraga has been supported by the Spanish State Secretary of Education and Universities of the Ministry of Education, Culture and Sports (Grant SAB 2000-0048) and by Social Fund of the European Community.

References

Alsina, C., Trillas, E. & Valverde, L. (1983). On Some Logical Connectives for Fuzzy Set Theory. *Journal of Mathematical Analysis and its Applications* **93**(1): 15–26.

Assilian, S. & Mamdani, E. H. (1974). Learning Control Algorithms in Real Dynamic Systems. In *Proceedings 4th Int. IFAC/IFIP Conference on Digital Computer Application to Process Control*. Zürich.

Baldwin, J. F. (1985). A Model of Fuzzy Reasoning Through Multi-Valued Logic and Set Theory. *International Journal of Man-Machine Studies* **11**(3): 351–380.

Castro, J. L. (1995). Fuzzy Logic Controllers Are Universal Approximators. *IEEE Transactions on System, Man and Cybernetics* **25**(4): 629–635.

Campo, C. del & Trillas, E. (2000). On Mamdani-Larsen s Type Fuzzy Implications. In *Proceeding 8th IPMU, Madrid*, 712–716.

Cross, V. V. & Sudkamp, Th. A. (2002). *Similarity and Compatibility in Fuzzy Set Theory*. Heidelberg: Physica Verlag.

Driankov, D., Hellendoorn, H. & Reinfrank, M. (1996). *An Introduction to Fuzzy Control*. Heidelberg: Springer.

Dubois, D. & Prade, H. (1985). A Review of Fuzzy Set Aggregation Connectives. *Information Sciences* **36**: 85–121.

Dunn, J. M. & Epstein, G. (eds.) (1977). *Modern Uses of Multiple-Valued Logic*. Boston: D. Reidel.

Gottwald, S. (1993). *Fuzzy Sets and Fuzzy Logic*. Braunschweig: Vieweg & Sohn.

Guadarrama, S., Trillas, E., Gutierréz, J. & Fernández, F. (2002). A Step Towards Conceptually Improving Takagi-Sugeno's Approximation. In *Proceedings IPMU'2002 3, 1789–1794*.

Holmblad, L. P. & Østergaard, J. J. (1982). Control of a Cement Kiln by Fuzzy Logic. In Gupta, M. & Sánchez, E. (eds.) *Fuzzy Information and Decision Processes*, 389–399. New York: North Holland.

Klir, G. J. & Yuan, B. (1995). *Fuzzy Sets and Fuzzy Logic: Theory and Applications*. Prentice Hall.

Huertas, J. L., Sánchez S., Barriga A. & Baturone I. (1994). A Hardware Implementation of a Fuzzy Controller Using Analog/Digital Techniques. *International Journal of Computers and Electrical Engineering* **20**(5): 409–419.

Klement, P., Mesiar R., Pap E. (2000). *Triangular Norms*. Dordrecht: Kluwer Academic Publisher.

Lee, C. C. (1990). Fuzzy Logic in Control Systems: Fuzzy Logic Controller. Part I and II. IEEE Trans. *Systems, Man and Cybernetics* **20**(2): 404–418 and **20**(2): 419–435.

Lukasiewicz, J. (1923). On 3-valued Logic. *Ruch Filozoficzny* **5**: 169–171.

Mamdani, E. H. (1974). Application of Fuzzy Algorithms for Control of a Simple Dynamic Plant. *Proceedings IEE* **121**(12): 1585–1588.

Mamdani, E. H. (1976). Advances in the Linguistic Synthesis of Fuzzy Controllers. *International Journal Man-Machine Studies* **8**: 669–678.

Menger, K. (1942). Statistical Metric Spaces. In *Proceedings National Academy of Science USA* **28**.

Moraga, C., Pradera, A. & Trillas, E. (2003). Evolutionary Tunning of Fuzzy If-then Rules at the Level of Operations: A Proposal. *Actas III Congreso Español sobre Metaheurísticas, Algoritmos evolutivos y bioinspirados MAEB'2003*, 530–537.

Proceedings 32nd International IEEE Symposium on Multiple-valued Logic. University of Massachusetts at Boston, May 2002. IEEE-CS-Press.

Nguyen, H. T. & Walker, E. A. (1999). *First Course in Fuzzy Logic*. Bocar Raton, FL: CRC Press.

Pedrycz, W. (1992). *Fuzzy Control and Fuzzy Systems*. Research Studies Publisher.

Pradera, A., Trillas, E. & Cubillo, S. (2000). On Modus Ponens Generating Functions. *International Journal of Uncertainty, Fuzziness and Knowledge Based Systems* **8**(1): 7–19.

Rescher, N. (1969). *Many-valued Logics*. New York: McGraw Hill.

Rine, D. (ed.) (1977). *Multiple-valued Logic and Computer Science*. Amsterdam: North Holland. Second Edition, 1984.

Schweizer, B. & Sklar, A. (1983). *Probabilistic Metric Spaces*. Amsterdam: North-Holland.

Skala, H. J. (1978). On Many-valued Logics, Fuzzy Sets, Fuzzy Logics and Their Applications. *Fuzzy Sets and Systems* **1**(2): 129–149.

Skala, H. J., Termini, S. & Trillas, E. (1984). *Aspects of Vagueness*. Boston: D. Reidel.

Smet, P., Mamdani, A., Dubois, D. & Prade, H. (1988). *Non-Standard Logics for Automated Reasoning*. New York: Academic Press.

Sugeno, M. (1999). Development of an Intelligent Unmanned Helicopter. In Nguyen, H. T. & Prasad, N. R. (eds.) *Fuzzy Modeling and Control*, 13–43. Boca Raton FL: CRC Press LLC.

Sugeno, M. & Murofushi, T. (1991). Helicopter Flight Control Based on Fuzzy Logic. In *Proceedings 1st International Fuzzy Engineering Symposium*, 1120–1121. Yokohama.

Sugeno, M., Guriffin, M. & Bastian, A. (1993). Fuzzy Hierarchical Control of an Unmanned Helicopter. In *Proceeding 5th IFSA Congress*, 179–182. Seoul.

Tanaka, K. (1996). *An Introduction to Fuzzy Logic for Practical Applications*. New York: Springer-Verlag.

Teodorescu, H. N., Yamakawa, T. & Rascanu, A. (1991). *Fuzzy Systems and Artificial Intelligence*. Iasi, Romania: Iasi University Publishing House.

Trillas, E. (1998). Sobre funciones de negación en la teoría de subconjuntos difusos. *Stocastica* **III-1**: 47–59 (in Spanish). Reproduced (English version) in *Advances of Fuzzy Logic* (S. Barro et al. (eds.)), 31–43. Spain: Press of the Universidad de Santiago de Compostela.

Trillas, E. & Alsina, C. (2001). On the Joint Verification of Modus Ponens and Modus Tollens in Fuzzy Logic. In *Proceedings EUSFLAT'01*, 257–260.

Trillas, E., Campo, C. del & Cubillo, S. (2000a). When QM-operators Are Implication Functions and Condicional Fuzzy Relations. *International Journal of Intelligent Systems* **15**: 647–655.

Trillas, E., de Soto, A. R. & Cubillo, S. (2000b). A Glance at Implitation and T-Conditional Functions. In Novák, V. & Perfiliera, I. (eds.), *Discovering the World with Fuzzy Logic*, 126–147. Heidelberg: Physic-Verlag.

Trillas, E. & Guadarrama, S. (2001). On Logical Aspects of Numerical Fuzzy Inference. In *Proceedings IFSA 2001*, 998–1002. IEEE-CS-Press.

Trillas, E. & Pradera, A. (2002). Non-functional Fuzzy Connectives: The Case of Negations. In *Proceedings ESTYLF 2002*, 527–532. Spain: Press of the University of Leon.

Trillas, E. & Valverde, L. (1985a). On Implications and Indistinguishability in the Setting of Fuzzy Logic. In Kacprzyk, J. & Yager, R. R. (eds.) *Management of Decision Support Systems Using Fuzzy Sets and Possibility Theory*, 198–212. Köln: Verlag TUV Rheinland.

Trillas, E. & Valverde, L. (1985b). On Mode and Implication in Aproximate Reasoning. In Gupta, M.M. et al. (eds.) *Approximate Reasoning in Expert Systems*, 157–166. Amsterdam: North Holland.

Watanabe, H., Dettloff, W. & Yount, K. E. (1990). A VLSI Fuzzy Logic Controller with Reconfigurable, Cascadable Architecture. *IEEE Jr. Solid State Circuits* **25**(2): 367–382.

Whalen, T. & Schott, B. (1985). Alternative Logics for Approximate Reasoning in Expert Systems: A Comparative Study. *International Journal of Man-Machine Studies* **22**(3): 327–346.

Yagishita, O., Itoh, O. & Sugeno, M. (1999). Application of Fuzzy Reasoning to the Water Purification Process. In Nguyen, H. T. & Prasad, N. R. (eds.) *Fuzzy Modeling and Control*, 98–104. Boca Raton, FL: CRC Press LLC.

Yamakawa, T. (1988). High Speed Fuzzy Controller Hardware System: The Mega FIPS Machine. *Information Sciences* **45**: 113–128.

Yasunobu, S. et al (1985). Automatic Train Operation System by Predictive Fuzzy Control. In M. Sugeno (ed.) *Industrial Applications of Fuzzy Control*, 1–18. Amsterdam: Elsevier Science.

Zadeh, L. A. (1965). Fuzzy Sets. *Information and Control*.

Zadeh, L. A. (1972). A Rationale for Fuzzy Control. *Journal of Dynamic Systems, Measurements and Control* **34**, 3–4.

Evolutionary Synthesis of Arithmetic Circuit Structures

TAKAFUMI AOKI[1], NAOFUMI HOMMA[1,2] and TATSUO HIGUCHI[3]

[1]*Graduate School of Information Sciences, Tohoku University, Aoba-yama 05, Sendai 980-8579, Japan (E-mail: egg@aoki.ecei.tohoku.ac.jp);* [2]*PRESTO, Japan Science and Technology Corporation;* [3]*Department of Electronics, Tohoku Institute of Technology, Sendai 982-8577, Japan*

Abstract. This paper presents an efficient graph-based evolutionary optimization technique called Evolutionary Graph Generation (EGG), and its application to arithmetic circuit synthesis. Key features of EGG are to employ a graph-based representation of individuals and to manipulate the graph structures directly by evolutionary operations. The potential capability of EGG is demonstrated through experimental synthesis of arithmetic circuits with different levels of abstraction. Design examples include (i) combinational multipliers using word-level arithmetic components (such as parallel counters and parallel shifters), (ii) bit-serial multipliers using bit-level arithmetic components (such as 1-bit full adders and 1-bit registers), and (iii) multiple-valued current-mode arithmetic circuits using transistor-level components (such as current sources and current mirrors).

Keywords: arithmetic circuits, circuit design, evolutionary computation, evolutionary design, genetic algorithms, genetic programming, multiple-valued logic

1. Introduction

Electronic design automation (EDA) has matured as a technology to be universally accepted for producing highly integrated VLSI systems. However, there still remain many difficult design problems to be addressed in the advanced SoC (System on Chip) era. For example, further research and development are needed to establish systematic automated techniques for synthesizing high-level processor architectures, high-performance arithmetic datapath circuits, transistor-level circuits for high-performance logic, analog circuits and analog/digital-mixed circuits, etc. Design of these structures still requires the knowledge and experience of the experts who had trained in a particular way to understand basic characteristics of hardware algorithms and circuit/device technologies.

We have proposed a new approach to designing circuit structures using an evolutionary optimization technique called Evolutionary Graph Generation (EGG) (see Aoki et al. (1999a), Homma et al. (2000a, b, 2001) for earlier discussions on this topic). Key features of the proposed EGG system are to employ general graph structures as individuals and to introduce new evolutionary operations for manipulating the individual graph structures directly without encoding them into other indirect representations, such as bit strings used in Genetic Algorithm (GA) and trees used in Genetic Programming (GP). A major challenge for implementing the EGG system is related to the ability to manipulate complex graph structures efficiently. The proposed EGG system transforms the individual graph structures by exchanging their compatible subgraphs. We carefully designed the data structure for individual graphs so as to make possible fast search of compatible subgraph pairs between specified individuals. Our experimental observation indicates that the subgraph-based *crossover* and *mutation* operations, newly introduced in the EGG system, are useful for many circuit synthesis problems; these operations could naturally acquire useful circuit building blocks during evolution process.

There have been already some approaches to the evolutionary design of hardware structures (Miller et al. 1997; Koza et al. 1997; Lohn and Colombano 1999), and another new field of research has come to be known as *evolvable hardware* for real-world applications (Higuchi et al. 1997, 1999; Huelsbergen et al. 1999; Haddow and Tufte 2000; Tyrrell et al. 2001). Their main motivation is based on the hypothesis that the evolutionary algorithms would inevitably explore a much richer set of possibilities in the design space that are beyond the scope of conventional methods (Thompson et al. 1999). Reference (Miller et al. 1997), for example, presents a GA-based circuit synthesis technique that is capable of evolving 100% functional arithmetic circuits and is closely related to this work. The reported design method, however, is based on direct evolution with gate-level primitive components such as logic gates and flip-flops. Due to its limited capability of modeling arithmetic algorithms, it seems to be difficult to apply the method to the design problems of practical size.

Our approach solves this problem by introducing graph-based abstraction of hardware algorithms, which can cover a wide range of structures appeared in high-level hardware design problems. In stead of creating bit-level circuits directly, the EGG system described in this paper generates arithmetic data-flow graphs that can be transformed into actual bit-level circuit configurations. The proposed graph-based abstraction is applicable to various arithmetic algorithms including those using unconventional number systems, such as redundant number systems and high-radix number systems (Avizienis 1961; Koren 1993; Omondi 1994; Aoki and Higuchi 2000). The graph-based modeling will be effective not only for high-level hardware algorithms but also for low-level circuit structures such as logic gate networks and analog/digital circuits. The EGG system will be also applicable to such design tasks.

In this paper, we demonstrate the potential capability of EGG system through some case studies. Our focus is on the design of arithmetic circuit structures in various different levels of abstraction. Design examples discussed in this paper are (i) combinational constant-coefficient multipliers using word-level arithmetic components such as parallel counters and parallel shifters, (ii) bit-serial constant-coefficient multipliers using bit-level arithmetic components such as 1-bit full adders and 1-bit registers, and (iii) multiple-valued current-mode arithmetic circuits using transistor-level components such as current sources and current mirrors. This paper is organized as follows. Section 2 describes basic idea of EGG (Evolutionary Graph Generation) and its mathematical formulation. Section 3 discusses the application of EGG system to the synthesis of combinational constant-coefficient multipliers. Section 4 describes the synthesis of bit-serial constant coefficient multipliers. Section 5 deals with the synthesis of multiple-valued current-mode arithmetic circuits. We conclude in Section 6.

2. EGG System for Arithmetic Circuit Synthesis

The Evolutionary Graph Generation (EGG) technique can be regarded as a unique variation of evolutionary computation techniques (Back et al. 1997). In general, evolutionary methods mimic the process of natural evolution, the driving process for emergence of complex structures well-adapted to the given environment. The better an individual performs under the conditions the greater is the chance for the individual to live for a longer while and generate offspring. As a result, the individuals are transformed to the suitable forms on the designer's defined constraint. In the EGG system, a graph representing a specific circuit structure is modeled as an individual, and a population of individual graphs is evolved through evolutionary operations. The evolutionary operations are designed to manipulate the graph structures directly by exchanging their subgraphs. This section summarizes the framework of EGG for automated arithmetic circuit synthesis.

2.1. *Basic concept*

The EGG system employs *circuit graphs* (Figure 1) to represent circuit structures. A circuit graph consists of nodes and directed edges. Nodes are of two classes: functional nodes and input/output nodes. Every node has its own name, the function type and input/output terminals. We assume that every directed edge must connect one output terminal (of a node) and one input terminal (of another node), and that each terminal has one edge connection at most. A circuit graph is said to be *complete* if all the terminals have an edge connection. In order to guarantee valid circuit structures, all the circuit graphs used in the EGG system are complete circuit graphs.

More precisely, a circuit graph is defined as follows:

$$G = (N^G, T_O^G, T_I^G, v_O^G, v_I^G, \epsilon^G), \tag{1}$$

where

N^G: the set of nodes,

T_O^G: the set of output terminals,

T_I^G: the set of input terminals,

v_O^G: mapping from T_O^G to N^G;
  $n = v_O^G(u)$ means that the output terminal u ($\in T_O^G$) belongs to the node n ($\in N^G$),

v_I^G: mapping from T_I^G to N^G;
  $n = v_I^G(v)$ means that the input terminal v ($\in T_I^G$) belongs to the node n ($\in N^G$),

ϵ^G: bijection from S_O^G to S_I^G, where $S_O^G \subseteq T_O^G$, $S_I^G \subseteq T_I^G$ and $|S_O^G| = |S_I^G|$; $v = \epsilon^G(u)$ means that the output terminal u ($\in S_O^G$) and the input terminal v ($\in S_I^G$) have a directed edge connection.

Note here that S_O^G (or S_I^G) is the set of output (or input) terminals having edge connections. For example, the circuit graph $G = (N^G, T_O^G, T_I^G, v_O^G, v_I^G, \epsilon^G)$ illustrated in Figure 1 can be defined as follows:

$$N^G = \{n_1, n_2, n_3, n_4, n_5, n_6, n_7, n_8, n_9, n_{10}\}, \tag{2}$$

$$T_O^G = \{u_1, u_2, u_3, u_4, u_5, u_6, u_7, u_8, u_9, u_{10}, u_{11}, u_{12}\}, \tag{3}$$

$$T_I^G = \{v_1, v_2, v_3, v_4, v_5, v_6, v_7, v_8, v_9, v_{10}, v_{11}, v_{12}\}, \tag{4}$$

$$v_O^G = \{ (u_1, n_1), (u_2, n_2), (u_3, n_3), (u_4, n_4), (u_5, n_4), (u_6, n_4),$$
$$(u_7, n_5), (u_8, n_5), (u_9, n_6), (u_{10}, n_7), (u_{11}, n_8), (u_{12}, n_8) \}, \tag{5}$$

$$v_I^G = \{ (v_1, n_4), (v_2, n_5), (v_3, n_5), (v_4, n_6), (v_5, n_6), (v_6, n_6),$$
$$(v_7, n_7), (v_8, n_8), (v_9, n_8), (v_{10}, n_8), (v_{11}, n_9), (v_{12}, n_{10}) \}, \tag{6}$$

$$\epsilon^G = \{ (u_1, v_2), (u_2, v_5), (u_3, v_1), (u_4, v_3), (u_6, v_{10}), (u_7, v_{11}),$$
$$(u_8, v_4), (u_9, v_7), (u_{10}, v_8), (u_{11}, v_6), (u_{12}, v_{12}) \}, \tag{7}$$

where the domain and the range of ϵ^G, i.e., S_O^G (the set of output terminals having edge connections) and S_I^G (the set of input terminals having edge connections), are given by

$$S_O^G = \{u_1, u_2, u_3, u_4, u_6, u_7, u_8, u_9, u_{10}, u_{11}, u_{12}\}, \tag{8}$$

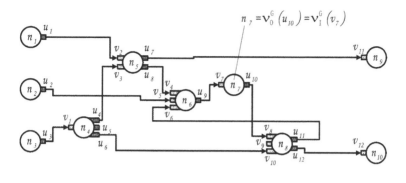

$n_7 = \nu_0^G\left(u_{10}\right) = \nu_I^G\left(v_7\right)$

○ : Node   ▭ : Input terminal

⟶ : Directed edge   ▭ : Output terminal

Figure 1. Circuit graph G.

$$S_I^G = \{v_1, v_2, v_3, v_4, v_5, v_6, v_7, v_8, v_{10}, v_{11}, v_{12}\}. \tag{9}$$

The circuit graph G is said to be complete if and only if $S_O^G = T_O^G$ and $S_I^G = T_I^G$. For example, the graph G of Figure 1 becomes a complete circuit graph if we add the edge from u_5 to v_9. The EGG system is designed to generate only complete circuit graphs during evolution process.

Given a complete circuit graph G defined by Equation (1), its *sub-circuit graph* G' induced by the node subset $N^{G'}(\subseteq N^G)$ is defined as

$$G' = (N^{G'}, T_O^{G'}, T_I^{G'}, \nu_O^{G'}, \nu_I^{G'}, \epsilon^{G'}), \tag{10}$$

where the four items $T_O^{G'} \sim \nu_I^{G'}$ are defined as

$$T_O^{G'} = \{u \in T_O^G | \nu_O^G(u) \in N^{G'}\}, \tag{11}$$

$$T_I^{G'} = \{v \in T_I^G | \nu_I^G(v) \in N^{G'}\}, \tag{12}$$

$$\nu_O^{G'} = \nu_O^G \backslash T_O^{G'}, \tag{13}$$

$$\nu_I^{G'} = \nu_I^G \backslash T_I^{G'}. \tag{14}$$

In the above expressions, $f \backslash A$ (where f is a function and A is a set) denotes the same function as f, but with its domain restricted to A. The fifth item in (10), the bijection $\epsilon^{G'}$, which defines the directed edges in G', has the domain $S_O^{G'}$ and the range $S_I^{G'}$ given below:

$$S_O^{G'} = \{u \in T_O^{G'} | \nu_I^G(\epsilon^G(u)) \in N^{G'}\}, \tag{15}$$

$$S_I^{G'} = \{v \in T_I^{G'} | \nu_O^G(\epsilon^{G^{-1}}(v)) \in N^{G'}\}, \tag{16}$$

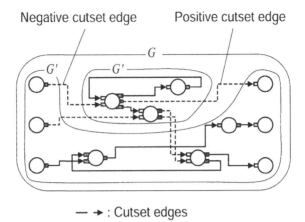

Negative cutset edge   Positive cutset edge

— ➔ : Cutset edges

Figure 2. Partitioning the graph G into the sub-circuit graph G' and its complement $\overline{G'}$.

where $\epsilon^{G^{-1}}$ is the inverse mapping of ϵ^G. Thus, we have

$$\epsilon^{G'} = \epsilon^G \backslash S_O^{G'}. \tag{17}$$

Given a sub-circuit graph G' of a circuit graph G, its complement denoted by

$$\overline{G'} = (N^{\overline{G'}}, T_O^{\overline{G'}}, T_I^{\overline{G'}}, v_O^{\overline{G'}}, v_I^{\overline{G'}}, \epsilon^{\overline{G'}}) \tag{18}$$

is defined as the sub-circuit graph of G induced by $N^{\overline{G'}}(= N^G - N^{G'})$. Figure 2 illustrates the partition of the complete circuit graph G into the sub-circuit graph G' and its complement $\overline{G'}$. The set of the directed edges of G that do not belong to either G' or $\overline{G'}$ is called the *cutset* with respect to the partition of G into G' and $\overline{G'}$. Every edge in the cutset has either *positive* or *negative* direction with respect to the sub-circuit graph G', where "positive" implies the direction from the nodes in G' to the nodes in $\overline{G'}$ and "negative" implies the reverse direction. The positive edges and negative edges in the cutset are defined by the bijections:

$$\epsilon^{pc} = \epsilon^G \backslash (T_O^{G'} - S_O^{G'}), \tag{19}$$

$$\epsilon^{nc} = \epsilon^G \backslash (T_O^{\overline{G'}} - S_O^{\overline{G'}}), \tag{20}$$

respectively.

2.2. *EGG system flow*

Figure 3 shows the overall procedure of the EGG system. At first, the system generates embryonic circuit graphs randomly to form the initial population

```
program EGG_System_Flow
  begin
    t := 0;   { t: number of generations }
    initialize(P(t));   { P(t): population }
    evaluate(P(t));
    while t ≤ Max. num. of generations do
      begin
        C(t) := crossover(P(t));   { C(t): offsprings generated by crossover }
        M(t) := mutation(P(t));   { M(t): offsprings generated by mutation}
        evaluate(C(t) ∪ M(t));
        P(t + 1) := select(C(t) ∪ M(t) ∪ P(t));
        t := t + 1
      end
  end.
```

Figure 3. EGG system flow.

$P(0)$. At this point, we try to reduce the search space of graph structures by restricting the class of generated graph structures according to the target application. For example, if the target of synthesis is in the class of combinational circuits, graphs to be generated are restricted to directed acyclic graphs. The next step is to evaluate fitness of the circuit graphs in the population $P(t)$. The evaluation is defined as a combination of functional validity and performance. After the evaluation, the system selects a set of circuit graphs having higher scores to perform evolutionary operations: *crossover* and *mutation*. The probability of selecting an individual for crossover and mutation is proportional to its fitness value. The offsprings generated by these evolutionary operations form the populations $C(t)$ and $M(t)$, where $C(t)$ and $M(t)$ are obtained by crossover and mutation operations, respectively. The individuals for the next generation $P(t + 1)$ are selected from the current population $C(t) \cup M(t) \cup P(t)$.

The crossover, illustrated in Figure 4 (a), recombines two parent graphs into two new graphs. When a pair of parent graphs G_{p1} and G_{p2} is selected from the population, the crossover determines a pair of subgraphs G_{p1}' and G_{p2}' to be exchanged between the parents, and generates offsprings G_{c1} and G_{c2} by replacing the subgraph of one parent by that of the other parent. In this process, the system selects a subgraph G_{p1}' randomly from the mother circuit graph G_{p1}, and selects a *compatible* subgraph G_{p2}' from the father circuit graph G_{p2}, where "compatible" means that the cutsets for these subgraphs contain the same number of edges for both negative and positive directions. Note here that "positive" implies the direction from the nodes in G_{p1}' (or G_{p2}') to the nodes in the complement of G_{p2}' (or G_{p1}') and "negative" implies the

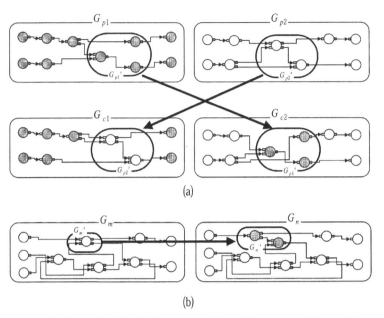

Figure 4. Examples of evolutionary operations: (a) crossover, (b) mutation.

reverse direction. The compatibility condition ensures the completeness of the generated offsprings.

The offspring G_{c1} generated by the crossover operation is formally represented as

$$G_{c1} = (N^{G_{c1}}, T_O^{G_{c1}}, T_I^{G_{c1}}, \nu_O^{G_{c1}}, \nu_I^{G_{c1}}, \epsilon^{G_{c1}}), \tag{21}$$

where $N^{G_{c1}}, T_O^{G_{c1}}, T_I^{G_{c1}}, \nu_O^{G_{c1}}, \nu_I^{G_{c1}}$ and $\epsilon^{G_{c1}}$ are defined as

$$N^{G_{c1}} = N^{\overline{G'_{p1}}} \cup N^{G'_{p2}}, \tag{22}$$

$$T_O^{G_{c1}} = T_O^{\overline{G'_{p1}}} \cup T_O^{G'_{p2}}, \tag{23}$$

$$T_I^{G_{c1}} = T_I^{\overline{G'_{p1}}} \cup T_I^{G'_{p2}}. \tag{24}$$

$$\nu_O^{G_{c1}} = (\nu_O^{G_{p1}} \backslash T_O^{\overline{G'_{p1}}}) \cup (\nu_O^{G_{p2}} \backslash T_O^{G'_{p2}}), \tag{25}$$

$$\nu_I^{G_{c1}} = (\nu_I^{G_{p1}} \backslash T_I^{\overline{G'_{p1}}}) \cup (\nu_I^{G_{p2}} \backslash T_I^{G'_{p2}}), \tag{26}$$

$$\epsilon^{G_{c1}} = \epsilon^{\overline{G'_{p1}}} \cup \epsilon^{G'_{p2}} \cup \epsilon^{pc} \cup \epsilon^{nc}, \tag{27}$$

respectively. In Equation (27), ϵ^{pc} denotes a bijection from $(T_O^{G'_{p2}} - S_O^{G'_{p2}})$ to $(T_I^{\overline{G'_{p1}}} - S_I^{\overline{G'_{p1}}})$, which corresponds to positive-cutset edges from G'_{p2} to $\overline{G'_{p1}}$.

On the other hand, ϵ^{nc} denotes a bijection from $(T_O^{\overline{G'_{p1}}} - S_O^{\overline{G'_{p1}}})$ to $(T_I^{G'_{p2}} - S_I^{G'_{p2}})$, which corresponds to negative-cutset edges from $\overline{G'_{p1}}$ to G'_{p2}. In the crossover operation, the system adds the cutset edges ϵ^{pc} and ϵ^{nc} randomly to connect $\overline{G'_{p1}}$ and G'_{p2}, which creates a new offspring G_{c1}. The other offspring G_{c2} can be defined in a similar way. Note that the crossover operation requires the compatibility of sub-circuit graphs G'_{p1} and G'_{p2}. This condition is written as

$$\left| T_O^{G'_{p2}} - S_O^{G'_{p2}} \right| = \left| T_O^{G'_{p1}} - S_O^{G'_{p1}} \right|,$$

$$\left| T_I^{G'_{p2}} - S_I^{G'_{p2}} \right| = \left| T_I^{G'_{p1}} - S_I^{G'_{p1}} \right|. \tag{28}$$

This ensures that the offsprings G_{c1} and G_{c2} are complete circuit graphs if the parents G_{p1} and G_{p2} are complete circuit graphs. For efficient implementation of the crossover operation, the data structure for individual circuit graphs should be carefully designed to allow fast checking of the compatibility condition (28).

The mutation operation illustrated in Figure 4 (b), on the other hand, partially reconstructs the given circuit graph. When a parent graph G_m is selected from the population, the mutation operation determines a sub-circuit graph G_m' randomly, and generates a new offspring G_n by replacing G_m' with a randomly generated sub-circuit graph G_n'. In this process, G_n' must be compatible with the original sub-circuit graph as shown in Figure 4 (b). The system generates G_n' within the range of a specific number of nodes. (The generation process is also employed for generating embryonic circuit graphs.)

The offspring G_n generated by the mutation operation is formally represented as

$$G_n = (N^{G_n}, T_O^{G_n}, T_I^{G_n}, v_O^{G_n}.v_I^{G_n}, \epsilon^{G_n}), \tag{29}$$

where $N^{G_n}, T_O^{G_n}, T_I^{G_n}, v_O^{G_n}, v_I^{G_n}$ and ϵ^{G_n} are defined as

$$N^{G_n} = N^{\overline{G_m}} \cup N^{G_n'}, \tag{30}$$

$$T_O^{G_n} = T_O^{\overline{G_m}} \cup T_O^{G_n'}, \tag{31}$$

$$T_I^{G_n} = T_I^{\overline{G_m}} \cup T_I^{G_n'}, \tag{32}$$

$$v_O^{G_n} = (v_O^{G_m} \backslash T_O^{\overline{G_m}}) \cup v_O^{G_n'}, \tag{33}$$

$$v_I^{G_n} = (v_I^{G_m} \backslash T_I^{\overline{G_m}}) \cup v_I^{G_n'}, \tag{34}$$

$$\epsilon^{G_n} = \epsilon^{\overline{G_m}} \cup \epsilon^{G_n'} \cup \epsilon^{pc}, \cup \epsilon^{nc} \tag{35}$$

respectively. In Equation (35), ϵ^{pc} denotes a bijection from $(T_O^{G'_n} - S_O^{G'_n})$ to $(T_I^{\overline{G'_m}} - S_I^{\overline{G'_m}})$, which corresponds to positive-cutset edges from G'_n to $\overline{G'_m}$. On the other hand, ϵ^{nc} denotes a bijection from $(T_O^{\overline{G'_m}} - S_O^{\overline{G'_m}})$ to $(T_I^{G'_n} - S_I^{G'_n})$, which corresponds to negative-cutset edges from $\overline{G'_m}$ to G'_n. In the mutation operation, the system adds the cutset edges ϵ^{pc} and ϵ^{nc} randomly to connect $\overline{G'_m}$ and G'_n, which creates a new offspring G_n. Note that the randomly generated sub-circuit graph G'_n is satisfied with the following compatibility condition:

$$
\begin{aligned}
\left| T_O^{G'_m} - S_O^{G'_m} \right| &= \left| T_O^{G'_n} - S_O^{G'_n} \right|, \\
\left| T_I^{G'_m} - S_I^{G'_m} \right| &= \left| T_I^{G'_n} - S_I^{G'_n} \right|.
\end{aligned}
\tag{36}
$$

This ensures that the offsprings G_n is a complete circuit graph if the parent G_m is a complete circuit graph.

An important characteristic of the two evolutionary operations is that both operations are designed to reconfigure circuit graphs by modifying their sub-circuit graphs. The underlying hypothesis here can be stated as follows: every functional circuit is a collection of many useful sub-circuits operating as basic building blocks, and the total functionality of the circuit emerges from a collective behavior of its sub-circuits. In other words, we consider the sub-circuits as basic units of circuit construction.

3. Synthesis of Combinational Constant-coefficient Multipliers

3.1. *Motivation and method of experiment*

In the following, we demonstrate the potential capability of EGG system through some case studies. Our approach solves various design problem by introducing graph-based abstraction of hardware algorithms, which can cover a wide range of structures appeared in high-level hardware design problems. In stead of creating bit-level circuits directly, the EGG system described in this section generates arithmetic data-flow graphs that can be transformed into actual bit-level circuit configurations. The proposed graph-based abstraction is applicable to various arithmetic algorithms including those using unconventional number systems, such as redundant number systems and high-radix number systems.

This section addresses the problem of synthesizing combinational circuits for multiplication in the form: $Y = KX$, where K is an integer coefficient, and X and Y are the integer input and output. Main reasons for choosing the constant-coefficient multiplier as a target function are as follows: (i) there are many possible choices for multiplier structure for a specific coefficient K, and (ii) the complexity of the multiplier structure significantly varies with the coefficient value K. The goal of this experiment is to demonstrate that the EGG system can generate efficient multiplier structures whose performance and complexity are comparable with those designed by experienced designers.

One of the most important techniques in constant-coefficient multiplier design is to encode the target coefficient K by Canonic Signed-Digit (CSD) number representation (Hwang 1979). The CSD number representation is defined as a specific binary Signed-Digit (SD) number representation that contains the least number of non-zero digits. This encoding technique makes possible to reduce the number of partial products, which is equal to the number of non-zero digits. The CSD encoding combined with the fast partial product accumulation technique using parallel counter trees is widely used in practical DSP applications, such as high-frequency FIR filter architectures (Wong and Samueli 1991; Khoo et al. 1996).

As for compact counter tree design for partial product accumulation, the authors' group has recently proposed the Signed-Weight (SW) arithmetic (Aoki et al. 1999b). The use of SW arithmetic instead of conventional two's complement arithmetic makes possible the construction of compact counter trees without using irregular arithmetic operations, such as sign extension and two's complementation. As a result, the combination of the CSD encoding technique with counter tree architectures consisting of the least number of SW counter stages (e.g., Wallace trees (Wallace 1964)) seems to provide the best possible approach to the practical hardware implementation of fast constant-coefficient multipliers at present. Thus, we will compare the multipliers generated by the EGG system with the multipliers designed by hand employing the knowledge of the above techniques (CSD encoding plus SW Wallace tree).

Figure 5 shows the seven types of nodes used in this experiment. The input and output of these nodes are vectors of binary digits corresponding to specific SW operand encoding. The EGG system interprets a circuit graph consisting of these seven types of nodes as a data-flow graph representing arithmetic computation process (see Figure 6(a)). A directed edge in the data-flow graph represents the dependence of operands. Also, two attributes are assigned to each edge: (i) the definition of SW number system used for operand encoding and (ii) the activated operand digits. Each node, on the other hand, represents a word-level arithmetic component such as parallel counters and parallel shifters, and has a rule for generating the corresponding bit-level circuit interpretation. Thus, the node itself has no circuit details at first. It can be transformed into a set of bit-level circuit elements only when

Name	Symbol	Function	Mathematical representation
Operand inputs	X (IN)→Y	Input signal	$Y = X$
	X (IN)→Y	Inverse input signal	$Y = -X$
Operand output	X →(OUT) Y	Output signal	$Y = X$
3-2 counter	X_1 X_2 →3-2→ λS X_3 →λC	3-input 2-output carry-free addition with output branches	$\lambda C + \lambda S = X_1 + X_2 + X_3$ $\lambda \in \{-1, 1\}$
Final-stage adder	X_1 X_2 →FSA→Y	Carry-propagate addition with a bias canceling stage	$Y = X_1 + X_2$
1-bit shifter	X→(1-S)→Y	1-bit arithmetic shifter	$Y = 2X$
2-bit shifter	X→(2-S)→Y	2-bit arithmetic shifter	$Y = 4X$
4-bit shifter	X→(4-S)→Y	4-bit arithmetic shifter	$Y = 16X$

Figure 5. Functional nodes.

the attributes of all the input operands are determined. Therefore, the actual interpretation of a node depends on the overall structure of the data-flow graph.

In this example, the EGG system is designed to generate only complete circuit graphs without any feedback loops, called Complete Circuit Trees (CCTs), in the evolution process since the target circuit to be synthesized belongs to the class of combinational circuits. Figure 6(a) gives an example of CCT considered in this experiment. Assuming 4-bit two's complement data to be given for the input node, the graph is translated into the bit-level circuit configuration as shown in Figure 6(b). In general, given a CCT, its function is represented as $Y = \hat{K}X + f(X)$, where $\hat{K}$ is an integer constant and $f(X)$ is a nonlinear function of X which is represented by intermediate variables. When $f(X) = 0$, the circuit graph can be recognized as a specific constant-coefficient multiplier. In this experiment, the use of the graph theoretic constraint of CCTs reduces the number of possible graph structures by about 10^{-17} times.

The generated structures are evaluated by a combination of two different evaluation functions, *functionality* and *performance*. The functionality measure F evaluates the validity of the logical function compared with the target function. The performance measure P, on the other hand, is assumed to be inversely proportional to the product of circuit delay D and number of inter-module interconnections A (Brent and Kung 1981). First, we describe the functionality measure F in detail. Let K be the target coefficient given by the following representation:

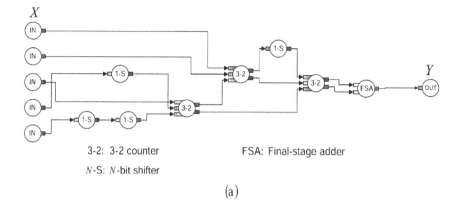

3-2: 3-2 counter   FSA: Final-stage adder

N-S: N-bit shifter

(a)

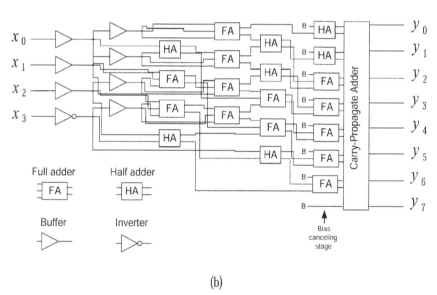

(b)

Figure 6. Circuit graph in the experiment: (a) an example, (b) actual circuit interpretation of the graph (a).

$$K = k_0 2^0 + k_1 2^1 + k_2 2^2 + \cdots = \sum_{j=0}^{||K||-1} k_j 2^j, \qquad (37)$$

where $||K||$ denotes the string length of the coefficient K, and $k_j \in \{-1, 0, 1\}$. The system checks the function of a circuit graph by symbolic verification (Homma et al. 2000a) and obtains the estimated coefficient $\hat{K}$. Note that the function of a CCT can be represented by $Y = \hat{K} X + f(X)$. Let assume that $\hat{K}$ is written as

$$\hat{K} = \hat{k}_0 2^0 + \hat{k}_1 2^1 + \hat{k}_2 2^2 + \cdots = \sum_{j=0}^{||\hat{K}||-1} \hat{k}_j 2^j. \qquad (38)$$

The similarity between K and $\hat{K}$ are evaluated by digit-coincidences for all the digit positions of the given two strings. Using the difference of the string lengths $d \ (= ||\hat{K}|| - ||K||)$, the correlation M_s of the two coefficient strings at the shift amount s $(0 \leq s \leq |d|)$ is defined by

$$M_s = \begin{cases} \dfrac{1}{||\hat{K}||} \displaystyle\sum_{j=0}^{||\hat{K}||-1} \delta(\hat{k}_j - k_{j-s}) & d \geq 0 \\[2ex] \dfrac{1}{||K||} \displaystyle\sum_{j=0}^{||K||-1} \delta(\hat{k}_{j-s} - k_j) & d < 0, \end{cases}$$

where $\delta(x)$ is defined as

$$\delta(x) = \begin{cases} 1 & x = 0 \\ 0 & x \neq 0. \end{cases}$$

In the above calculation, we assume that the values of the undefined digit positions make $\delta(x)$ 0 for both coefficient strings. Using this correlation function, the functionality measure F is defined as

$$F = \max_{0 \leq s \leq |d|} [100 M_s - C_1 s], \qquad (39)$$

where $C_1 = 2$ in this section.

On the other hand, the performance measure P is defined as

$$P = \frac{C_2}{DA}, \qquad (40)$$

where D is the number of counter stages, and A is the total number of inter-module interconnections in the translated bit-level circuit. We use $F + P$ as a total fitness function, where the ratio P_{max}/F_{max} is adjusted about 5/100 by tuning the constant C_2.

3.2. Experimental result

In this experiment, we assume that the population size is 1000, the number of generations is 1000, the maximum number of nodes is 50, the crossover rate is 0.7 and the mutation rate is 0.1. The parameter values are optimized through a set of experiments. The EGG system with the above parameter values are used to generate 50 distinct multipliers whose coefficients are selected

Table 1. DA product of multipliers: (a) the CSD multipliers, (b) the multipliers generated by EGG system

Index	Coefficient (Wordlength)	DA (a)	(b)	Index	Coefficient (Wordlength)	DA (a)	(b)
K_1	147(8)	756	747	K_{26}	−2429(12)	1312	1312
K_2	−166(8)	744	729	K_{27}	1796(12)	420	418
K_3	159(8)	396	396	K_{28}	−1477(12)	1312	1292
K_4	−169(8)	756	744	K_{29}	2118(12)	810	795
K_5	98(8)	388	376	K_{30}	−1759(12)	822	810
K_6	−135(8)	396	396	K_{31}	7574(14)	1552	1516
K_7	87(8)	756	723	K_{32}	−9293(14)	1600	1568
K_8	−117(8)	756	744	K_{33}	6068(14)	1320	1292
K_9	109(8)	756	744	K_{34}	−9500(14)	1304	1300
K_{10}	−124(8)	142	141	K_{35}	9152(14)	408	406
K_{11}	474(10)	774	759	K_{36}	−8180(14)	440	438
K_{12}	−405(10)	1256	1256	K_{37}	6362(14)	1596	1536
K_{13}	642(10)	408	406	K_{38}	−10346(14)	1604	1604
K_{14}	−448(10)	136	135	K_{39}	10075(14)	1824	1600
K_{15}	343(10)	1248	1220	K_{40}	−6635(14)	1612	1556
K_{16}	−526(10)	408	406	K_{41}	26204(16)	2275	2080
K_{17}	392(10)	392	380	K_{42}	−23649(16)	1652	1604
K_{18}	−669(10)	1256	1256	K_{43}	34000(16)	1304	1300
K_{19}	520(10)	148	147	K_{44}	−23336(16)	1604	1604
K_{20}	−559(10)	786	768	K_{45}	38009(16)	1632	1632
K_{21}	1895(12)	1288	1288	K_{46}	−37507(16)	1656	1648
K_{22}	−2314(12)	822	807	K_{47}	30187(16)	1640	1632
K_{23}	2679(12)	1304	1304	K_{48}	−39752(16)	1612	1596
K_{24}	−1985(12)	436	436	K_{49}	39767(16)	2345	2300
K_{25}	2261(12)	1528	1504	K_{50}	−27846(16)	2330	2280

randomly. The results of the 50 evolutionary runs are listed in Table 1. The target coefficients are randomly selected in the following ranges: $K_1 \sim K_{10}$: $-256 \sim 255$, $K_{11} \sim K_{20}$: $-1024 \sim 1023$, $K_{21} \sim K_{30}$: $-4096 \sim 4095$, $K_{31} \sim K_{40}$: $-16384 \sim 16383$, $K_{41} \sim K_{50}$: $-65536 \sim 65535$. Thus, the wordlength of randomly selected coefficients is increased from 8 digits to 16 digits at every 10 coefficient number. We have performed a single evolutionary run for every coefficient.

In Table 1, we compare the DA product of the corresponding 16-bit CSD multipliers using the optimal Wallace tree architecture (*a*) and that of the

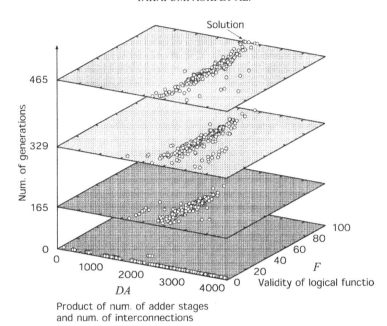

Figure 7. Example of evolution process for the case of $K_{41} = 26204$.

16-bit multipliers generated by the EGG system (b). Note here that smaller DA product means higher performance. From this table, we can confirm that all the generated solutions exhibit same or higher performance compared with the corresponding optimal CSD multipliers. The success rate of a set of evolutionary trials is 100% in this experiment.

Figure 7 shows some snapshots of evolution process for the case of $K_{41} = 26204_{10}$. The vertical axis indicates the number of generations, and the horizontal axes indicate the functionality measure F and the DA complexity (the number of counter stages × the number of inter-module interconnections). Given the initial random population, the evolution is mainly driven towards better functionality. In the evolution process, each individual shows a tendency to keep a specific level of DA product corresponding to the given coefficient. The first individual achieving 100% functionality appears in the 329th generation with the DA product of 2345. This multiplier has lower performance than the optimal CSD multiplier ($DA = 2275$). In the 465th generation, we obtain a better solution, where DA complexity is reduced to 2080.

Figure 8 (a) is the best solution generated by the EGG system for the target coefficient $K_{39} = 10075$. Figure 8 (b), on the other hand, shows the conventional CSD multiplier using Wallace tree architecture consisting of the least stages of 3-2 counters. We can observe that the solution (a) obtained by

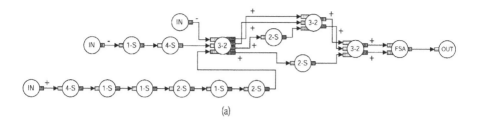

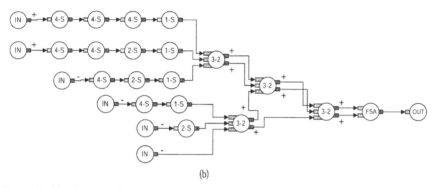

(b)

Figure 8. Circuit graphs for the multiplier of $K_{39} = 10075$: (a) the best solution generated by EGG system, (b) the CSD multiplier using Wallace tree architecture.

the EGG system employs a 3-2 counter with two-way carry-save branches at the first stage, and this hardware resource is shared by the successive stages. Compared with the structure of Figure 8 (b), this feature significantly reduces the complexity of the corresponding bit-level circuit configurations. Figure 9 shows the actual circuit interpretations corresponding to the circuit graphs of Figure 8 with 16-bit two's complement data. Recently, some papers have presented the method of reducing the complexity of the constant-coefficient multipliers by inserting such branches at adders (Hartley 1996; Dempster and Macleod 1994). Note here that the EGG system can derive the optimal structure without using the knowledge of such techniques.

4. Synthesis of Bit-Serial Constant-Coefficient Multipliers

4.1. *Motivation and method of experiment*

In this section, we demonstrate that the graph-based modeling can be effective not only for high-level hardware algorithms but also for low-level circuit structures such as logic gate networks. The target function considered here is bit-serial constant-coefficient multiplication which is frequently appeared

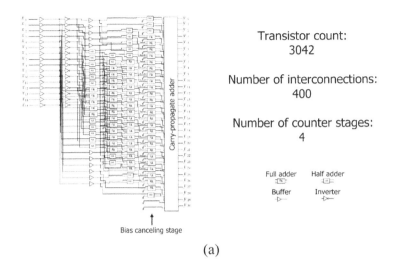

Transistor count:
3042

Number of interconnections:
400

Number of counter stages:
4

Full adder   Half adder

Buffer   Inverter

(a)

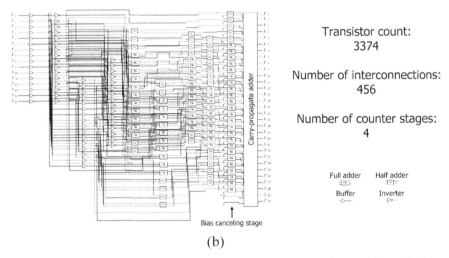

Transistor count:
3374

Number of interconnections:
456

Number of counter stages:
4

Full adder   Half adder

Buffer   Inverter

(b)

Figure 9. Multiplier configurations corresponding to the graphs of Figure 8: (a) the best solution generated by EGG system, (b) the CSD multiplier using Wallace tree architecture.

in signal processing applications. The target function $Y = KX$ is the same as that used in the previous section. We first summarize some techniques for designing the bit-serial constant-coefficient multipliers using bit-level arithmetic components. As is the case with the combinational constant-coefficient multipliers, one of the important techniques is to encode the target coefficient by the CSD number representation. As a result, the known best possible

Name	Symbol	Schematic	Mathematical representation
Operand inputs	X (IN) $\rightarrow Y$	X —— Y	$Y = X$
	X (IN) $\rightarrow Y$	X —▷∘— Y	$Y = -X$
Operand output	$X \rightarrow$ (OUI) Y	X —— Y	$Y = X$
Full adder	$X_1 \rightarrow$ FA $\rightarrow S$ $X_2 \rightarrow$ $\rightarrow C$ $X_3 \rightarrow$	X_1 FA S X_2 C X_3	$2C + S = X_1 + X_2 + X_3$
Final-stage adder	$X \rightarrow$ (FSA) $\rightarrow Y$	X — FA S C $-Y$ Bias — R	$Y = X$
1-bit registers	$X \rightarrow$ (1-R) $\rightarrow Y$	X -R- Y	$Y = 2X$
	$X \rightarrow$ (2-R) $\rightarrow Y$	X -R-R- Y	$Y = 4X$
	$X \rightarrow$ (4-R) $\rightarrow Y$	X -R-R-R-R- Y	$Y = 16X$

Figure 10. Functional nodes.

approach to the design of bit-serial constant-coefficient multipliers is to use a one-dimensional systolic array of m processing elements (i.e., full adders), where m is the number of non-zero digits in the given CSD coefficients. Also, the bias canceling technique of SW arithmetic is useful for reducing circuit complexity. In this experiment, the EGG system synthesizes the multiplier structure without using the knowledge of the above techniques.

Figure 10 shows the five types of nodes in this experiment. Each node represents a bit-level arithmetic component such as 1-bit full adders and 1-bit registers. The EGG system is designed to generate complete circuit graphs with feedback loops in the evolution process since the target circuit to be synthesized belongs to the class of sequential circuits. Thus, the variation of possible graph structures is larger than the previous example (which assumes CCTs: Complete Circuit Trees).

As described in the previous section, the generated structures are evaluated by a combination of *functionality* and *performance*. Both measures are defined similarly to the previous section. First, we describe the functionality measure F in detail. In the following discussion, we assume the use of LSB-first bit-serial arithmetic based on two's complement binary number system, where the first bit has the weight 2^0, the second has 2^1, the third has 2^2, and so on. The functionality measure F is defined similarly to (39) with few modification, and is given by

$$F = \max_{0 \le s \le |d|} [100 M_s - C_1 s] - C_2 q, \qquad (41)$$

where a new term $-C_2 q$ is added to the functionality measure F. In this equation, q is the number of delay-free loops in the evolved circuit. We set the constants as $C_1 = 10$ and $C_2 = 5$ in this experiment. On the other hand, the performance measure P is defined as

$$P = \frac{C_3}{DA}, \tag{42}$$

where D is the maximum register-to-register delay estimated with an inverter gate as a unit delay element, and A is the total number of transistors. We use $F + P$ as a total fitness function, where the ratio P_{max}/F_{max} is adjusted about 5/100 by tuning the constant C_3.

4.2. *Experimental result*

Table 2 shows the result of a set of evolutionary runs, in which the EGG system generates bit-serial multipliers with 8-bit integer coefficients. In this experiment, we assume that the population size is 300, the number of generations is 10000, the maximum number of nodes is 100, the crossover rate is 0.7, and the mutation rate is 0.1.

Total 50 coefficients are selected randomly in the range of -128 to 127. We have performed 10 distinct evolutionary runs for each coefficient. As an example, Figure 11 shows the transition of the best individual fitness for the case of $Y = -105X$. We can see the staircase improvements of the best individual fitness for every trial. We have obtained the solutions achieving 100% functionality with a 100% success rate in 10 trials. The calculation time for each evolutionary run is about 15 minutes on a Linux PC with 1 GHz Pentium III and 1 GB memory. Percent distribution of the calculation time is approximately as follows: selection-37%, evolutionary operations-30%, evaluation-28%, others-5%.

In Table 2, we show the DA product (the maximum register-to-register delay × the total number of transistors) of the corresponding CSD multipliers (a) and that of the multipliers generated by the EGG system (b). From this table, we can confirm that the EGG system can generate efficient multiplier structures whose performances are comparable or sometimes superior to those of systolic CSD multipliers. Figure 12 shows an example of the solution for the case of $Y = 103X$. Note here that the EGG system can synthesize efficient multiplier structure without using the knowledge of arithmetic algorithms. The EGG system can also generate the structural Hardware Description Language (HDL) code for the evolved multipliers, and this may provide a good initial description for state-of-the-art logic synthesis tools in some applications.

Table 2. *DA* product of multipliers: (a) the CSD multipliers, (b) the multipliers generated by the EGG system

Index	Coefficient (Wordlength)	DA (a)	(b)	Index	Coefficient (Wordlength)	DA (a)	(b)
K_1	−123	1088	1088	K_{26}	−4	180	180
K_2	−115	1368	1352	K_{27}	13	968	912
K_3	−111	1088	1088	K_{28}	26	630	476
K_4	−105	1557	1530	K_{29}	27	1152	1152
K_5	−101	1557	1530	K_{30}	31	837	837
K_6	−91	1352	1328	K_{31}	35	1134	992
K_7	−87	1352	1328	K_{32}	43	1512	1256
K_8	−82	675	675	K_{33}	50	655	496
K_9	−79	1064	1064	K_{34}	52	655	496
K_{10}	−78	665	552	K_{35}	55	1197	1197
K_{11}	−77	1512	1328	K_{36}	58	655	544
K_{12}	−58	665	524	K_{37}	66	384	384
K_{13}	−54	524	524	K_{38}	70	655	516
K_{14}	−47	1048	1024	K_{39}	73	1032	1032
K_{15}	−44	524	524	K_{40}	77	1312	1296
K_{16}	−40	475	400	K_{41}	84	516	516
K_{17}	−38	640	532	K_{42}	99	1539	1296
K_{18}	−35	1024	1024	K_{43}	103	1539	1476
K_{19}	−34	475	475	K_{44}	104	680	516
K_{20}	−29	1152	1008	K_{45}	105	1352	1296
K_{21}	−26	640	504	K_{46}	108	690	592
K_{22}	−25	1152	1125	K_{47}	114	680	564
K_{23}	−23	1008	984	K_{48}	120	515	432
K_{24}	−13	1107	968	K_{49}	123	1242	1242
K_{25}	−10	425	425	K_{50}	127	927	927

The EGG system can be applied not only to the synthesis of constant-coefficient multipliers but also to other design problems by changing the target function. For example, we can easily synthesize the constant-coefficient multiply-adder such as $Y = 58X_1 + 43X_2$. Figure 13 shows the best solution obtained in the 10000th generation. Our observation suggests the potential capability of EGG for handling generic synthesis tasks for bit-serial arithmetic.

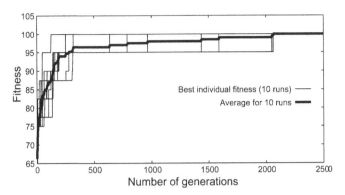

Figure 11. Transition of the best individual fitness for the case of $Y = -105X$.

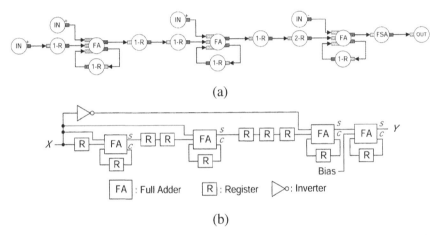

(b)

Figure 12. Example solution of bit-serial constant-coefficient multiplier ($Y = 103X$): (a) the circuit graph generated by the EGG system, (b) the multiplier structure corresponding to graph (a).

5. Synthesis of Current-Mode Arithmetic Circuits

5.1. *Motivation and method of experiment*

In this section, we apply the EGG system to more general class of circuit synthesis problems. The circuit structures considered in the previous experiments consist of only limited kinds of functional nodes such as 3-2 counters and arithmetic shifters. The synthesized circuits are, in a sense, *homogeneous* networks of a limited variety of components. However, there are many important design problems in which a variety of building blocks with different functions are to be used to synthesize the target function. Typical examples of such problems include synthesis of *heterogeneous* networks of analog/digital components. The search space of heterogeneous circuit config-

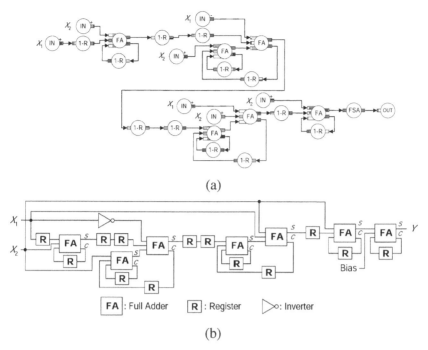

Figure 13. Example solution of bit-serial multiply-adder ($Y = 58X_1 + 43X_2$): (a) the circuit graph generated by EGG system, (b) the multiply-addition structure corresponding to graph (a).

urations increases rapidly as the variety of available components increases. This section presents a new variation of EGG system that can be used to synthesize heterogeneous architectures.

The modified EGG system reduces the search space of heterogeneous circuit structures by introducing constraint in terms of connectivity among circuit components. For example, if a component produces current signals, then its output cannot be connected to components accepting only voltage inputs. We generalize this idea and obtain a new model of a circuit graph using *terminal-color constraint*, where every input/output terminal of functional nodes has its own attribute called *color*. In this model, if and only if two terminals have compatible *colors*, they can be connected together.

The *color* of a terminal is an abstraction of signal attributes, such as voltage and current, associated with the terminal. The terminal-color compatibility defines the constraint on the connectivity between the terminals, and can be defined formally as follows. Let the set of terminal colors be given by $C = \{c_1, c_2, \cdots, c_m\}$. We define the *terminal-color compatibility* R as a binary relation: $R \subseteq C \times C$. Assuming that an output terminal of a node has the color c_i and an input terminal of another node has the color c_j,

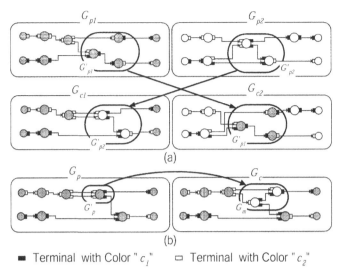

■ Terminal with Color "c_1"   □ Terminal with Color "c_2"

Figure 14. Evolutionary operation considering the terminal-color constraint: (a) crossover, (b) mutation.

this (output, input)-terminal pair can have a directed edge, if and only if $(c_i, c_j) \in R$.

According to the terminal-color constraint, the modified EGG system introduces evolutionary operations considering terminal-color compatibility of the generated offsprings as well as their completeness property. Figure 14(a) illustrates the crossover operation considering the terminal-color compatibility: $R = \{(c_1, c_1), (c_2, c_2)\}$ as an example. The system determines a pair of subgraphs G'_{p1} and G'_{p2} to be exchanged between the parents, and generates offsprings by replacing the subgraph of one parent by that of the other parent. In this process, the subgraphs should be carefully selected so that their cut sets contain the same number of positive/negative edges for each color c_1 or c_2. This ensures terminal-color compatibility of the generated offsprings as well as their completeness. We can define the mutation operation (Figure 14(b)) in a similar way.

The performance of the EGG system is demonstrated through experimental synthesis of CMOS current-mode logic circuits. The CMOS current-mode logic (Kameyama et al. 1988; Kawahito et al. 1988; Kawahito et al. 1990) is particularly useful for implementing arithmetic functions, since it can represent each digit of arithmetic number systems with a current on a wire. Another important property of current-mode logic is that a simple wire connection can perform linear summation of multiple digits without using any active devices. This significantly reduces the wiring complexity, chip area and power consumption of arithmetic circuits (Degawa et al. 2003). The effective utilization of the current-mode circuit technology, however, requires the knowledge of transistor-level design of analog-digital-mixed circuits, which is not an easy task for non-specialists. This is the major motivation to introduce EGG system for design automation.

In the following discussion, we mainly focus on the experimental synthesis of the full adder circuit based on radix-4 Signed-Digit (SD) arithmetic employing the digit set $L = \{-3, -2, -1, 0, 1, 2, 3\}$. The same discussion can be applied also to the synthesis of other circuit structures. Let the two SD numbers be given by $X = (x_{n-1} \cdots x_i \cdots x_0)$ and $Y = (y_{n-1} \cdots y_i \cdots y_0)$, where $x_i, y_i \in L$ for all i. The carry-propagation-free addition of the two SD numbers is performed by the following three successive steps for each digit:

$$z_i = x_i + y_i \tag{43}$$
$$4c_i + w_i = z_i \tag{44}$$
$$s_i = w_i + c_{i-1} \tag{45}$$

where $z_i \in \{-6, \cdots, 0, \cdots, 6\}$, $w_i \in \{-2, -1, 0, 1, 2\}$, and $c_i \in \{-1, 0, 1\}$. Each signal is represented by current on a wire: the amount of current represents digit magnitude and the direction of current flow represents sign $(+/-)$. The linear addition steps (43) and (45) can be performed by wired summation without active devices. The step (44) is performed by an SD full adder core, which is the target of synthesis in this section.

Figure 15 (a) shows the corresponding block diagram of the radix-4 SD full adder. The two input operands x_i and y_i ($\in \{-3, -2, -1, 0, 1, 2, 3\}$) are added by current summation node at the first stage to form the linear sum z_i ($\in \{-6, \ldots, 0, \ldots, 6\}$) in (43). The full adder core for the step (44) – the target of synthesis in this experiment – accepts the linear sum z_i and converts it into a pair intermediate sum (inverted form) $-w_i$ ($\in \{-2, -1, 0, 1, 2\}$) and carry c_i ($\in \{-1, 0, 1\}$). The signal $-w_i$ enters the inverted quantizer to produce the level-restored signal w_i ($\in \{-2, -1, 0, 1, 2\}$), which is added to the carry c_i ($\in \{-1, 0, 1\}$) from the lower digit position to form the final sum s_i ($\in \{-3, -2, -1, 0, 1, 2, 3\}$) in the step (44). Figure 15 (b) defines the target functions $-w_i(z_i)$ and $c_i(z_i)$ of the full adder core.

For modeling the current-mode circuits, we consider four different signal types: positive current, negative current, bi-directional current, and voltage. These four types are used as "colors" for terminals in circuit graphs. We define terminal-color compatibility as $R = \{(c_p, c_p), (c_n, c_n), (c_b, c_b), (c_p, c_v), (c_n, c_v), (c_b, c_v), (c_v, c_v)\}$, where c_p, c_n, c_b and c_v denote positive current, negative current, bi-directional current and voltage, respectively. This constraint is illustrated in Figure 16, where compatible colors (signal types) are indicated by edge connections. Note that the positive/negative/bi-directional currents can be easily converted into voltage signals at high-

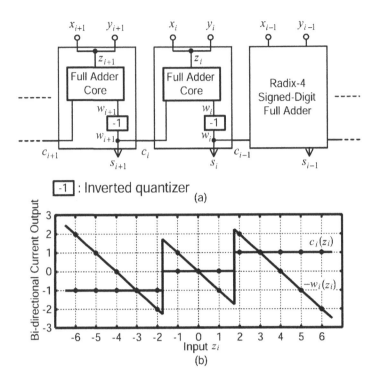

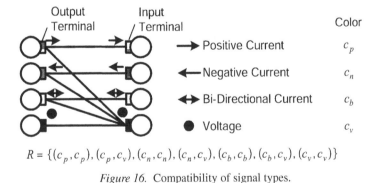

Figure 15. Target of synthesis: (a) block diagram of a radix-4 SD adder, and (b) transfer functions for carry $c_i(z_i)$ and intermediate sum $-w_i(z_i)$.

Figure 16. Compatibility of signal types.

impedance terminals connected to gate electrodes of transistors. Thus, a voltage input terminal (high-impedance terminal) can accept any signal type.

Figure 17 shows the functional nodes with colored terminals that are used in our experiment. The bi-directional current input circuit is used to convert a bi-directional current into a set of positive/negative currents. If the

Name	Symbol	Schematic	Mathematical representation
Operand input			$Y = X$
Operand output			$Y = X$
Bi-directional current input			$Y_1 = 0, Y_2 = -X \; (X>0)$ $Y_1 = -X, Y_2 = 0 \; (X<0)$
Current source			$Y = a$
nMOS Current mirrors			
pMOS Current mirrors			$Y = -X$
nMOS transistor switches			$Y = 0 \; (X_2 = \text{low})$ $Y = X_1 \; (X_2 = \text{high})$
pMOS transistor switches			$Y = 0 \; (X_2 = \text{high})$ $Y = X_1 \; (X_2 = \text{low})$
Current summation nodes			$Y = X_1 + X_2$
Voltage detection node			$Y = \text{high} \; (X_1 + X_2 > 0)$ $Y = \text{low} \; (X_1 + X_2 < 0)$
Voltage branch node			$Y_1 = X$ $Y_2 = X$

Figure 17. Functional nodes with colored terminals.

input current X has positive direction ($X > 0$), only the negative-output terminals labeled by Y_2 are activated. On the other hand, the positive-output terminals labeled by Y_1 are activated only when $X < 0$. The current sources are of two types (positive and negative) depending on the direction of the produced current. We also assume the use of current mirrors to make copies of positive/negative currents. We have four types of transistor models since there are two different transistor switches (nMOS and pMOS) and two different directions of current flow. There are passive elements, called current summation nodes, a voltage detection node and a voltage branch node. Actually, all these passive elements are simple wiring points; their difference comes from their usage. The voltage terminals of the voltage detection/branch nodes are high-impedance terminals (which may be located on open circuit paths connected to gate terminals of transistors).

The evolved circuit structures are evaluated by a combination of *functionality F* and *performance P*. The functionality measure F evaluates the validity of the transfer functions compared with the target functions $-w_i(z_i)$ and $c_i(z_i)$. The EGG system evaluates the functions of evolved circuit graphs by using mathematical definitions of the functional nodes shown in Figure 17. Let the estimated transfer functions of the evolved circuit graphs be denoted by $-\tilde{w}_i(z_i)$ and $\tilde{c}_i(z_i)$, and let the forward differences of the functions $-w_i(z_i)$, $c_i(z_i)$, $-\tilde{w}_i(z_i)$ and $\tilde{c}_i(z_i)$ be denoted by $-w_i'(z_i)$, $c_i'(z_i)$, $-\tilde{w}_i{}'(z_i)$ and $\tilde{c}_i{}'(z_i)$, respectively. We first calculate the following quantities, which estimate the distance between the target function and the evolved function:

$$d = \sum_{z_i} |w_i(z_i) - \tilde{w}_i(z_i)| + \sum_{z_i} |c_i(z_i) - \tilde{c}_i(z_i)|,$$

$$d' = \sum_{z_i} |w_i'(z_i) - \tilde{w}_i{}'(z_i)| + \sum_{z_i} |c_i'(z_i) - \tilde{c}_i{}'(z_i)|.$$

We define the functionality measure by

$$F = C_1(1 - \frac{d}{d_{max}}) + C_2(1 - \frac{d'}{d'_{max}}) + C_3(1 - \frac{n}{n_{max}}),$$

where n is the number of output terminals whose values cannot be determined uniquely through functional verification. We assume that $0 \le d \le d_{max}$, $0 \le d' \le d'_{max}$ and $0 \le n \le n_{max}$, where $d_{max} = 20$, $d'_{max} = 400$ and $n_{max} = 100$ in our experiments. The constants C_1, C_2 and C_3 are selected so as to keep the range of F as $0 \le F \le 100$. In our experiments, $C_1 = 45$, $C_2 = 45$ and $C_3 = 10$.

On the other hand, the performance measure P is defined as

$$P = \frac{C_4}{DA},$$

where D is the relative delay time of the generated circuit graph, A is the total number of transistors, and C_4 is a constant. We use $F + P$ as a synthetic fitness value of the graph, where the ratio P_{max}/F_{max} is adjusted around $10/100$ by tuning the constant as $C_4 = 1000$. This implies that the evolution in earlier generations is mainly driven by functionality measure F toward 100% functionality, and then eventually the population evolves to better performance.

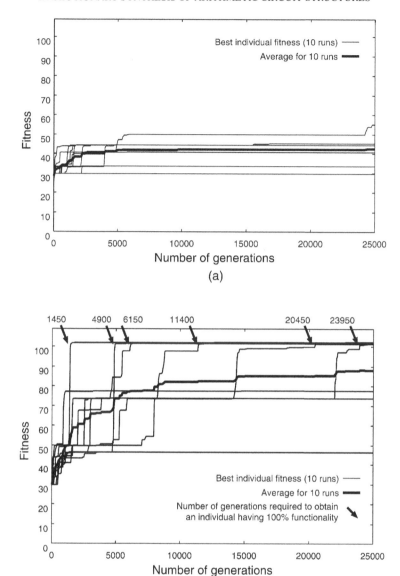

Figure 18. Best fitness transition for (a) the original EGG (10 runs), and (b) the modified EGG adopting terminal-color constraint (10 runs).

5.2. *Experimental result*

In this experiment, we assume that the population size is 300, the number of generations is 25000, the maximum number of nodes is 50, the crossover rate is 0.8, and the mutation rate is 0.1. Figure 18 compares the best fitness

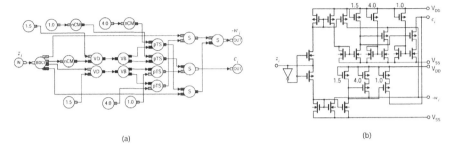

Figure 19. Evolved radix-4 SD full adder circuit: (a) graph structure synthesized by EGG, (b) circuit structure corresponding to the graph (a).

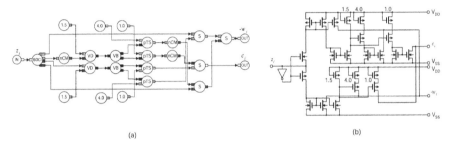

Figure 20. Radix-4 SD full adder circuit by human design: (a) graph representation, (b) circuit structure corresponding to the graph (a).

transition of the original EGG system (a) and that of the modified EGG system adopting terminal-color constraint (b). The modified EGG system shows the tendency to improve the fitness more rapidly, and successfully finds the optimal solution in 6 trials out of 10. Figure 19 shows the evolved circuit structure and Figure 20 shows the human design reported in (Kawahito et al. 1988). Both circuits have complete functionality for the radix-4 SD full adder core, but have different performance values: $P = 2.39$ for the evolved structure and $P = 1.96$ for the reported structure. The reason for this is that the evolved structure has shorter critical path compared with the critical path in the human design, where the critical path delay is evaluated by using the relative delay. These delay values are roughly estimated in advance by HSPICE simulation. For more detailed performance evaluation, careful circuit simulation is required using the optimized L/W parameters. Note that the current version of EGG system can create only the netlist of circuits, and cannot optimize the circuit parameters for HSPICE simulation. The simultaneous optimization for circuit structures and parameters are being left for future studies.

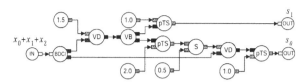

Figure 21. Evolved 3-2 counter structure.

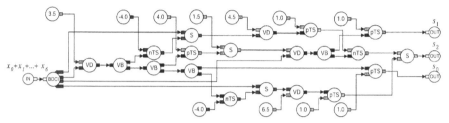

Figure 22. Evolved 7-3 counter structure.

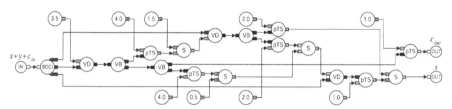

Figure 23. Evolved radix-4 non-redundant full adder.

5.3. *Other examples*

The above experiment focuses on a specific target function (i.e., a radix-4 SD full adder) simply for the purpose of comparing the quality of the evolved circuit structure with that of human design. Note that the proposed system can synthesize other current-mode logic circuits without any modification. For example, Figures 21, 22 and 23 show the evolved current-mode logic circuits for a 3-2 counter, a 7-3 counter and a radix-4 non-redundant full adder. The functions of these arithmetic circuits are given as follows:

- 3-2 counter
  - Inputs: $x_0, x_1, x_2 \in \{0, 1\}$
  - Outputs: $s_0, s_1 \in \{0, 1\}$
  - Function: $s_0 + 2s_1 = x_0 + x_1 + x_2$
- 7-3 counter
  - Inputs: $x_0, x_1, \cdots, x_6 \in \{0, 1\}$
  - Outputs: $s_0, s_1, s_2 \in \{0, 1\}$
  - Function: $s_0 + 2s_1 + 4s_2 = x_0 + x_1 + \cdots + x_6$

- — Radix-4 non-redundant full adder
  - Inputs: $x, y \in \{0, 3\}$ $c_{in} \in \{0, 1\}$
  - Outputs: $s \in \{0, 3\}$, $c_{out} \in \{0, 1\}$
  - Function: $s + 4c_{out} = x + y + c_{in}$

We have confirmed, through HSPICE simulation, that the evolved circuits realize the complete target functions. Note that the target functions considered here are multiple output functions. The simplest way of implementing a multiple-output function is to realize every single-output function independently. However, the EGG system finds more compact implementations naturally, where common circuit components are shared by multiple functions. For example, in the 3-output circuit of Figure 22, 22% of basic components are shared by two functions and 15% of basic components are shared by three functions. The EGG system can create such efficient circuit structures with limited knowledge of the circuit design. These results indicate that the EGG system can become a powerful tool for various heterogeneous circuit synthesis problems, for which no systematic design techniques have been developed yet.

6. Conclusion

In this paper, we have presented an efficient graph-based evolutionary optimization technique called Evolutionary Graph Generation (EGG), and its application to the synthesis of combinational multipliers using word-level arithmetic components, bit-serial multipliers using bit-level arithmetic components, and multiple-valued current-mode arithmetic circuits using transistor-level components. Our initial observation suggests the potential capability of the graph-based evolutionary synthesis technique to solve various circuit design problems with limited knowledge of underlying design principle. The key to success is to reduce the search space of possible graph structures by employing adequate graph-theoretic constraint on the evolved structures. For example, we have employed the "CCT (Complete Circuit Tree) constraint" in Section 3 and "terminal-color constraint" in Section 5 to eliminate needless computation.

Further investigations are now being conducted to develop a generic object-oriented framework for EGG system, which can be systematically modified for different design problems by inheriting the framework class templates. The current version of EGG framework used in this paper is open to the public at *http://www.aoki.ecei.tohoku.ac.jp/egg/*.

References

Aoki, T. & Higuchi, T. (2000). Beyond-Binary Arithmetic – Algorithms and VLSI Implementations –. *Interdisciplinary Information Sciences* **6**(1): 75–98.

Aoki, T., Homma, N. & Higuchi, T. (1999a). Evolutionary Design of Arithmetic Circuits. *IEICE Trans. Fundamentals* **E82-A**(5): 798–806.

Aoki, T., Sawada, Y. & Higuchi, T. (1999b). Signed-Weight Arithmetic and Its Application to a Field-Programmable Digital Filter Architecture. *IEICE Trans. Electronics* **E82-C**(9): 1687–1698.

Avizienis, A. (1961). Signed-Digit Number Representations for Fast Parallel Arithmetic. *IRE Trans. Electronic Computers* **10**: 389–400.

Back, T., Hammel, U. & Schwefel, H. P. (1997). Evolutionary Computation: Comments on the History and Current State. *IEEE Trans. Evolutionary Computation* **1**(1): 3–13.

Brent, P. R. & Kung, H. T. (1981). The Area-Time Complexity of Binary Multiplication. *Journal of the Association for Computing Machinery* **28**(3): 521–534.

Degawa, K., Aoki, T. & Higuchi, T. (2003). A Field-Programmble Digital Filter Chip Using Multiple-Valued Current-Mode Logic. *Proceedings 33rd IEEE Int'l Symp. Multiple-Valued Logic*, 213–220.

Dempster, A. G. & Macleod, D. M. (1994). Constant Integer Multiplication Using Minimum Adders. *IEE Proceedings Circuits Devices Syst.* **141**(5): 407–413.

Haddow, C. P. & Tufte, G. (2000). An Evolvable Hardware FPGA for Adaptive Hardware. *Proceedings of the 2000 Congress on Evolutionary Computation*, 553–560.

Hartley, R. I. (1996). Subexpression Sharing in Filters Using Canonic Signed Digit Multipliers. *IEEE Trans. Circuits Syst. II, Analog Digit. Signal Process.* **43**(10): 677–688.

Higuchi, T., Iwata, M., Keymeulen, D., Sakanashi, H., Murakawa, M., Kajitani, I., Takahashi, E., Toda, K., Salami, M., Kajihara, N. & Otsu, N. (1999). Real-World Applications of Analog and Digital Evolvable Hardware. *IEEE Trans. Evolutionary Computation* **3**(3): 220–235.

Higuchi, T., Murakawa, M., Iwata, M., Kajitani, I. & Weixin, L. (1997). Evolvable Hardware at Function Level. *Proceedings of the 1997 IEEE International Conference on Evolutionary Computation (ICEC '97)*, 187–192.

Homma, N., Aoki, T. & Higuchi, T. (2000a). Evolutionary Graph Generation System with Symbolic Verification for Arithmetic Circuit Design. *IEE Electronics Letters* **36**(11): 937–939.

Homma, N., Aoki, T. & Higuchi, T. (2000b). Evolutionary Synthesis of Fast Constant-coefficient Multipliers. *IEICE Trans. Fundamentals* **E83-A**(9): 1767–1777.

Homma, N., Aoki, T. & Higuchi, T. (2001). Evolutionary Graph Generation System with Transmigration Capability for Arithmetic Circuit Design. *Proceedings of The 2001 IEEE International Symposium on Circuits and Systems* **5 of 5**, 171–174.

Huelsbergen, L., Rietman, E. & Slous, R. (1999). Evolving Oscillators in Silico. *IEEE Trans. Evolutionary Computation* **3**(3): 197–204.

Hwang, K. (1979). *Computer Arithmetic: Principles, Architecture, and Design*. John Wiley & Sons.

Kameyama, M., Kawahito, S. & Higuchi, T. (1988). A Multiplier Chip with Multiple-Valued Bidirectional Current-Mode Logic Circuits. *IEEE Computer* **21**(4): 43–56.

Kawahito, S., Kameyama, M. & Higuchi, T. (1990). Multiple-Valued Radix-2 Signed-Digit Arithmetic Circuits for High-Performance VLSI Systems. *IEEE J. Solid-State Circuits* **25**(1): 125–131.

Kawahito, S., Kameyama, M., Higuchi, T. & Yamada, H. (1988). A 32×32-Bit Multiplier Using Multiple-Valued MOS Current-Mode Circuits. *IEEE J. Solid-State Circuits* **23**(1): 124–132.

Khoo, K., Kwentus, A. & Willson, N. A. (1996). A Programmable FIR Digital Filter Using CSD Coefficients. *IEEE J. Solid-State Circuits* **31**(6): 869–874.

Koren, I. (1993). *Computer Arithmetic Algorithms*. Prentice Hall.

Koza, R. J., Bennett, F. H., Andre, D., Keane, M. A. & Dunlap, F. (1997). Automated Synthesis of Analog Electrical Circuits by Means of Genetic Programming. *IEEE Trans. Evolutionary Computation* **1**(2): 109–128.

Lohn, D. J. & Colombano, S. P. (1999). A Circuit Representation Technique for Automated Circuit Design. *IEEE Trans. Evolutionary Computation* **3**(3): 205–219.

Miller, F. J., Thomson, P. & Fogarty, T. (1997). Designing Electronic Circuits Using Evolutionary Algorithms. Arithmetic Circuits: A Case Study. *Genetic Algorithms and Evolution Strategies in Engineering and Computer Science*, 105–131.

Omondi, A. R. (1994). *Computer Arithmetic Systems: Algorithms, Architecture and Implementations*. Prentice Hall.

Thompson, A., Layzell, P. & Zebulum, R. S. (1999). Explorations in Design Space: Unconventional Electronics Design Through Artificial Evolution. *IEEE Trans. Evolutionary Computation* **3**(3): 167–196.

Tyrrell, M. A., Hollingworth, G. & Smith, S. L. (2001). Evolutionary Strategies and Intrinsic Fault Tolerance. *Proceeding of the 3rd NASA/DoD Workshop on Evolvable Hardware*, 98–106.

Wallace, C. S. (1964). A Suggestion for a Fast Multiplier. *IEEE Trans. Electron. Computer* **13**: 14–17.

Wong, B. C. & Samueli, H. (1991). A 200-MHz All-Digital QAM Modulator and Demodulator in 1.2-μm CMOS for Digital Radio Applications. *IEEE Journal Solid-State Circuits* **26**(12): 1970–1979.

Bi-Decomposition of Function Sets in Multiple-Valued Logic for Circuit Design and Data Mining

CHRISTIAN LANG[1] and BERND STEINBACH[2]

[1] *Institute for Microelectronic and Mechatronic Systems, Konrad-Zuse-Str. 14, 99099 Erfurt, Germany (E-mail: christian.lang@imms.de);* [2] *TU Bergakademie Freiberg, Institute of Computer Science, Germany (E-mail: steinb@informatik.tu-freiberg.de)*

Abstract. This article presents a theory for the bi-decomposition of functions in multi-valued logic (MVL). MVL functions are applied in logic design of multi-valued circuits and machine learning applications. Bi-decomposition is a method to decompose a function into two decomposition functions that are connected by a two-input operator called gate. Each of the decomposition functions depends on fewer variables than the original function. Recursive bi-decomposition represents a function as a structure of interconnected gates. For logic synthesis, the type of the gate can be chosen so that it has an efficient hardware representation. For machine learning, gates are selected to represent simple and understandable classification rules.

Algorithms are presented for non-disjoint bi-decomposition, where the decomposition functions may share variables with each other. Bi-decomposition is discussed for the min- and max-operators. To describe the MVL bi-decomposition theory, the notion of incompletely specified functions is generalized to function intervals. The application of MVL differential calculus leads to particular efficient algorithms. To ensure complete recursive decomposition, separation is introduced as a new concept to simplify non-decomposable functions. Multi-decomposition is presented as an example of separation.

The decomposition algorithms are implemented in a decomposition system called YADE. MVL test functions from logic synthesis and machine learning applications are decomposed. The results are compared to other decomposers. It is verified that YADE finds decompositions of superior quality by bi-decomposition of MVL function sets.

Keywords: bi-decomposition, data mining, differential calculus, logic synthesis, machine learning, multi-level circuit design, multiple-valued logic

Abbreviations: BDC – Boolean Differential Calculus; BDD – Binary Decision Diagram; BEMDD – Binary Encoded Multi-valued Decision Diagram; DFC – Discrete Function Cardinality; ISF – Incompletely Specified Function; MDD – Multi-valued Decision Diagram; MFI – Multi-valued Function Interval; MFS – Multi-valued Function Set; MM – Max and Min; MVL – Multiple-Valued Logic; MVR – Multi-Valued Relation

1. Introduction

The divide and conquer principle has been recognized as an important and powerful tool in many areas of computer science. Applied to the area of

digital circuits, this strategy is implemented by the decomposition of functions. Given a function as a behavioral description of the circuit, the function
is decomposed into interconnected smaller functions which reveals the structure of the circuit. Although the idea of *multi-valued logic* circuits is not
new (Smith 1981), the overwhelming majority of discrete circuits today is
two-valued. Two-valued systems are best described by the Boolean logic.
Synthesis and analysis of two-valued systems have been research topics for
more than fifty years. Methods for decomposition of Boolean functions are
well developed.

Advances in the field of semi-conductors indicate increasing interest
in circuits with more than two values. The number of interconnections is
becoming an increasing problem in modern VLSI chips (Kameyama 1999).
Transmission systems in telecommunications rely on multi-valued signals.
Three- and four-valued signals are typical for coding on expensive access
lines. MVL signals can carry more information than two-valued signals, and
hence, the number of interconnections can be reduced. In modern memories
the size of a memory cell reaches physical limits, and the density of the
memory can be increased by storing more than one bit per cell, which implies
more than two states.

MVL circuits can be processed in a variety of technologies including
current-mode CMOS (Teng and Bolton 2001) and neuron MOS (Inaba et
al. 2001; Han et al. 2001). It was shown that MVL circuits improve circuit
performance in terms of chip area, operation speed and power consumption
(Hanyu and Kameyama 1995; Dhaou et al. 2001). Future quantum electronic
circuits may exploit tunneling effects (Uemura and Baba 2001; Waho et al.
2001). Research on MVL circuits has led to commercial applications. Four-
valued flash memories and DRAMs are commercially available (Ricco et al.
1998). Intel designed a RAMBUS interface that doubles the bandwidth of
the communication between the processor and the memory by application
of four-valued signals that transmit 2 bits per clock cycle over a single wire
(Rambus Inc.).

The application of MVL circuits requires design methods for MVL
systems. *Bi-decomposition* decomposes a given function $f(A, B, C)$ into
two decomposition functions $g(A, C)$ and $h(B, C)$ that are combined by an
operator function $\pi(x, y)$ according to

$$f(A, B, C) = \pi\big(g(A, C), h(B, C)\big),$$

where both decomposition functions depend on fewer variables than the
original function. Bi-decomposition has a direct realization as a circuit as
shown in Figure 1. Bi-decomposition of Boolean functions was shown to be
superior to other methods of decomposition, such as Curtis decomposition

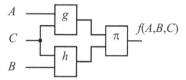

Figure 1. Bi-decomposition of a function represented by a circuit.

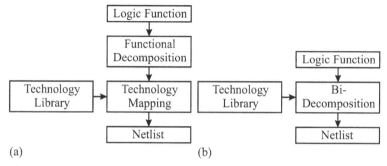

Figure 2. Design flow in logic synthesis. a. Design by functional decomposition with technology mapping. b. Design by bi-decomposition without technology mapping.

and factorization (Mishchenko et al. 2001a). Therefore, bi-decomposition of MVL functions should also be investigated.

Synthesis of logic by functional decomposition involves a decomposition step and a technology mapping, see Figure 2(a). A logic function is decomposed into smaller blocks by functional decomposition. Then a mapping with a technology library synthesizes each block using the gates of the library. The result is a netlist of gates that is realized in the chosen technology.

Bi-decomposition gives complete control over the type of gates that is selected during decomposition. The bi-decomposition process can be controlled by the technology library of available gates directly, and no technology mapping is necessary, see Figure 2(b).

Synthesis of MVL functions was also suggested as a preprocessing step in the synthesis of Boolean circuits (Brayton and Khatri 1999; Dubrova et al. 2001). Many signals from real-world applications are multi-valued. For instance, the direction of an elevator has the three values "up", "down" and "stop". Decomposition of a logic function on this multi-valued level can help to find a good encoding of the problem in Boolean variables for a subsequent synthesis of the Boolean circuit.

Recently, MVL decomposition found applications in the field of *data mining* (Zupan 1997). Over the past decades much data has been stored digitally and huge databases have been created. There arises the question of how to retrieve important information and how to restructure this large amount of data.

Table 1. Lenses problem: Codes for the variables of the lens prescription rules

Variable	Type	Meaning	Code	Value
Age	input	age of the patient	0	young
			1	pre-presbyopic
			2	presbyopic
Prescription	input	spectacle prescription	0	myope
			1	hypermetrope
Tear	input	tear production rate	0	reduced
			1	normal
Lens	output	type of contact lenses	0	hard
			1	soft
			2	no-contacts

EXAMPLE 1. *Consider a simplified version of the lenses problem from the UCI database (*Cendrowska 1987; *Blake and Merz 1998). There are guidelines for a doctor, which type of contact lenses should be prescribed for a patient. In this example the prescription depends on three independent parameters, the age of the patient, the spectacle prescription and the tear production rate. The parameters are encoded by the input variables age, prescription and tear respectively. The codes and values of these variables are shown in Table 1. The type of contact lenses, which is the goal variable, is encoded by the variable lens, which is also shown in the table. The guidelines for the prescription are given in Table 2. For each patient the doctor can determine the values of the variables age, prescription and tear. Then, with help of Table 2, the doctor can look up the type of contact lenses that should be prescribed.*

Consider a patient of presbyopic age, with myope spectacle prescription and normal tear production rate. According to Table 1, this case corresponds to the set of input variables {age = 2, prescription = 0, tear = 1}. It can be derived from Table 2 (boldfaced row) that lens = 2, and hence, this patient should not be prescribed contact lenses.

However, there are difficulties when looking up a solution from a table:
— The data has an arbitrary structure. Each case is treated independently, no relations between similar cases are immediately apparent from the table.
— The solution is difficult to understand by humans. The table just tells the solution without explanation.

Table 2. Lenses problem: Rules for prescription of contact lenses

Age	Prescription	Tear	Lens
0	0	0	2
0	0	1	0
0	1	0	2
0	1	1	1
1	0	0	2
1	0	1	1
1	1	0	2
1	1	1	1
2	0	0	2
2	**0**	**1**	**2**
2	1	0	2
2	1	1	1

- For many problems of the real world there is a great number (up to several hundred) input variables. Because the number of parameter combinations grows exponentially with the number of variables, it is impossible to list all combinations in a table. The reason is not primarily the storage capacity of computers, but the fact that knowledge is often incomplete. In many cases the value of the goal variable is not known for some input combinations. In fact, there are many examples, where the value of the goal variable is not known for more than 99.9 % of the input combinations (Blake and Merz 1998).

To solve these problems, an appropriate structure should be introduced into the data. The structure should reveal hidden relationships between cases and should be understandable by humans. Furthermore, it should be possible to correctly extrapolate from the known values of the goal variable to the unknown cases. In the machine learning theory there are many ways to introduce such a structure into a database. Data mining includes pattern recognition and statistical analysis as methods of data analysis.

Given a set of training examples, there are two major goals for data mining. The first one is to partition new examples with a small classification error. The second one is to derive a simple model of the data that can easily be understood by humans.

Usually, data analysis starts with a model developed by a human domain expert. Then the parameters of the model are adjusted according to the given training examples. This method requires an expert with detailed knowledge

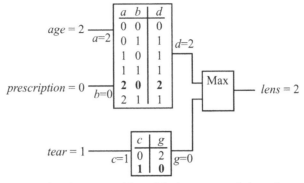

Figure 3. Bi-decomposition of the lenses prescription rules.

about the laws and dependencies of the data. The structure of the model is induced from the knowledge of the expert rather than from the data.

To be independent from (difficult to find) experts and to make it possible to find new structures, which were not previously known by experts, machine learning systems were developed that derive a model of the data directly from the set of training examples by data analysis algorithms. Functional decomposition of MVL functions has successfully been applied to solve machine learning problems (Zupan 1997). The set of training examples is transformed into a discrete function. The input variables are the attributes of the examples, and the output variable is the desired classification of the examples. The function is recursively decomposed into small blocks. The result is a structure of interconnected blocks, where each block is associated with a function that represents a rule.

For MVL functions, there are many discrete functions with two input variables. Only a few functions (e.g., sum, product, min and max) can easily represented by natural language. Many other functions require complicated explanations or tables to be understood. Therefore, if the solution should be simple from a human point of view, there should be some way to control the decomposition functions.

EXAMPLE 2. *Figure 3 shows a bi-decomposition of the data from Table 2. The operator function is the max-gate. The figure also shows the values of the intermediate and goal variables for the set of input variables $\{age = 2, prescription = 0, tear = 1\}$. Note that the original table is split into two smaller tables that are connected by a max-gate. The structure shows that the value $tear = 0$ implies the value $lens = 2$. This translates to the rule "If a patient has a reduced tear production rate then she should not be prescribed contact lenses". Recursive decomposition of the upper table results in a network that consists of only min- and max-gates, which reveals further rules.*

This article suggests bi-decomposition as a way to control the decomposition functions, where the operator function can be chosen so that simple to understand decompositions are produced.

Bi-decomposition was shown to give solutions of low complexity in the Boolean case. Therefore, it is expected that bi-decomposition is well suitable to solve machine learning problems. For this application it is necessary to develop efficient bi-decomposition algorithms for large MVL functions.

The remainder of this paper is organized as follows. Section 2 defines MVL functions and extends Boolean differential calculus (BDC) to the MVL case. Section 3 introduces sets of functions as a generalization of MVL functions. Bi-decomposition algorithms for the min- and max-operators are developed in Section 4. To ensure complete decomposition of all MVL functions, bi-decomposition is generalized to separation in Section 5. The algorithms are implemented in a decomposition program called YADE, which is presented in Section 6. Section 7 presents the decomposition results of test functions and compares these with the results obtained by other decomposers. The article concludes with a summary and ideas for further work in Section 8.

2. MVL Functions and Operations

2.1. *Definition of MVL functions*

Signals in binary circuits have exactly one of two values. Therefore, Boolean functions are ideal for modeling such systems. In some applications, such as machine learning or *multi-valued logic (MVL)* synthesis, functions with more than two values are applied. To model these systems, MVL functions, where more than two values can be assigned to the variables, are necessary.

It is always possible to employ a vector of Boolean functions instead of an MVL function by a proper encoding of the values. However, there are some problems with this method:

— The Boolean representation introduces operations that cannot be interpreted in the MVL domain. Decomposition is an example. Boolean variables that represent a single MVL variable can be separated by Boolean decomposition. This decomposition cannot be translated into a corresponding MVL decomposition.

— Which encoding of the values to Boolean vectors should be chosen? Some coding schemes will introduce unnecessary complexity into the system.

— The number of values of MVL variables is, in general, not a power of two. Therefore, some values will not be used and there is the question of what to do with these values.

— In some cases, as in machine learning, it is very important to produce solutions that can easily be understood by humans. The solution produced by replacing MVL by Boolean functions will be in terms of Boolean operators. These operators may not be easily understood in the context of the MVL system.

To avoid these problems, the theory of Boolean functions has been extended to MVL functions, which are a generalization of two-valued Boolean functions. Some properties of Boolean functions do not hold for MVL functions. However, as this article will show, many properties of Boolean functions can directly be extended to the MVL case. Furthermore, investigation of MVL systems emphasizes some properties that were not very apparent in Boolean systems.

A Boolean variable can only assume one of two values. For an *MVL variable* the number of possible values must be defined.

DEFINITION 1. *An* MVL variable *a can assume any value from the set of integers* $\{0, \ldots, m - 1\}$, *where the integer* $m \geq 2$ *is called the* cardinality *of the variable a. A set* $A = \{a_1, \ldots, a_n\}$ *of MVL variables, with the cardinalities* $m_1, \ldots, m_n$ *respectively, can assume any element of*

$$\mathbb{M}_A = \{0, \ldots, m_1 - 1\} \times \{0, \ldots, m_2 - 1\} \times \ldots \{0, \ldots, m_n - 1\}, \quad (1)$$

where × *denotes the Cartesian product of sets. The set* $\mathbb{M}_A$ *is the* MVL space *of the set of variables A. An element of* $\mathbb{M}_A$ *is an* MVL minterm. *The MVL space* $\mathbb{M}_A$ *has* $|\mathbb{M}_A| = m_1 * \ldots * m_n$ *minterms.*

Sometimes the cardinalities of all variables are assumed to be equal and constant. This assumption is called *fixed cardinality* and enables elegant theoretical formulations. However, this assumption is only valid for some applications, like MVL circuits. In machine learning the cardinalities of attributes typically differ from each other. Therefore, in this article the cardinalities of the variables are assumed to be different, which is called *variable cardinality*. However, for a single variable, its cardinality is assumed to be constant. The case of fixed cardinality can be treated as a special case of variable cardinality.

MVL functions are defined similar to Boolean functions.

DEFINITION 2. *Let* $A = \{a_1, \ldots, a_n\}$ *be a set of MVL variables that have the cardinalities* $m_1, \ldots, m_n$ *respectively. An* MVL function $f(A)$ *is many-to-one mapping*

$$\mathbb{M}_A \rightarrow \{0, \ldots, m_o - 1\} \quad (2)$$

from the set of minterms $\mathbb{M}_A$ *into the set of integers* $\{0, \ldots, m_o - 1\}$. *The integers* $m_i, i = 1, \ldots, n$ *are called the* input cardinalities *and* m_o *is called the* output cardinality *of the function* $f(A)$.

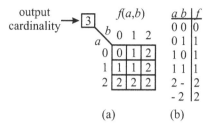

(a)   (b)

Figure 4. MVL-functions. a. Display as a map. b. Display as a table.

The set of all MVL functions that depend on the set of MVL variables A and have the output cardinality m_o is denoted by $\mathbb{F}_{m_o}(A)$. If the set A contains $|A| = n$ variables, with the cardinalities $m_1, \ldots, m_n$, there are

$$\left|\mathbb{F}_{m_o}(A)\right| = m_o^{|\mathbb{M}_A|} = m_o^{m_1 * \ldots * m_n} \tag{3}$$

MVL functions with the output cardinality m_o that depend on A. Like Boolean functions, MVL functions can be visualized by maps and tables, see Figure 4. The output cardinality of the function is indicated by the number shown in the square box at the upper left corner of the map. Similar to tables for Boolean functions, a dash '–' in the column of the variable a stands for any value in the range $[0, m_a - 1]$, where m_a is the cardinality of the variable a. For instance, the last row in Figure 4(b), "-2 2", abbreviates the three rows "02 2", "12 2" and "22 2". Note, that a dash can only be used to abbreviate minterms that exist for all values of the variable a. For instance, two minterms "02 2" and "22 2" cannot be replaced by a row containing a dash.

A rough measure of the complexity of a Boolean function is its number of variables. For MVL functions, the cardinality of the variables must be taken into account. A simple measure of the complexity of an MVL function is the discrete function cardinality.

DEFINITION 3.   *The* discrete function cardinality *(DFC)* $D(f)$ *of an MVL function* $f(a_1, \ldots, a_n)$ *is the product*

$$D(f) = m_1 * \ldots * m_n \tag{4}$$

of the input cardinalities m_i, $i = 1, \ldots, n$.

The DFC of the function $f(a, b)$ from Figure 4 is $D(f) = 3 * 3 = 9$ because cardinality of the variables a and b is '3'.

2.2. *MVL composition operators*

New MVL functions can be created from existing functions by composition. For MVL functions, the number of functions depends on the input and output

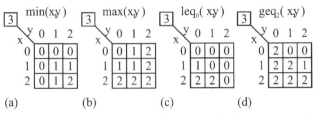

Figure 5. Maps of binary MVL functions. a. Min-operator. b. Max-operator. c. leq0-operator. d. geqm-operator.

cardinalities. In this article, unary MVL functions are called *literals*. Note that there are different definitions of MVL literals. Dubrova and Brayton defined literals to be two-valued functions (Dubrova et al. 2001). Here, literals have a multi-valued output. Examples of binary functions are:

— the *min-operator* $\min(x, y)$,
— the *max-operator* $\max(x, y)$,
— the *leq0-operator* $leq_0(x, y)$ defined to be 0 for $x \leq y$ and x otherwise,
— the *geqm-operator* $geq_m(x, y)$ defined to be m for $x \geq y$ and x otherwise.

The maps of these MVL functions are displayed in Figure 5 for the input and output cardinalities $m_x = m_y = m_o = 3$.

For the special case of input and output cardinalities of 2, the min-operator becomes the Boolean AND, the max-operator becomes the Boolean OR. The leq0- and geqm-operators are applied in the bi-decomposition algorithms shown in Section 4.

All of the associative binary operators can be extended to n-ary operators by repeated application of the binary operator. For instance, there is

$$\max(x_1, x_2, \ldots, x_n) = \max\left(x_1, \max\left(x_2, \ldots \max(x_{n-1}, x_n) \ldots\right)\right)$$

for the max-operator.

2.3. *MVL derivative operators*

There are several possibilities to extend the Boolean differential calculus (BDC) to MVL functions (Yanushkevich 1998). Decomposition problems can easily be described in terms of the min- and max-operators over variables.

DEFINITION 4. *The* minimum *of the MVL function* $f(A, b)$ *over the single variable b with the cardinality* m_b *is defined as the MVL function*

$$\min_b f(A, b) := \min\left(f(A, b = 0), f(A, b = 1), \ldots, f(A, b = m_b - 1)\right). \quad (5)$$

The k-times minimum *of the MVL function* $f(A, B)$ *over the set of variables* $B = \{b_1, \ldots, b_k\}$ *is defined as the MVL function*

$$\min_B^k f(A, B) = \min_{b_1}\left(\min_{b_2}(\ldots \min_{b_k} f(A, B)\ldots)\right). \tag{6}$$

DEFINITION 5. *The* maximum *of the MVL function* $f(A, b)$ *over the single variable* b *with the cardinality* m_b *is defined as the MVL function*

$$\max_b f(A, b) = \max(f(A, b = 0), f(A, b = 1), \ldots, f(A, b = m_b - 1)). \tag{7}$$

The k-times maximum *of the MVL function* $f(A, B)$ *over the set of variables* $B = \{b_1, \ldots, b_k\}$ *is defined as the MVL function*

$$\max_B^k f(A, B) = \max_{b_1}\left(\max_{b_2}(\ldots \max_{b_k} f(A, B)\ldots)\right). \tag{8}$$

The variables of the derivative operators are centered below the operator, such as in (8) if the formula is displayed on a separate line, and the variables are set as an index of the operator, such as $\max_B^k f(A, B)$, if the operator is embedded into text. The meaning of both notations is the same.

Note that the minimum and maximum over a variable do not depend on this variable anymore, and the minimum and maximum over a set of variables do not depend on this set of variables. The function $\min_B^k f(A, B)$ is the projection of the smallest value of the function $f(A, B)$ over the MVL space $\mathbb{M}_B$. Similarly, $\max_B^k f(A, B)$ is the projection of the largest value of function $f(A, B)$ over the MVL space $\mathbb{M}_B$.

EXAMPLE 3. *Figure 6 shows the function* $f(a, b, c, d)$ *and the functions*

$$g(a, b, c) = \min_d f(a, b, c, d), \tag{9}$$

$$h(a, b) = \min_c g(a, b, c) = \min_{\{c,d\}}^k f(a, b, c, d). \tag{10}$$

The values of the first column of the function $g(a, b, c)$ *are the minimum values of the first three columns of the function* $f(a, b, c, d)$. *The second column of* $g(a, b, c)$ *consists of the minimum values of the last three columns of* $f(a, b, c, d)$. *The shaded values of the function* $f(a, b, c, d)$ *indicate the values of the function* $g(a, b, c)$.

The values of the function $h(a, b)$ *are the minimum values of the two columns of the function* $g(a, b, c)$ *(indicated by darkly shaded values), which is the same value as the minimum value of all six columns of the function* $f(a, b, c, d)$ *(also indicated by darkly shaded values).*

The min-operator for integers is characterized by two properties.

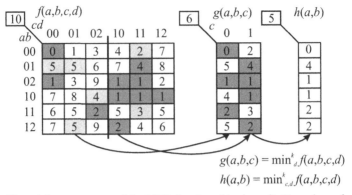

$$g(a,b,c) = \min^k_d f(a,b,c,d)$$
$$h(a,b) = \min^k_{c,d} f(a,b,c,d)$$

Figure 6. The minimum operator of the MVL function $f(a, b, c, d)$. The values of $g(a, b, c)$ are indicated by shaded boxes in $f(a, b, c, d)$. The values of $h(a, b)$ are indicated by darkly shaded boxes in $f(a, b, c, d)$ and $g(a, b, c)$.

1. The minimum is smaller than or equal to its arguments,

$$\min(a_1, \ldots, a_n) \leq a_i, \quad i = 1, \ldots, n.$$

2. There exists at least one argument of the min-operator that equals the minimum,

$$\exists i : a_i = \min(a_1, \ldots, a_n), \quad 1 \leq i \leq n.$$

These properties can be extended to the $\min^k$ and $\max^k$ operators. Property 1 translates into the observation that the values of the function $h(a, b)$ are smaller than or equal to the values of the function $f(a, b, c, d)$ in the same row. According to property 2, the value of the function $h(a, b)$ appears at least once in the same row in the chart of the function $f(a, b, c, d)$. Both properties can be summarized as shown in the theorem below.

THEOREM 1. *Let $f(A, B)$ be an MVL function, where A and B are disjoint sets of variables. Every function $g(A)$, that does not depend on the set of variables B, satisfies*

$$g(A) \leq f(A, B) \text{ iff } g(A) \leq \min_B^k f(A, B) \quad \text{and} \qquad (11a)$$

$$g(A) \geq f(A, B) \text{ iff } g(A) \geq \max_B^k f(A, B). \qquad (11b)$$

Theorem 1 provides a test criteria for a partial order between functions, when these functions depend on different sets of variables. Such comparisons are frequently required for bi-decomposition tests.

3. Sets of MVL Functions

3.1. *Incompletely specified MVL functions*

Incompletely specified MVL functions (MVL ISFs) are frequently used in logic synthesis and machine learning. If a certain combination of inputs cannot occur in an application, the value of the synthesized function for this input combination can be any value, which is modeled by a don't care.

In machine learning there is a mapping from the attributes to the goal for a given set of examples. Examples are given only for a small subset of all possible combinations of attributes. The aim of machine learning is to determine values of the goal for attribute combinations that are not given. The set of examples can be modeled by MVL ISFs, which are a direct extension of Boolean ISFs.

DEFINITION 6. *An MVL ISF $F(A)$ is a mapping $\mathbb{C} \to \{0, \ldots, m_o - 1\}$ from a subset $\mathbb{C} \subseteq \mathbb{M}_A$ of all minterms into the set of integers $\{0, \ldots, m_o - 1\}$. The set $\mathbb{C}$ is called* care set, *the complement set $\mathbb{D} = \mathbb{M}_A \setminus \mathbb{C}$ is the* don't care set *(DC-set).*

MVL ISFs can be visualized by maps and tables similar to MVL functions, see Section 2.1. The elements of the care set are denoted by their respective values, the elements of the DC-set by Φ.

Although ISFs are defined as partially defined functions, it is sometimes useful to view ISFs as a multiple-valued function set (MFS), where all MVL functions of the set have the same values as the ISF for its cares. Sections 3.2 and 3.3 show further examples of MFSs. The MFS defined by an ISF is called its characteristic function set.

DEFINITION 7. *The* characteristic function set $\mathbf{F}(A)$ *of an MVL ISF $F(A)$ with the care set $\mathbb{C}$ is the set of all MVL functions $f(A)$ with $F(A_i) = f(A_i)$ for all minterms $A_i \in \mathbb{C}$:*

$$\mathbf{F}(A) = \big\{ f(A) \mid \forall A_i \in \mathbb{C} : f(A_i) = F(A_i) \big\}. \qquad (12)$$

EXAMPLE 4. *The ISF $F(a, b)$ (see Figure 7(a)) has the characteristic function set $\mathbf{F}(a, b) = \big\{ f_1(a, b), f_2(a, b), f_3(a, b) \big\}$ (see Figure 7(b)).*

3.2. *Multi-valued function intervals*

A Boolean ISF can be described as an interval of functions $\mathbf{F}(A) = [f_l, f_u]$. An interval of functions can also be defined for MVL functions. In this section

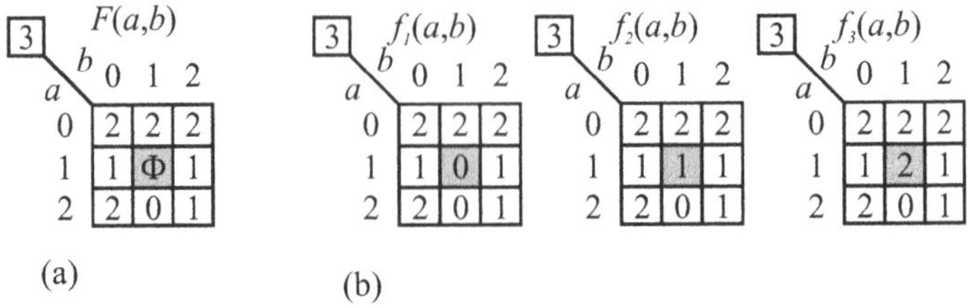

Figure 7. Characteristic function set of an ISF. a. ISF $F(a, b)$. b. Characteristic function set $\mathbf{F}(a, b) = \left\{ f_1(a, b), f_2(a, b), f_3(a, b) \right\}$.

multi-valued function intervals are defined as a generalization of MVL ISFs (Takagi and Nakashima 2000; Lang and Steinbach 2001; Mishchenko et al. 2001b). MFIs form the basis of max- and min-bi-decomposition (MM-bi-decomposition) of MVL functions, see Section 4.

DEFINITION 8. *A multi-valued function interval (MFI)* $\mathbf{F}(A) = [f_l, f_u]$ *with* $f_l(A) \leq f_u(A)$ *is the set of MVL functions*

$$\mathbf{F}(A) = \left\{ f(A) \mid f_l(A) \leq f(A) \leq f_u(A) \right\}. \tag{13}$$

The function $f_l(A)$ *is the* lower bound, *and the function* $f_u(A)$ *is the* upper bound *of the MFI.*

An MFI $\mathbf{F}(A) = [f_l, f_u]$ can be visualized by a map similar to ISFs. For a minterm A_i, the bounds $f_l(A_i)$ and $f_u(A_i)$ are shown as an interval $[f_l(A_i), f_u(A_i)]$ in the field A_i of the map. If $f_l(A_i) = f_u(A_i)$ then the minterm A_i specifies a *care* and only the value $f_l(A_i)$ of the care is shown in the field A_i.

EXAMPLE 5. *Figure 8(a) shows the map of the MFI* $\mathbf{F}(a, b) = [f_1, f_4]$, *where the functions* $f_1(a, b)$ *and* $f_4(a, b)$ *are displayed in Figure 8(b). The element functions* $\mathbf{F}(a, b) = \left\{ f_1(a, b), f_2(a, b), f_3(a, b), f_4(a, b) \right\}$ *are also displayed in Figure 8(b).*

3.3. *Multi-valued relations*

MVL ISFs either specify the value of a function for a particular minterm completely, or the value is undefined. MFIs give more control over the don't cares by specifying intervals of values. There are several reasons to extend this control over the don't cares to arbitrary subsets of values as is the case in multi-valued relations (MVR).

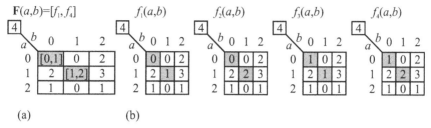

(a)   (b)

Figure 8. Characteristic function set of an MFI. a. MFI $\mathbf{F}(a, b) = [f_1, f_4]$. b. Characteristic function set $\mathbf{F}(a, b) = \Big\{ f_1(a, b),\ f_2(a, b),\ f_3(a, b),\ f_4(a, b) \Big\}$.

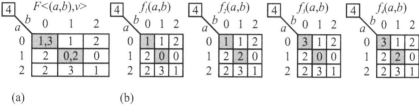

(a)   (b)

Figure 9. Characteristic function set of an MVR. a. MVR $F \langle (a, b), v \rangle$. b. Characteristic function set $\mathbf{F}(a, b) = \Big\{ f_1(a, b),\ f_2(a, b),\ f_3(a, b),\ f_4(a, b) \Big\}$ of the MVR $F \langle (a, b), v \rangle$.

DEFINITION 9. *A* multi-valued relation (MVR) $R \langle A, v \rangle$ *is a many-to-many mapping, or relation, between the set of minterms* $\mathbb{M}_A$ *of the variables A and the set of output values* $\mathbb{M}_v = \{0, \ldots, m_o - 1\}$

$$R \langle A, v \rangle \subseteq \mathbb{M}_A \times \mathbb{M}_v, \tag{14}$$

where at least one value is specified for each minterm A_i,

$$\forall A_i : \exists v_j : \langle A_i, v_j \rangle \in R \langle A, v \rangle. \tag{15}$$

Similar to ISFs and MFIs, MVRs define a set of MVL functions.

DEFINITION 10. *The* characteristic function set *of the MVR $R \langle A, v \rangle$ is the set* $\mathbf{R}(A)$ *of all functions* $f(A)$, *whose function values agree with the values specified by the MVR $R \langle A, v \rangle$:*

$$\mathbf{R}(A) = \big\{ f(A) \mid \forall A_i \in \mathbb{M}_A : \langle A_i, f(A_i) \rangle \in R \langle A, v \rangle \big\}. \tag{16}$$

EXAMPLE 6. *Consider the MVR $F \langle (a, b), v \rangle$ from Figure 9(a). For every member function $f(a, b)$ of the characteristic function set* $\mathbf{F}(a, b)$ *there is $f(ab = 00) \in \{1, 3\}$ and $f(ab = 11) \in \{0, 2\}$. The combination of these values results in the characteristic function set* $\mathbf{F}(a, b) = \big\{ f_1(a, b), \ldots, f_4(a, b) \big\}$ *shown in Figure 9(b). Because $F \langle (a, b), v \rangle$ is an*

MVR and not an MFI, the values in the map of Figure 9(a) are listed without brackets, and "0, 2" in the cell $ab = 11$ represents the values '0' and '2'. In contrast, the interval $[0, 2]$ represents the value '1' in addition to '0' and '2'.

The main reason that MVRs are introduced in this article is that there is an extremely efficient data structure called binary encoded multi-valued decision diagram (BEMDD) to store and process MVRs on computers (Mishchenko et al. 2000). Because MVRs generalize ISFs and MFIs, BEMDDs can be applied to process these structures too.

The BEMDD of the relation $R \langle A, v \rangle$ encodes the MVL variables A and v by the sets of Boolean variables B_A and B_v respectively. A Boolean function $f(B_A, B_v)$ is defined as $f(B_A, B_v) = 1$ if $\langle A, v \rangle \in R \langle A, v \rangle$ and $f(B_A, B_v) = 0$ otherwise. The Boolean function $f(B_A, B_v)$ is stored as a BDD and can be processed with standard BDD packages.

4. Max- and Min-Bi-Decomposition

4.1. *Max- and min-bi-decomposition problem*

This section discusses bi-decomposition of MFIs with respect to the max- and min-operators, see Figure 1 on page 235 with the max- and min-operator for the operator π. Given an MFI $\mathbf{F}(A, B, C) = [f_l, f_u]$ the max-bi-decomposition problem consists in finding pairs of functions $\langle g(A, C), h(B, C) \rangle$ that satisfy

$$\max(g(A, C), h(B, C)) \geq f_l(A, B, C) \quad \text{and} \tag{17a}$$

$$\max(g(A, C), h(B, C)) \leq f_u(A, B, C). \tag{17b}$$

The sets A and B are called the *dedicated sets*. The set C is called the *common set*. If there exists a pair $\langle g(A, C), h(B, C) \rangle$ that satisfies (17), the MFI $\mathbf{F}(A, B, C)$ is called *max-bi-decomposable*. Additional terms must be defined to describe the relation between the functions $g(A, C)$ and $h(B, C)$ if there is more than one pair that satisfies (17).

– The *free set* is the set $\mathbf{G}(A, C)$ of all functions $g(A, C)$ for which there exists at least one function $h(B, C)$ that satisfies (17).
– The *bound set* $H_{g0}(B, C)$ bound to the function $g_0(A, C)$ is the set of functions $h(B, C)$ so that $\langle g_0(A, C), h(B, C) \rangle$ satisfies (17).

Similar terms can be defined for the min-bi-decomposition by replacing the max-operator with the min-operator.

The properties developed in this section are mostly generalizations of the bi-decomposition theory for Boolean ISFs using the BDC. The results for

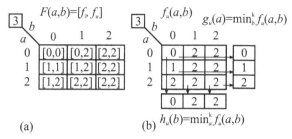

Figure 10. Max-bi-decomposition of an MFI. a. Function interval $\mathbf{F}(a, b)$. b. Upper bounds $g_u(a)$ and $h_u(b)$ of the max-bi-decomposition functions of $\mathbf{F}(a, b)$.

the OR- and AND-bi-decomposition of Boolean ISFs are generalized to the max- and min-bi-decomposition of MFIs respectively (Lang and Steinbach 2001; Mishchenko et al. 2001b). Instead of the BDC, multi-valued derivative operators are applied.

4.2. *Bounds on the max-decomposition functions*

An upper bound on the max-bi-decomposition functions can be developed from the upper bound of the given MFI. A max-bi-decomposition $\langle g(A, C), h(B, C) \rangle$ of the MFI $\mathbf{F}(A, B, C) = [f_l, f_u]$ must satisfy inequality (17b). Because there is $g(A, C) \leq \max\big(g(A, C), h(B, C)\big)$, it can be concluded that $g(A, C) \leq f_u(A, B, C)$.

EXAMPLE 7. *Consider the max-bi-decomposition of the MFI $\mathbf{F}(a, b)$ shown in Figure 10(a) with respect to dedicated sets $A = \{a\}$ and $B = \{b\}$. To keep the example simple, the common set C is empty. The values of the function $g(a_i)$ should not exceed the values of $f_u(a_i, b)$ as shown in Figure 10(b). Therefore, an upper limit on the function values of $g(a)$ is the minimum value of each row of the chart of $f_u(a, b)$. Because the max-operator is commutative, a similar observation holds for $h(b)$.*

Note that function $g(A, C)$ does not depend on the set of variables B. Therefore, Theorem 1 can be applied to compare the functions $g(A, C)$ and $f_u(A, B, C)$ as shown in Lemma 1.

LEMMA 1. *Let the MFI $\mathbf{F}(A, B, C) = [f_l, f_u]$ be max-bi-decomposable with respect to the dedicated sets A and B. If $\langle g(A, C), h(B, C) \rangle$ is a max-bi-decomposition of $\mathbf{F}(A, B, C)$ then there is*

$$g(A, C) \leq g_u(A, C) \leq f_u(A, B, C) \quad \text{and} \tag{18a}$$

$$h(B, C) \leq h_u(B, C) \leq f_u(A, B, C), \tag{18b}$$

where the functions $g_u(A, C)$ and $h_u(B, C)$ are defined by

$$g_u(A, C) = \min_B{}^k f_u(A, B, C) \quad \text{and} \tag{19a}$$

$$h_u(B, C) = \min_A{}^k f_u(A, B, C). \tag{19b}$$

Lemma 1 shows that there are upper limits for the decomposition functions. It remains the questions are these limits tight, i.e., are there decompositions that reach these limits? Consider a max-bi-decomposition $\langle g(A, C), h(B, C) \rangle$. This pair satisfies (17a). Certainly this inequality remains satisfied if $g(A, C)$ is increased to $g_u(A, C)$. Now consider (17b). Because of $g_u(A, C) \leq f_u(A, B, C)$, this inequality is satisfied for $g_u(A, C)$ if it is satisfied for $g(A, C)$. Hence, the pair $\langle g_u(A, C), h(B, C) \rangle$ is also a max-bi-decomposition as stated in Lemma 2.

LEMMA 2. *Let the MFI* $\mathbf{F}(A, B, C) = [f_l, f_u]$ *be max-decomposable, and let the pair* $\langle g(A, C), h(B, C) \rangle$ *be a max-bi-decomposition of* $\mathbf{F}(A, B, C)$. *Then* $\langle g_u(A, C), h(B, C) \rangle$ *and* $\langle g(A, C), h_u(B, C) \rangle$ *are also max-bi-decompositions of* $\mathbf{F}(A, B, C)$, *where*

$$g_u(A, C) = \min_B{}^k f_u(A, B, C) \quad \text{and} \tag{20a}$$

$$h_u(B, C) = \min_A{}^k f_u(A, B, C). \tag{20b}$$

Lemma 2 leads to an important consequence. In general, there are many max-bi-decomposition functions, say $g_1, \ldots, g_m$ and $h_1, \ldots, h_n$. Not every pair of functions $\langle g_i, h_j \rangle$, $i = 1, \ldots, m$, $j = 1, \ldots, n$ is a max-bi-decomposition, but Lemma 2 shows that all pairs $\langle g_u, h_j \rangle$, $j = 1, \ldots, n$ and $\langle g_i, h_u \rangle$, $i = 1, \ldots, m$ are a max-bi-decompositions.

The development of the lower bound on the decomposition functions is complicated by the fact that the lower bound for the functions $g(A, C)$ depends on the other decomposition function $h(B, C)$. A decomposition must satisfy inequality (17a) for all minterms (A_i, B_j, C_k). If there is $h(B_1, C_1) \geq f_l(A_1, B_1, C_1)$ for the minterm (A_1, B_1, C_1) then inequality (17a) is satisfied for (A_1, B_1, C_1) independently of the value of $g(A_1, C_1)$, and no restriction results from this minterm for $g(A_1, C_1)$. Otherwise, if $h(B_2, C_2) < f_l(A_2, B_2, C_2)$, there must be $g(A_2, C_2) \geq f_l(A_2, B_2, C_2)$ to satisfy (17a). Because $g(A, C)$ must satisfy (17a) for all minterms B, the maximum value over B must be taken.

EXAMPLE 8. *Consider the max-bi-decomposition of the MFI* $\mathbf{F}(a, b)$ *from Figure 10(a) with respect to the function* $h_u(b)$ *shown below the map of* $f_l(a, b)$ *in Figure 11(a). For the minterm* $ab = 11$ *there is* $h_u(b = 1) =$

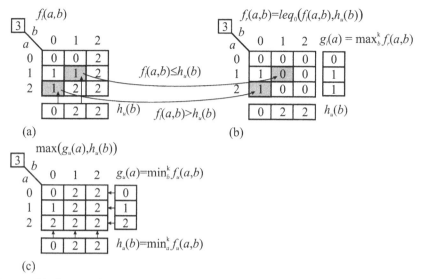

Figure 11. Computation of the max-bi-decomposition function $g_l(a)$ for the MFI $\mathbf{F}(a, b) = [f_l, f_u]$ from Figure 10. a. Lower bound $f_l(a, b)$. b. Computation of the lower bound $g_l(a)$ of the free set. c. Test for max-decomposability by comparison of $\max\big(g_u(a), h_u(b)\big)$ with $f_l(a, b)$.

$2 \geq 1 = f_l(ab = 11)$ *and* $\max\big(g(a = 1), h(b = 1)\big) \geq f_l(ab = 11)$ *is satisfied independently of the value of* $g(a = 1)$. *For the minterm* $ab = 20$ *there is* $h_u(b = 0) = 0 < 1 = f_l(ab = 20)$. *Therefore, the minimum value of* $g(a = 2)$ *must be* $f_l(ab = 20) = 1$. *The restrictions on* $g(a)$ *for all minterms* b_j *are shown in the map of function* $f_r(a, b)$ *in Figure 11(b). The lower bound* $g_l(a)$, *after taking the maximum over the variable b, is shown to the right of the map.*

The restrictions for the lower bound can be expressed using the operator $leq_0(x, y)$, see Section 2.2. There is $leq_0\big(f_l(A_i, B_j, C_k), h(B_j, C_k)\big) = 0$, which implies no restriction if there is $f_l(A_i, B_j, C_k) \leq h(B_j, C_k)$, and there is $leq_0\big(f_l(A_i, B_j, C_k), h(B_j, C_k)\big) = f_l(A_i, B_j, C_k)$ as restriction otherwise. Lemma 3 shows how to compute the lower bound $g_l(A, C)$ using the operator leq_0.

LEMMA 3. *Let the MFI* $\mathbf{F}(A, B, C) = [f_l, f_u]$ *be max-decomposable with respect to the dedicated sets A and B. If* $\langle g(A, C), h(B, C) \rangle$ *is a max-bi-decomposition of* $\mathbf{F}(A, B, C)$ *then there is*

$$g(A, C) \geq g_l(A, C) \tag{21}$$

$$g_l(A, C) = \max_B{}^k leq_0\big(f_l(A, B, C), h(B, C)\big). \tag{22}$$

Again, there remains the question if the lower bound is tight. Observe that equation (22) restricts $g_l(A, C)$ so that (17a) is satisfied for $\langle g_l(A, C), h(B, C) \rangle$. Furthermore, function $g_l(A, C)$ is a lower bound on the decomposition functions because there is $g_l(A, C) \neq 0$ only where necessary to satisfy (17a). Therefore, there is $g_l(A, C) \leq g(A, C)$ if $\langle g(A, C), h(B, C) \rangle$ is a max-bi-decomposition that satisfies (17b). It follows that (17b) remains satisfied if $g(A, C)$ is decreased to $g_l(A, C)$. Lemma 4 summarizes this argument.

LEMMA 4. *Let the MFI* $\mathbf{F}(A, B, C) = [f_l, f_u]$ *be max-decomposable with respect to the dedicated sets A and B, and let $\langle g(A, C), h(B, C) \rangle$ be a max-bi-decomposition of* $\mathbf{F}(A, B, C)$. *Then the pair of functions $\langle g_l(A, C), h(B, C) \rangle$ is also a max-bi-decomposition of* $\mathbf{F}(A, B, C)$, *where*

$$g_l(A, C) = \max_{B}{}^k leq_0 \big(f_l(A, B, C), h(B, C) \big). \tag{23}$$

4.3. *Max-decomposition test*

A test condition for the max-decomposability of MFIs can now be derived from the upper bounds of the decomposition functions. On the one hand, Lemmas 1 and 2 showed tight upper bounds $g_u(A, C)$ and $h_u(B, C)$ for the decomposition functions, see (20). Therefore, a decomposition can exist only if these bounds satisfy (17a). On the other hand if (17a) is satisfied by the upper bounds, the functions $g_u(A, C)$ and $h_u(B, C)$ are a decomposition and the MFI is decomposable. Theorem 2 formalizes this argument.

THEOREM 2. *The MFI* $\mathbf{F}(A, B, C) = [f_l, f_u]$ *is max-decomposable with respect to the dedicated sets A and B iff*

$$f_l(A, B, C) \leq \max \big(g_u(A, C), h_u(B, C) \big), \tag{24}$$

where

$$g_u(A, C) = \min_{B}{}^k f_u(A, B, C) \tag{25a}$$

$$h_u(B, C) = \min_{A}{}^k f_u(A, B, C). \tag{25b}$$

EXAMPLE 9. *Consider the max-bi-decomposition of the MFI* $\mathbf{F}(a, b)$ *shown in Figure 10(a). The functions $g_u(a)$ and $h_u(b)$ are displayed together with $\max \big(g_u(a), h_u(b) \big)$ in Figure 11(c). Comparison with $f_l(a, b)$ in Figure 11(a) shows that (24) is satisfied. Additional comparison with $f_u(a, b)$ from Figure 10(a) reveals that $\langle g_u(a), h_u(b) \rangle$ is indeed a max-bi-decomposition of $F(a, b) = [f_l, f_u]$. The equality $\max \big(g_u(a), h_u(b) \big) = f_u(a, b)$ is a coincidence.*

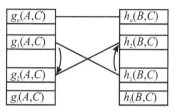

Figure 12. Relation between free and bound set for the max-bi-decomposition.

4.4. *Computation of the max-decomposition functions*

To apply max-bi-decomposition in decomposers, it is necessary to compute the free and bound sets. First, a formula for the bound set is derived in Theorem 3. Then the free set is computed in Theorem 4. The bound set can be computed directly from the tight upper and lower bounds of the decompositions functions shown in Lemmas 1 and 3 (A and $g(A, C)$ is exchanged with B and $h(B, C)$).

THEOREM 3. *Let the MFI* $\mathbf{F}(A, B, C) = [f_l, f_u]$ *be max-decomposable with respect to the dedicated sets A and B, and let the function $g_0(A, C)$ be a decomposition function of* $\mathbf{F}(A, B, C)$. *Then the bound set of the max-bi-decomposition of* $\mathbf{F}(A, B, C)$ *with respect to the function $g_0(A, C)$ is the MFI* $\mathbf{H}_{g_0}(B, C) = [h_l, h_u]$, *where*

$$h_l(B, C) = \max_A{}^k leq_0\big(f_l(A, B, C), g_0(A, C)\big) \qquad (26)$$

$$h_u(B, C) = \min_A{}^k f_u(A, B, C). \qquad (27)$$

The free set is computed similarly. Theorem 3 can be applied, where $g_0(A, C)$ is substituted with the upper bound of Lemma 1, $h_u(B, C) = \min_A{}^k f_u(A, B, C)$ to make $g_l(A, C)$ as small as possible.

THEOREM 4. *Let the MFI* $\mathbf{F}(A, B, C) = [f_l, f_u]$ *be max-decomposable with respect to the dedicated sets A and B. Then the free set $G'(A, C)$ of the max-bi-decomposition of* $\mathbf{F}(A, B, C)$ *is equal to the MFI* $\mathbf{G}(A, C) = [g_l, g_u]$, *where*

$$g_l(A, C) = \max_B{}^k leq_0\big(f_l(A, B, C), \min_A{}^k f_u(A, B, C)\big) \qquad (28)$$

$$g_u(A, C) = \min_B{}^k f_u(A, B, C). \qquad (29)$$

The relation between the free and bound set of max-bi-decomposition is shown in Figure 12. Every function of the MFI $\mathbf{G}_1(A, C) = [g_1, g_u]$ is a decomposition with every function of the MFI $\mathbf{H}_1(B, C) = [h_1, h_u]$. The functions $g_u(A, C)$ and $h_u(B, C)$ are computed by (20), and $h_1(B, C)$ is computed by (26). If the function $g_1(A, C)$ is decreased to $g_2(A, C)$ the upper bounds $g_u(A, C)$ and $h_u(B, C)$ do not change. However, the lower bound of $\mathbf{H}(B, C)$ increases to $h_2(B, C)$, i.e. the bound set becomes smaller. The smallest possible value for $g(A, C)$ with a nonempty bound set $\mathbf{H}(B, C)$ is the function $g_l(A, C)$ computed by (28).

It is possible to exchange the sizes of the sets $\mathbf{G}(A, C)$ and $\mathbf{H}(B, C)$. Increasing the MFI $\mathbf{G}(A, C)$ (so that it contains more functions), reduces the size of the MFI $\mathbf{H}(B, C)$ and vice versa.

4.5. *Min-decomposition*

Min-bi-decomposition is dual to max-bi-decomposition. The results for max-bi-decomposition are adapted and summarized for the case of min-bi-decomposition.

THEOREM 5. *The MFI* $\mathbf{F}(A, B, C) = [f_l, f_u]$ *is min-decomposable with respect to the dedicated sets A and B iff*

$$f_u(A, B, C) \geq \min\bigl(g_l(A, C), h_l(B, C)\bigr), \tag{30}$$

where

$$g_l(A, C) = \max_{B}^{k} f_l(A, B, C) \tag{31a}$$

$$h_l(B, C) = \max_{A}^{k} f_l(A, B, C). \tag{31b}$$

Theorem 6 applies the geqm-operator (see Section 2.2) as dual operator to the leq0-operator.

THEOREM 6. *Let the MFI* $\mathbf{F}(A, B, C) = [f_l, f_u]$ *be min-decomposable with respect to the dedicated sets A and B, and let the function* $g_0(A, C)$ *be a decomposition function of* $\mathbf{F}(A, B, C)$. *Then the bound set of the min-bi-decomposition of* $\mathbf{F}(A, B, C)$ *with respect to the function* $g_0(A, C)$ *is the MFI* $\mathbf{H}_{g_0} = [h_l, h_u]$, *where*

$$h_l(B, C) = \max_{A}^{k} f_l(A, B, C) \tag{32}$$

$$h_u(B, C) = \min_{A}^{k} geq_{m_F}\bigl(f_u(A, B, C), g_0(A, C)\bigr). \tag{33}$$

where m_F *is the output cardinality of the MFI* $\mathbf{F}(A, B, C)$.

THEOREM 7. *Let the MFI* $\mathbf{F}(A, B, C) = [f_l, f_u]$ *be min-decomposable with respect to the dedicated sets A and B. Then the free set of the min-bi-decomposition of* $\mathbf{F}(A, B, C)$ *is the MFI* $\mathbf{G}(A, C) = [g_l, g_u]$, *where*

$$g_l(A, C) = \max_B^k f_l(A, B, C) \tag{34}$$

$$g_u(A, C) = \min_B^k geq_{m_F}\left(f_u(A, B, C), \max_A^k f_l(A, B, C)\right) \tag{35}$$

5. Separation of Non-Decomposable MFSs

5.1. *Separation methods*

There are MFSs that are not bi-decomposable with respect to a given set of operators. Weak bi-decomposition is applied to simplify non-bi-decomposable Boolean functions. Unfortunately, there are MVL functions that are neither bi-decomposable nor weakly bi-decomposable (Mishchenko et al. 2000). Therefore, new decomposition principles are necessary to ensure complete decomposition of all MVL functions.

Bi-decomposition decomposes a given MFS $\mathbf{F}(A, B, C)$ into the sets $\mathbf{G}(A, C)$ and $\mathbf{H}(B, C)$, where each set depends on fewer variables. In order to simplify non-bi-decomposable MFSs, these requirements must be relaxed.

Separation is a generalization of bi-decomposition. An non-decomposable MFS is split into two or more separation MFSs, which depend on the same set of variables as the original MFS. To ensure simplification, the separation MFSs are either bi-decomposable or supersets of the original MFS. Two separation methods implement this principle:

1. *Multi-decomposition* splits an MFS into n bi-decomposable separation sets, where the number n is minimized.
2. *Set separation* splits an MFS into a bi-decomposable MFS and a superset of the original MFS.

Multi-decomposition of MFIs with respect to the min- and max-operators are discussed in the next section. See (Lang 2003) for an example of set separation.

5.2. *Multi-decomposition*

5.2.1. *Definition of multi-decomposition*
It will be shown, that it is always possible to split an MFS $\mathbf{F}(A, B, C)$ into a set of n bi-decomposable functions $f_1(A, B, C), \ldots, f_n(A, B, C)$. Below, there is a precise definition of this concept.

DEFINITION 11. *A multi-decomposition of an MFS* $\mathbf{F}(A, B, C)$ *with respect to the dedicated sets A and B, and the operators* $\mu(x_1, \ldots, x_n)$ *and* $\pi(y, z)$ *is a vector* $\vec{f}(A, B, C) = [f_1(A, B, C), \ldots, f_n(A, B, C)]$ *of n functions so that*

$$\mu\big(f_1(A, B, C), \ldots, f_n(A, B, C)\big) \in \mathbf{F}(A, B, C), \tag{36}$$

where the functions $f_1(A, B, C), \ldots, f_n(A, B, C)$ *are* π*-decomposable with respect to the dedicated sets A and B.*

To obtain decompositions of low complexity, the size n of the vector $\vec{f}(A, B, C)$ should be as small as possible.

5.2.2. *Max-min multi-decomposition*
In the following, multi-decomposition of MFIs with respect to the MM-operators is considered.

DEFINITION 12. *A max-min multi-decomposition of an MFI* $\mathbf{F}(A, B, C)$ *with respect to the dedicated sets A and B is a vector of n functions*

$$\vec{f}(A, B, C) = [f_1(A, B, C), \ldots, f_n(A, B, C)]$$

so that
$$\max\big(f_1(A, B, C), \ldots, f_n(A, B, C)\big) \in \mathbf{F}(A, B, C), \tag{37}$$

where the functions $f_1(A, B, C), \ldots, f_n(A, B, C)$ *are min-decomposable with respect to the dedicated sets A and B.*

To compute max-min multi-decompositions, another criterion for min-decomposability of MFIs is developed in extension of Section 4.5. Then it is shown how this criterion can be used to determine the minimum number of functions for max-min multi-decomposition.

The min-decomposability of an MFI can be checked locally by the inspection of selected values of the interval bounds.

EXAMPLE 10. *Consider the MFI* $\mathbf{F}(a, b) = [f_l, f_u]$ *in Figure 13(a). To see that the MFI is not min-decomposable, it is sufficient to consider only three values of the interval bounds, printed in bold type in Figure 13(a). For every min-decomposition function* $g(a)$ *there is* $g(a = 0) \geq f_l(ab = 00) = 2$, *and for every min-decomposition function* $h(b)$ *there is* $h(b = 2) \geq f_l(ab = 22) = 3$. *It follows that*

$$\min\big(g(a = 0), h(b = 2)\big) \geq 2. \tag{38}$$

This restriction contradicts the condition $f_u(ab = 02) = 0 < 2$. *Therefore, the MFI* $\mathbf{F}(a, b)$ *is not min-decomposable.*

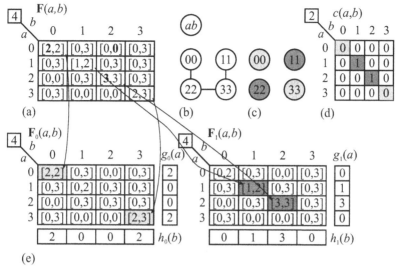

Figure 13. Multi-decomposition of MFIs. a. Non-decomposable MFI $\mathbf{F}(a, b)$. Values indicating that $\mathbf{F}(a, b)$ is not min-decomposable are printed in bold type. b. Min-incompatibility graph of $\mathbf{F}(a, b)$. c. Coloring of the incompatibility graph. d. Coloring function $\xi(a, b)$. e. Multi-decomposition of $\mathbf{F}(a, b)$ into $\mathbf{F}_0(a, b)$ and $\mathbf{F}_1(a, b)$.

Min-decomposability of an MFI can be checked by considering all applicable triple of function values. Theorem 8 formulates a min-decomposition criterion based on this observation.

THEOREM 8. *An MFI $\mathbf{F}(A, B, C) = [f_l, f_u]$ is min-decomposable iff for all 5-tuples of minterms A_i, A_j, B_m, B_n, C_k there is*

$$\min\left(f_l(A_i, B_m, C_k), f_l(A_j, B_n, C_k)\right) \le f_u(A_i, B_n, C_k). \qquad (39)$$

5.2.3. *Incompatibility graphs*
Properties of max-min multi-decompositions can be shown with the help of an *incompatibility graph*, which is a graph with one node for each minterm and an edge between minterms for every pair of minterms that does not satisfy (39).

DEFINITION 13. *The min-incompatibility graph of the MFI $\mathbf{F}(A, B, C)$ with respect to the dedicated sets A and B is a graph with one node for each minterm (A_i, B_m, C_k). There is an edge between all pairs of minterms (A_i, B_m, C_k) and (A_j, B_n, C_k) with*

$$\min\left(f_l(A_i, B_m, C_k), f_l(A_j, B_n, C_k)\right) > f_u(A_i, B_n, C_k), \qquad (40)$$

where $\mathbf{F}(A, B, C) = [f_l, f_u]$.

EXAMPLE 11. *The min-incompatibility graph of the MFI* $\mathbf{F}(a, b)$ *from Figure 13(a) is shown in Figure 13(b). The nodes are labeled with the value ab of the minterms. To simplify the graph, nodes with $f_l(a, b) = 0$ are omitted because for these nodes (39) is always satisfied and these nodes are not connected to any edge. Therefore, only four nodes of the incompatibility graph of* $\mathbf{F}(a, b)$ *are shown. For instance, there is an edge between the nodes $ab = 00$ and $ab = 22$ because there is*

$$\min\big(f_l(ab = 00), f_l(ab = 22)\big) \; > \; f_u(ab = 02) \tag{41}$$

$$\min(2, 3) \; > \; 0. \tag{42}$$

A direct consequence of Definition 13 and Theorem 8 is Lemma 5.

LEMMA 5. *Let* $\mathbf{F}(A, B, C) = [f_l, f_u]$ *be an MFI. Then* $\mathbf{F}(A, B, C)$ *is min-decomposable iff its min-incompatibility graph does not contain any edges.*

5.2.4. *Coloring incompatibility graphs*
Lemma 5 shows that max-min multi-decomposition problem is equivalent to partitioning the incompatibility graph of an MFI into subsets so that there are no edges between the nodes of the same subset. This is the *graph coloring problem*, which assigns a color to each node so that no pair of nodes with the same color is connected by an edge.

THEOREM 9. *An MFI* $\mathbf{F}(A, B, C) = [f_l, f_u]$ *is max-min multi-decomposable into n functions iff its min-incompatibility graph is colorable with n colors.*

EXAMPLE 12. *A coloring of the incompatibility graph for the MFI* $\mathbf{F}(a, b)$ *from Figure 13(a) is drawn in Figure 13(c). Two colors are needed to color the graph. The minterms that are not shown in the graph are isolated and do not have any edges. Therefore, they can be colored by any color. Color 0 is assigned to the lightly shaded and the isolated nodes. Color 1 is assigned to the darkly shaded nodes. The coloring function that maps the minterms to the number of its assigned color is shown in Figure 13(d). The max-min multi-decomposition of* $\mathbf{F}(a, b)$ *into* $\mathbf{F}_0(a, b)$ *and* $\mathbf{F}_1(a, b)$ *that corresponds to this coloring function is shown in Figure 13(e). The lower bound of* $\mathbf{F}_0(a, b)$ *is set to '0' for all minterms (a_i, b_j) with the coloring function $\xi(a_i, b_j) \neq 0$. The min-bi-decomposition* $\min\big(g_0(a), h_0(b)\big) \in \mathbf{F}_0(a, b)$ *is shown in Figure 13(e). Similarly, the lower bound in* $\mathbf{F}_1(a, b)$ *is set to '0' for all minterms (a_i, b_j) with $\xi(a_i, b_j) \neq 1$. The min-bi-decomposition of* $\min\big(g_1(a), h_1(b)\big) \in \mathbf{F}_1(a, b)$ *is also shown Figure 13(e).*

5.2.5. *Number of subfunctions*

This section concludes with an upper bound on the number of subfunctions that can be produced by max-min multi-decomposition. The upper bound is computed from the maximum number of colors that are necessary to color the min-incompatibility graphs of MFIs.

EXAMPLE 13. *Consider the application of (39) to three minterms located in the same row a_i of the map in Figure 13(a), say $f_l(a_i, b_m)$, $f_l(a_i, b_n)$ and $f_u(a_i, b_n)$. Obviously, (39) is always satisfied for such a triple of function values because*

$$\min(f_l(a_i, b_m), f_l(a_i, b_n)) \leq f_l(a_i, b_n) \leq f_u(a_i, b_n) \tag{43}$$

follows from $f_l(a, b) \leq f_u(a, b)$.

Therefore, a coloring of the incompatibility graph is to color each row with a different color. Similarly a coloring of the incompatibility graph is to color each column with a different color. I can be concluded that the number of colors necessary to color the incompatibility graph is bounded by the minimum numbers of rows and columns.

THEOREM 10. *The min-incompatibility graph of an MFI $\mathbf{F}(A, B, C) = [f_l, f_u]$ with respect to the dedicated sets A and B is colorable with $n \leq \min(|\mathbb{M}_A|, |\mathbb{M}_B|)$ colors, where $|\mathbb{M}_A|$ and $|\mathbb{M}_B|$ are the numbers of different minterms of the variable sets A and B respectively.*

6. The Bi-Decomposer YADE

6.1. *Purpose of YADE*

The YADE decomposer is a program that decomposes MVRs into a network of interconnected MVL gates. Its primary application is MVL logic synthesis and data mining. The MVRs are decomposed by the bi-decomposition multi-decomposition algorithms shown in Sections 4 and 5. In addition to the algorithms shown in this article, bi-decomposition with respect to monotone, simple and modsum operators as well as set separation are implemented, see (Lang 2003).

6.2. *Input data format of YADE*

The input data of YADE is a so called *ML-file* containing all MVRs that are to be decomposed (Portland Logic and Optimization Group).

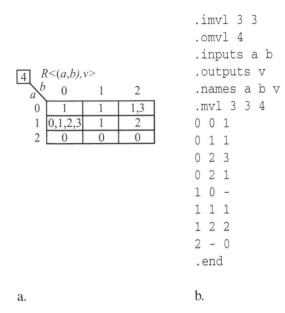

<table>
</table>

```
.imvl 3 3
.omvl 4
.inputs a b
.outputs v
.names a b v
.mvl 3 3 4
0 0 1
0 1 1
0 2 3
0 2 1
1 0 -
1 1 1
1 2 2
2 - 0
.end
```

a.   b.

Figure 14. Example input of YADE. a. Map of the MVR $R \langle (a, b), v \rangle$. b. ML-file of the MVR $R \langle (a, b), v \rangle$.

EXAMPLE 14. *The MVR $R \langle (a, b), v \rangle$ shown in Figure 14a. can be described by the ML file shown in Figure 14b. First, the cardinalities and names of the variables are defined. Then, the MVR is represented as a table.*

6.3. *Output data format of YADE*

The output data of YADE is a file in the *dotty-format* that describes the decomposed MVRs as a netlist of gates. A dotty-file is a plain ASCII file that describes graphs for viewing with the graph printing tool dotty (Koutsofios and North 1996). The dotty package also contains a description of the dotty-format. The inputs of the network are drawn at the top of the graph. The outputs are located at the bottom of the graph. Literal functions are labeled with the sequence of output values starting from the input value '0'. All other gates are labeled with their respective functions. Arrows indicate the direction of the data flow through the network.

EXAMPLE 15. *The decomposition of the MVR $R \langle (a, b), v \rangle$ from Figure 14 results in the graph shown in Figure 15.*

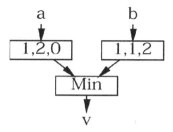

Figure 15. The graph drawn by dotty from the YADE output for the MVR from Figure 14.

Table 3. Comparison of the MDD-based decomposer (MDY) from (Lang and Steinbach 2001) to YADE. The table contains the DFC and the decomposition time. The column labeled M/Y contains the ratio MDY/YADE

Name	DFC			Decomposition time [s]		
	MDY	YADE	M/Y	MDY	YADE	M/Y
Balance	2012	**1470**	137%	18.00	**0.61**	2946%
Breastc	**634**	**634**	100 %	24.00	**0.46**	5206%
Flare1	932	**545**	171%	11.00	**0.64**	1716%
Flare2	8049	**2395**	336%	37.00	**1.40**	2639%
Hayes	128	**97**	132%	1.00	**0.04**	2500%
Average			175%			3001%

7. Decomposition Results

7.1. *Test setup*

Test functions from logic synthesis and machine learning applications were decomposed to evaluate the applicability of the proposed bi-decomposition approach to these problems. The results were compared to the results obtained by other decomposers. The test functions were take from the POLO Internet page that collects a variety of MVL functions from different applications (Portland Logic and Optimization Group).

7.2. *Comparison to the MDD-based version of YADE*

An earlier version of YADE, denoted by MDY, was based on multi-valued decision diagrams (MDDs) as central data structure (Lang and Steinbach 2001). A comparison of MDY to YADE is shown in Table 3. Better results are printed in bold type.

Table 4. Comparison of BI-DECOMP-MV (BDM) and YADE for fully decomposed functions by the BDFC and the number of gates (including literals)

Name	BDFC			Number of gates		
	BDM	YADE	BDM/YADE	BDM	YADE	BDM/YADE
Balance	1889	**1131**	167%	236	**192**	123%
Car	**372**	450	83%	**61**	64	95%
Employ1	309	**299**	103%	**49**	51	96%
Employ2	375	**321**	117%	63	**46**	137%
Hayes	**75**	76	99%	**17**	18	94%
Mushroom	**32**	42	76%	**11**	14	79%
Sleep	366	**280**	131%	34	**29**	117%
Sponge	**41**	44	93%	11	**9**	122%
Zoo	195	**147**	133%	17	**13**	131%
Average			111%			111%

MDY applies MM-decomposition of MFIs using derivative operators and set separation. The program is based on MDDs. Compared to MDY, YADE significantly decreases the DFC and the number of gates, which is mostly due to the improved decomposition algorithms and the larger variety of decomposition operators. The time for the decomposition is much larger for the decomposer MDY. Taken into account that a slower computer was used in (Lang and Steinbach 2001) (a 500 MHz Pentium III), YADE is still at least 10 times faster. This difference is caused by the application of BEMDDs instead of MDDs.

7.3. *Comparison to the BEMDD-based decomposer BI-DECOMP-MV*

The program BI-DECOMP-MV by Mishchenko is a decomposer that applies BEMDDs as the central data structure (Mishchenko et al. 2001b). The program decomposes MFIs by MM-bi-decomposition using the derivative operators. Non-decomposable functions are simplified by weak MM-bi-decomposition. A block is left undecomposed if neither MM- nor weak decomposition is possible. Therefore, the decomposition network may contain blocks with more than two inputs. The results are shown here only for test functions that could be fully decomposed by BI-DECOMP-MV (see (Lang 2003) for the results of the remaining test functions).

The results of the comparison are shown in Tables 4 and 5. Because BI-DECOMP-MV computes a modified DFC, this parameter (BDFC) is applied to compare BI-DECOMP-MV and YADE.

Table 5. Comparison of BI-DECOMP-MV (BDM) and YADE for fully decomposed functions by the number of levels and the decomposition time

Name	Number of levels			Time [s]		
	BDM	YADE	BDM/YADE	BDM	YADE	BDM/YADE
Balance	20	**11**	182%	**0.07**	0.61	11%
Car	10	**9**	111%	**0.02**	0.35	6%
Employ1	12	**10**	120%	**0.01**	0.82	1%
Employ2	8	**7**	114%	**0.01**	1.42	1%
Hayes	**5**	6	83%	**0.01**	0.04	25%
Mushroom	**6**	**6**	100%	**0.03**	2.52	1%
Sleep	8	**6**	133%	**0.04**	0.18	22%
Sponge	5	**4**	125%	**0.06**	1.16	5%
Zoo	6	**5**	120%	**0.01**	0.04	25%
Average			121%			11%

The networks produced by BI-DECOMP-MV have on average an 11% larger BDFC and contain 11% more gates, because YADE considers more bi-decomposition operators than BI-DECOMP-MV.

The number of levels is significantly larger for the program BI-DECOMP-MV because the program applies weak bi-decomposition. In the Boolean case there is a guarantee that a bi-decomposable function is obtained after at most two consecutive weak bi-decompositions. This fact is not true for MVL functions. There can be a long chain of weak bi-decompositions without guarantee that a bi-decomposable function is found, which increases the number of levels.

The improved performance of YADE comes at the cost of a significantly larger decomposition time. Both decomposition programs were executed on similar computers (YADE: 900 MHz Pentium III, BI-DECOMP-MV: 933 MHz Pentium III). Therefore, a direct comparison of the decomposition times is possible. The program BI-DECOMP-MV is faster than YADE because only MM-bi-decompositions are considered by BI-DECOMP-MV. In addition, YADE is implemented as an experimental platform, which can easily be extended by new decomposition algorithms. This flexibility causes significant overhead in YADE.

7.4. *Comparison to MVSIS*

MVSIS is the multi-valued extension of the well-known logic synthesis program SIS (Gao et al. 2001). MVSIS produces a multi-level network by

Table 6. Comparison of the decomposition results obtained by MVSIS and YADE

Name	Literal		
	MVSIS	YADE	MVSIS/YADE
monks1tr	173	**7**	2471%
monks2tr	155	**82**	189%
monks3tr	144	**29**	497%
pal3	30	**18**	167%
Average			831%

factoring expressions (Gao and Brayton 2001; Dubrova et al. 2001). The program represents an MVR by a set of m two-valued functions $F(A) = \{f_0(A), \ldots, f_{m-1}(A)\}$, where

$$f_i(A_k) = \begin{cases} 1 & \text{for } f(A) = i \text{ and} \\ 0 & \text{otherwise,} \end{cases} \qquad (44)$$

see Section 3.3. MVSIS produces a multi-level network for each function of $F(A)$ by the extraction of common factors. Therefore, the network produced by MVSIS consists of multi-valued input, two-valued output literals; AND- and OR-gates. Because YADE decomposes functions with an output cardinality of $m > 2$ into multi-valued gates with a single multi-valued output, the results of the two programs are comparable only for two-valued functions. In this case YADE can be configured to apply only MM-gates, which become AND- and OR-gates. The number of literals at the inputs can be counted if reuse of literals is disabled in YADE.

The number of literals obtained by MVSIS and YADE are shown in Table 6 for all two-valued MVL test functions from (Gao and Brayton 2001). The table shows that YADE produces significantly smaller networks. The networks found by MVSIS contain, on average, 8 times as many literals as the networks found by YADE. A direct comparison of the computation time is not possible. The run time of MVSIS was cited in (Gao and Brayton 2001) as "order of seconds" (hardware not specified), YADE needs less than 0.4 seconds per test function on a 900 MHz Pentium III PC.

8. Conclusion

This article introduced MFIs as a special case of MFSs to solve bi-decomposition problems in multi-valued logic. A closed solution was presented for bi-decomposition of MFIs with respect to the MM-operators. Operations from the MVL differential calculus were applied to generalize corresponding results for Boolean ISFs. The relation between the free and bound sets was shown. These results were applied to develop efficient bi-decomposition algorithms for MFIs.

There are MFSs that are not bi-decomposable. It was proposed to simplify these MFSs by separation, where the original MFS is decomposed into simpler MFSs. The results of separation are either supersets of the original MFS (and therefore simpler to decompose), or are bi-decomposable. Max-min multi-decomposition applies graph coloring to split an MFS into a minimal number of min-decomposable MFSs.

YADE is a decomposer for MVRs that can be applied to logic synthesis and data mining problems. The decomposer is well suited for very weakly specified MVRs. In contrast to existing decomposers the gates of the decomposition network can be selected from a library by the user. YADE produces decomposition networks by bi-decomposition of MFSs.

Several test functions from logic synthesis and data mining applications were decomposed. Even large test functions could be decomposed within a few minutes on a Pentium PC. The BEMDD based implementation of YADE is faster than its MDD based predecessor. The quality of the networks was also improved because of the additional decomposition algorithms. YADE finds better networks than the bi-decomposer BI-DECOMP-MV on average. However, BI-DECOMP-MV found better results for some test functions. This decomposer is also faster than YADE. For all test cases, YADE found better minimizations of the test functions than MVSIS.

Additional bi-decomposition algorithms, a more detailed description of the YADE decomposer and extensive test results can be found in Lang (2003).

In logic synthesis of MVL circuits, it is necessary to identify the basic gates of the chosen circuit technology. The algorithms of this article can be applied to produce networks that can be realized directly as gates of this technology. Further work is needed to extend bi-decomposition algorithms to the synthesis of sequential logic and to match the gates of given technology libraries.

This article introduced algorithms for the bi-decomposition of MFSs. Some of the algorithms, such as separation, are equally applicable to Boolean functions. It remains to investigate, whether these algorithms can improve current decomposition strategies for Boolean functions.

The theory of MFSs has its origins in the solution of Boolean differential equations. It can be expected that the solution of MVL differential equations will identify many other classes of MFSs. The authors of this article believes that the MM-decomposition and separation algorithms presented in this article are only the first step of this research. The extension of Boolean differential equations to multi-valued logic will form the theoretical basis for many problems in multi-valued combinatorial and sequential logic synthesis and data mining.

References

Blake, C. & Merz, C. (1998). UCI Repository of Machine Learning Databases. University of California, Irvine, Department of Information and Computer Sciences, Irvine, USA, URL http://www.ics.uci.edu/~mlearn/MLRepository.html.

Brayton, R. & Khatri, S. (1999). Multi-valued Logic Synthesis. In *12th International Conference on VLSI Design*, 196–206.

Cendrowska, J. (1987). PRISM: An Algorithm for Inducing Modular Rules. *International Journal of Man-Machine Studies* **27**: 349–370.

Dhaou, I. B., Dubrova, E. & Tenhunen, H. (2001). Power Efficient Inter-Module Communication for Digit-Serial DSP Architectures in Deep-Submicron Technology. In *31th IEEE International Symposium on Multiple-Valued Logic*, 61–66. Warsaw, Poland.

Dubrova, E., Jiang, Y. & Brayton, R. (2001). Minimization of Multiple-Valued Functions in Post Algebra. In *10th International Workshop on Logic & Synthesis*, 132–137. Granlibakken, USA.

Gao, M. & Brayton, R. (2001). Multi-Valued Multi-Level Network Decomposition. In *10th International Workshop on Logic & Synthesis*, 254–260. Granlibakken, USA.

Gao, M., Jiang, J., Jiang, Y., Li, Y., Sinha, S. & Brayton, R.: 2001, MVSIS. In *10th International Workshop on Logic & Synthesis*, 138–143. Granlibakken, USA.

Han, S., Choi, Y. & Kim, H. (2001). A 4-Digit CMOS Quaternary to Analog Converter with Current Switch and Neuron MOS Down-Literal Circuit. In *31th IEEE International Symposium on Multiple-Valued Logic*, 67–71. Warsaw, Poland.

Hanyu, T. & Kameyama, M. (1995). A 200 MHz Pipelined Multiplier Using 1.5 V-supply Multiple-valued MOS Current Mode Circuits with Dual-rail Source-coupled Logic. *IEEE Journal of Solid-State Circuits* **30**(11): 1239–1245.

Inaba, M., Tanno, K. & Ishizuka, O. (2001). Realization of NMAX and NMIN Functions with Multi-Valued Voltage Comparators. In *31th IEEE International Symposium on Multiple-Valued Logic*, 27–32. Warsaw, Poland.

Inc., R. Rambus Signaling Technologies – RSL, QRSL and SerDes Technology Overview. Rambus Inc. URL http://www.rdram.com/.

Kameyama, M. (1999). Innovation of Intelligent Integrated Systems Architecture – Future Challenge –. In *Extended Abstracts of the 8th International Workshop on Post-Binary Ultra-Large-Scale Integration Systems*, 1–4.

Koutsofios, E. & North, S. (1996). Drawing Graphs with Dot. AT&T Bell Labs, URL http://www.research.att.com/sw/tools/graphviz/.

Lang, C. (2003). Bi-Decomposition of Function Sets using Multi-Valued Logic. Dissertation thesis, Freiberg University of Mining and Technology, Germany.

Lang, C. & Steinbach, B. (2001). Decomposition of Multi-Valued Functions into Min- and Max-Gates. In *31th IEEE International Symposium on Multiple-Valued Logic*, 173–178. Warsaw, Poland.

Mishchenko, A., Files, C., Perkowski, M., Steinbach, B. & Dorotska, C. (2000). Implicit Algorithms for Multi-Valued Input Support Minimization. In *4th International Workshop on Boolean Problems*, 9–21. Freiberg, Germany.

Mishchenko, A., Steinbach, B. & Perkowski, M. (2001a). An Algorithm for Bi-Decomposition of Logic Functions. In *38th Design Automation Conference*, 18–22. Las Vegas, USA.

Mishchenko, A., Steinbach, B. & Perkowski, M. (2001b). Bi-Decomposition of Multi-Valued Relations. In *10th International Workshop on Logic & Synthesis*, 35–40. Granlibakken, USA.

Portland Logic and Optimization Group, 'POLO Web Page'. Portland State University. URL http://www.ee.pdx.edu/~polo/.

Ricco, B. et al. (1998). Non-volatile Multilevel Memories for Digital Applications. *Proceedings IEEE* **86**(12): 2399–2421.

Smith, K. (1981). The Prospects for Multivalued Logic: A Technology and Applications View. *IEEE Transactions on Computers* **C-30**(9): 619–634.

Takagi, N. & Nakashima, K. (2000). Some Properties of Discrete Interval Truth Valued Logic. In *30th IEEE International Symposium on Multiple-Valued Logic*, 101–106. Portland, USA.

Teng, H. & Bolton, R. (2001). The Use of Arithmetic Operators in a Self-Restored Current-Mode CMOS Multiple-Valued Logic Design Architecture. In *31th IEEE International Symposium on Multiple-Valued Logic*, 100–105. Warsaw, Poland.

Uemura, T. & Baba, T. (2001). A Three-Valued D-Flip-Flop and Shift Register Using Multiple-Junction Surface Tunnel Transistors. In *31th IEEE International Symposium on Multiple-Valued Logic*, 89–93. Warsaw, Poland.

Waho, T., Hattori, K. & Takamatsu, Y. (2001). Flash Analog-to-Digital Converter Using Resonant-Tunneling Multiple-Valued Circuits. In *31th IEEE International Symposium on Multiple-Valued Logic*, 94–99. Warsaw, Poland.

Yanushkevich, S. (1998). Logic Differential Calculus in Multi-Valued Logic Design. Habilitation thesis, Technical University of Szczecin, Poland.

Zupan, B. (1997). Machine Learning Based on Function Decomposition. Ph.D. thesis, University of Ljubljana, Ljubljana, Slowenia.

Advanced AI Search Techniques in Modern Digital Circuit Synthesis

LECH JÓŹWIAK

*Faculty of Electrical Engineering EH 9.24, Eindhoven University of Technology,
P.O. Box 513, 5600 MB Eindhoven, The Netherlands (E-mail: L.Jozwiak@tue.nl)*

Abstract. Progress in microelectronic technology is extremely fast and it is outstripping the designers' abilities to make use of the created opportunities. Development and application of new more suitable design methods and tools is therefore very important for the modern system industry. This paper shows the importance of the AI search techniques for the circuits and systems design space exploration, explains what sorts of search techniques are useful for this aim, and discusses the place, role and way of use of these techniques in circuit and system design. In particular, the paper explains the importance and usage of the heuristic search techniques for the automatic construction and selection of the most promising solutions to the circuit synthesis problems. The discussion and conclusions of the paper are illustrated with examples of three effective and efficient search algorithms, and experimental results from their application to two important circuit synthesis problems. The knowledge presented in the paper combines numerous valuable concepts of modern system engineering and artificial intelligence, and forms a base for further research and application of the AI search techniques to design of complex circuits and systems.

Keywords: artificial intelligence, automated design problem solving, circuit synthesis, double-beam search, genetic engineering algorithm, heuristic search, quick scan

1. Introduction

Microelectronic circuits and systems have important applications in virtually all fields of human activity and have a global influence on the performance of society. Modern microelectronic technology enables implementation of complete complex systems on a single chip. A new technology generation replaces the current one every two to three years, and the circuit complexity grows three to four times from generation to generation. Progress in microelectronic technology is extremely rapid and it is outstripping the designers' abilities to make use of the created opportunities. Therefore, development and application of new more suitable design paradigms, methods and tools is very important for the modern system industry.

What **system design** is about is a *definition of the required quality*, i.e. a satisfactory answer to the following two questions: what (new or modified) quality is required? and how can it be achieved? In (Jóźwiak 2001), I

109

proposed the following generic definition of **quality**: *quality of a purposive systemic solution is its total effectiveness and efficiency in solving the original real-life problem.* Here, the degree to which a systemic solution attains its goals is called its effectiveness, and the degree to which a solution uses resources to realize its aims is called its efficiency. Effectiveness and efficiency of a systemic solution together decide its grade of excellence. Their aggregation expresses quality. In turn, effectiveness and efficiency can be expressed in terms of measurable parameters, and in this way quality can be expressed and measured. The **system design** is an *evolutionary quality engineering process* that starts with an abstract, imprecise, incomplete and possibly contradictory initial model of the required quality (initial requirements) and tries to transform the initial model into a concrete, precise, complete, coherent and directly implementable final model (final design). In this process, analysis and modeling of the design problems, proposing of some tentative design solutions, analysis and evaluation of the proposed solutions, learning and adapting are all very important. Such an evolutionary quality-driven design process can be performed as an advanced heuristic search for solutions, involving combination of various appropriate AI search schemes and equipped with adequate multi-objective decision-making aids.

This paper focuses on the usage of the advanced AI search techniques in modern digital circuit synthesis. It aims at explanation why the AI search techniques are of primary importance for the modern digital circuit and system synthesis, discussing the place of the AI search techniques in the digital circuits and systems design space exploration, and showing what sorts of heuristic search techniques are useful for this aim, why are they useful, and how should they be developed and applied. In particular, the paper explains the importance and usage of the heuristic search techniques for the automatic construction and selection of the most promising solutions to the circuit synthesis problems. It discusses three effective and efficient search algorithms, representative to three different search algorithm classes, in their application to important circuit synthesis problems. The discussion of the paper is illustrated with some experimental results from the algorithms.

2. Design Space Exploration

System design is an evolutionary process that aims at a satisfactory definition of the required quality. Design space exploration with usage of models of the required quality is crucial in this process. The quality models can be seen as heuristic inventions that serve for setting and controlling the course of design. Creation of the well-structured models for the poor-structured design problems aims at enlarging the scope of the reasoning-based decision making with open and rational procedures that can be automated to a high degree. The design-time reduction due to design automation enables extensive exploration of the model and solution spaces, leading to designs of higher quality. The well-structured models of the design problems are used by search tools equipped in multi-objective decision-making aids (Jóźwiak and Ong 1996; Saaty 1980; Sakawa 1993; Vincke et al. 1992; Wierzbicki 1982, 1986; Yu 1985) in order to invent and select the most promising design solutions. The design intentions, goals, restrictions and their interrelations are modelled in the form of objectives, constraints and preferences expressed as functions of measurable performance variables. Analysis and evaluation tools extract information about values of the performance variables. Using this information, search tools equipped in multi-objective decision-making aids help the designers to invent and select the most promising design alternatives or to improve the previously found prototype solutions. If possible, the synthesis tools, using completely automated construction, modification, analysis, evaluation and decision making, perform fully automatic design transformations.

Each complex design problem is solved by its decomposition into a number of simpler sub-problems (design issues). For solving a particular issue, specific issue's models are constructed and used for the issue's solution space exploration. In this way, the design process breaks down a complex design problem into a network of interrelated simpler problems, but it also breaks down a complex system defined in abstract and non-precise terms into a structure of cooperating sub-systems defined in more concrete and precise terms. The sub-systems are in turn further broken down into some simpler sub-systems that can be directly implemented with the elements or sub-systems at the designer's disposal. The system is thus defined as an appropriate network of sub-systems that exposes the external aggregate behavior and characteristics that match the required behavior and characteristics.

This quality-driven design problem solving applies the framework of heuristic search for solutions and can be represented by a *design search tree*. The tree's nodes correspond to various design issues and branches to various alternative solutions possible for each issue. A *design decision* is a choice of a particular design option. Each option chosen can recursively raise new issues, expanding the search tree, until some final designs are found. To enable application of the multi-objective decision methods (Jóźwiak and Ong 1996; Saaty 1980; Sakawa 1993; Vincke et al. 1992; Wierzbicki 1982; Wierzbicki 1986; Yu 1985) for construction, improvement and selection of the most promising solutions, a *decision model* should be constructed for each issue, being an abstract model of the required quality, expressed in the decision-theoretical terms. For a certain design issue, a lot of alternative solu-

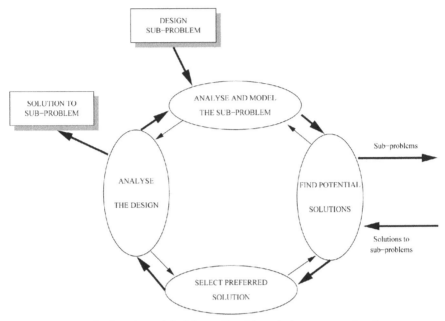

Figure 1. Generic model of the quality-driven design space exploration.

tions typically exist, but these alternatives are not known a priori. They must be constructed in the course of design and the process of their construction, improvement and selection has to be guided by the quality models. Since the quality models only give a temporary and restricted reflection of the quality really required, they have to be carefully validated prior to their usage. Also, posterior analysis of the design process and of the (partial) design results produced with use of those models should provide the information for their modification and for acceptation or rejection of the results. Consequently, the *quality-driven design space exploration* consists of the alternating phases of *exploration of the space of abstract models of the required quality* and *exploration of the space of the more concrete issues solutions obtained with the chosen quality models.* I proposed a generic model of the quality-driven design space exploration as represented in Figure 1 and briefly discussed below. More extensive discussion of this model can be found in (Jóźwiak 2001).

The design problem is first analysed and modelled. As a part of this modelling activity, a problem-specific decision model is constructed. Also, some general solution concepts (forms) are invented and the most promising of them are selected. In relation to these most promising general solution forms the problem-specific decision model is further refined. The design aim is expressed by a number of objectives (e.g. maximize the circuit or system speed, minimize the resource usage, minimize the inter-module communication, etc.), constraints (e.g. the minimum circuit or system speed, the maximum module dimensions, the maximum capacity of inter-module communication channels, etc.) and trade-off information. The objectives and constraints are expressed as functions of some basic parameters possible to estimate based on the module and communication channel models and performance characteristics, and they can be formulated as utility functions (Jóźwiak and Ong 1996; Saaty 1980; Sakawa 1993; Vincke et al. 1992; Wierzbicki, 1982, 1986; Yu 1985). The decision model constructed this way enables use of the multiple-objective decision methods and tools for invention and selection of the most promising solutions. The most promising of the general solution forms and their corresponding decision models that establish a preference relation on the set of solutions are then used in the framework of the heuristic search to find the most promising particular solutions (instantiations of the selected general solution forms). If the direct finding or construction of solutions is too difficult, the problem is decomposed into a number of sub-problems, and the same generic model of design space exploration is applied to solve each of these sub-problems. The composition of the sub-problems' solutions provides a solution for the original problem. The promising alternative solutions found this way are then evaluated and the most preferred of them are selected. Finally, the whole process is analysed to discover possible inconsistencies and start its improvement or to validate it and accept its results.

This paper focuses on the second and third stage of the design space exploration process outlined above, i.e. effective and efficient methods and tools for finding of the most promising solutions for the design sub-problems and for selection of the preferred solutions. It discusses the usage of the advanced AI search techniques for this aim.

3. Automatic Search for the Optimal Design Solutions

3.1. *Formulation and solution of the circuit and system design problems*

The complex and multi-aspect design problems can be solved by decomposing them into systems of simpler sub-problems, and then solving the systems of sub-problems by using certain systems of cooperating sub-problem specific methods. Typically, sub-problems in digital circuit and system design have very large and "irregular" discrete solution spaces. They can be modelled in the form of multi-objective constrained optimisation problems (MOCOP) (Jóźwiak and Kolsteren 1991; Jóźwiak and Volf 1992; Jóźwiak and Volf 1995). A MOCOP can be characterized by four sets of components: a set $V = (V_1, V_2, \ldots, V_n)$ of variables, a set $D = (D_1, D_2, \ldots, D_n)$ of domains for the variables, a set $C = (C_1, C_2, \ldots, C_k)$ of constraints, and a set $O = (O_1, O_2, \ldots O_p)$ of objectives. Each domain $D_i : D_i \in D$ represents a finite or infinite set of values. A constraint C_i: $C_i \in C$ is a m-ary relation on m of the n variables V_i: $V_i \in V$ and it can be viewed as a mapping from D to the set $\{T, F\}$ (T = "*true*", F = "*false*"). An objective $O_i : O_i \in O$ is a function on the variables from V with values being real numbers, i.e. a function from D to R (R – the set of real numbers). To solve a MOCOP is to find an assignment of values to variables V_j, $j = 1 \ldots n$, so that all constraints C_i, $i = 1 \ldots k$, are satisfied. Since a MOCOP involves several objectives, trade-off between them is possible. Therefore, the complete formulation of a MOCOP must include trade-off information. This trade-off information can be formulated in several different ways: as an order of the objectives, utility function, ranking information, local preference for small changes in values of the objectives etc. (Jóźwiak and Ong 1996; Saaty 1980; Sakawa 1993; Vincke et al. 1992; Wierzbicki 1982, 1986; Yu 1985). The choice of a specific formulation depends on the particular problem. To solve a MOCOP optimally is to solve it in such a way that the most preferred solution according to the objectives O_i, $i = 1 \ldots p$, and the actual trade-off information is found. When modelling the circuit synthesis problems, the particular variables and domains correspond to various structural attributes (dimensions) of the implementation architecture such as inputs, outputs, functional elements, interconnections, etc., and their possible values. The objectives and constraints are the functions and relations of the structural attributes, such as the number of inputs or outputs, area, speed, etc., or the limits imposed on such characteristics.

For complex, multi-aspect, large and "irregular" circuit synthesis problems, it is impractical or even impossible to generate and examine all the possible problem solutions. It is necessary to find only a limited number of the most promising solutions, when using heuristic search algorithms, and to choose the best of these. This section aims to describe the underlying principles of those heuristic algorithms.

One of the basic requirements for usage of the heuristic search schemes and algorithms in practical applications of circuit and system synthesis is the adequate effectiveness and efficiency of these schemes and algorithms. Effectiveness is understood here as the robustness of an algorithm in the sense of steadily delivering of the high-quality solutions in the whole range of its possible input data. This corresponds to construction of the high-quality circuits for the whole range of possible functions and values of the structural and parametric requirements. The "high quality" does not necessary mean strictly optimal, but steadily optimal or close to optimal. Efficiency is understood in the sense of a short enough computation time and practical memory usage. Since the practical circuit and system synthesis problems are solved typically by some systems of collaborating search algorithms, in which some particular algorithms may be repetitively used inside of some other algorithms, the computation time requirements are often extremely stringent. What precisely means "effective and efficient" depends of a particular application, as well as, the aim and way of usage of a particular algorithm in a certain system of collaborating algorithms. It is therefore incorrect to say that a particular search scheme or algorithm is the best one or not appropriate without referring to a particular aim and way of its use. What makes the search algorithms effective and efficient is the usage of a search scheme that closely matches the kind of problem to be solved, and usage of the problem-specific knowledge in the algorithm itself or in the information for the decision-making in the scope of the algorithm. Thus, an effective and efficient decision-making based on real facts.

The sequential search approaches go from one to another solution by making small local solution modifications and examine solutions sequentially one-by-one (e.g. best first search, simulated annealing or tabu search). Due to the intrinsic difficulty of the circuit synthesis problems, the usefulness of the probabilistic or local-search-based sequential search approaches to search the extremely large and "irregular" solution spaces of these circuit synthesis problems, is in general questionable. Since the solution space is huge and "irregular", its local optima can be isolated. Furthermore, there is often no clear information in the surroundings of a particular solution on the direction of improvement, because in the "irregular" solution space good, bad and average quality solutions can be in the neighbourhood in various configurations. Moreover, because of the discrete and "irregular" character of the solution space and tradeoffs non-linearity, a small change in a solution may result in a big change of its quality. Therefore, sequential search has usually a small chance of finding a global optimum or a very high-quality local optimum within an acceptable time. The quality of such a search decreases very quickly with the increase of the problem complexity and irregularity. Nevertheless, the probabilistic or local-search-based sequential search approaches can be used for solving some specific circuit synthesis sub-problems having reasonably regular solution spaces or sub-problems in which search of a reasonably small solution space or an extremely quick search is required (e.g. for the local improvement of the earlier found near-optimal solutions).

3.2. *Parallel constructive search*

For the really complex, multi-aspect, large and "irregular" circuit synthesis problems, it is necessary to apply a richer apparatus of the solution invention methodologies – not only the solution modification (as in the case of the local improvement), but the solution construction as well. It is necessary to construct only a limited set of the most promising solutions when using constructive heuristic search algorithms equipped in multi-objective decision making aids, and then to select the best of these solutions. At each stage of such constructive search, it is necessary to estimate and compare the partially constructed solutions and construction operators, to select and use only the most promising of them, and in this way to reduce the search space to a manageable size while keeping the high-quality solutions in the reduced space.The parallel search approaches that work with populations of solutions and consider a number of solutions in parallel (e.g. genetic algorithms (Beasley et al. 1993a, 1993b Davis 1991; Dill and Perkowski 2001; Goldberg 1975; Holland 1992, 1995; Jóźwiak et al. 1998; Jóźwiak and Postuła 2002; Louis and Rawlins 1992) or double-beam search (Jóźwiak and Kolsteren 1991; Jóźwiak 1992a; Jóźwiak and Volf 1992; Jóźwiak 1992b; Jóźwiak and Vol 1995)), have a good chance of performing this task adequately. They not only enable modification of solutions, but also interplay among various solutions, by combination of their parts, their comparison and selection. By enabling combination of partial solutions or solution elements, these search approaches make stepwise construction of solutions possible. When appropriately used, the parallel search approaches enable a constructive search in which either the most promising construction operators are applied to the most promising partial solutions (in the case of the deterministic search approaches) or the more promising construction operators have a higher chance of being applied to the more promising partial solutions (in the case of the probabilistic approaches).

To apply such a constructive search to a certain synthesis problem, after finding a generic solution form, building the solution's morphological model, and formulating the corresponding multi-objective constrained optimisation problem, the synthesis problem has to be represented in terms of a **space of states**, where each state corresponds to a particular (partially constructed) solution. The space of partial solutions is defined by means of an **implicit tree**, i.e. by an **initial state**, the **rules for generating the tree**, and **the termination criteria**. The **rules** describe how to generate successors to each partial solution, i.e. they define the construction or move operators that realize transitions between states and are partial mappings from states to states. Any state that meets a termination criterion is called a **goal state** and it represents a complete solution, which is a certain instantiation of the generic solution form. The partial solutions, and finally also complete solutions, are constructed by successively applying the construction operators to the initial state or to the previously constructed partial solutions. A heuristic construction algorithm based on the mechanism described above can be effective and efficien, if it is able to appropriately combine the exploration of the search space in many promising directions with the fast convergence to some optimal or near-optimal solutions (Davis 1991; Goldberg 1975; Holland 1995; Jóźwiak 1995; Jóźwiak and Postuła 2002). The divergent (construction) and convergent (selection) phases should alternate.

To select the most promising partial solutions and construction operators, they must be estimated and compared. The basic information for their estimation and selection is contained in the decision model of the problem at hand and must be extracted from this model to construct the adequate decision models for selection of construction operators and partial candidate solutions. One has to ensure that a solution that violates the hard constraints will never be constructed and one should try to satisfy the objectives (near-)optimally by limited expenditure of computation time and memory space. In order to fulfil the first task, the construction operators must be selected that do not lead to the violation of hard constraints if applied to a given state. In order to fulfil the second task, one has to develop approximate evaluation criteria and selection mechanisms (decision models and methods) for operators and partial solutions.

Generally, operators and partial solutions are estimated with some uncertainty. This uncertainty decreases with the progress of computations, because both the "sure" information contained in partial solutions and the quality of prediction grow with this progress. Therefore, usage of knowledge cumulated in the previous search steps for selecting the most promising partial solutions and move operators can result in faster convergence without sacrificing the result quality. In the first phase of the search, the choices of construction operators can be made with much more certainty than the choices of partial solutions. In this phase partial solutions almost do not exist or, in other words, they are far from being complete solutions and almost anything can happen to them on the way to complete solutions. The uncertainty of estimation decreases with the progress of computations. Therefore in the first phase, the search should be almost exclusively based on the choices of the most promising construction operators, and with the progress of computations, more and more based on the choices of the most promising partial solutions. Since the uncertainty of estimations, both for partial solutions and for construction operators, decreases with the progress of computations, the search has to be divergent to a high degree in the first phase (to achieve high effectiveness), and in the second phase it can become more and more convergent (for increasing efficiency).

The parallel constructive search schema described above was implemented by us in the form of a double-beam search and in the form of a modified genetic search in a number of tools for solving various synthesis problems in digital system design. When tested on benchmarks, it efficiently produced very good results in all considered applications (Jóźwiak and Kolsteren 1991; Jóźwiak 1992a; Jóźwiak and Volf 1992; Jóźwiak 1992b; Jóźwiak and Vol 1995; Jóźwiak et al. 2001; Jóźwiak and Chojnacki 2001).

In this section, we have only discussed some general principles of searching for the promising design solutions. For each particular design problem, the problem specific features should be used to find a generic solution concept and general morphologic model of potential solutions, to define construction operators, and to perform the construction processes effectively and efficiently. In this way the discussed generic parallel constructive search concept can be transformed into adequate problem-specific constructive search schemes and algorithms.

To illustrate this process, to show that the precise meaning of "effective and efficient" depends on a particular aim, context and way of usage of a particular algorithm, and to demonstrate that what makes the search algorithms effective and efficient is the usage of a search scheme closely matching the kind of problem to be solved, and usage of the problem-specific knowledge in the algorithm itself or in the information for the algorithm's decision-making, we will discuss below three effective and efficient algorithms that we developed for circuit synthesis: *Double-Beam Search*, *QuickScan* and *Genetic Engineering Algorithm*. We will show application of these algorithms to two important circuit synthesis problems: *decomposition* and *input support minimization* of Boolean Functions, Relations and Sequential Machines. These three algorithms were selected as being representative to three different search algorithm classes. *QuickScan* is a simple, but quite effective and extremely efficient, problem-specific deterministic sequential search. *Double-Beam Search* is a quite generally applicable deterministic parallel constructive search. *Genetic Engineering Algorithm* is a quite generally applicable mixed probabilistic/deterministic parallel constructive search.

4. Double-Beam Search

4.1. *Introduction to double-beam search*

The double beam-search scheme was developed by me to robustly and efficiently construct some optimal or near-optimal solutions for synthesis problems. The double beam-search is a representative example of a deterministic parallel constructive search that can be applied to a broad class of problems.

To apply a constructive search for finding the most promising solutions to a certain synthesis problem, the problem has to be represented in terms of a space of states, where each state corresponds to a particular (partially constructed) solution, being a particular (partial) instantiation of the generic solution form. The **state space of (partial) solutions** is defined as an implicit tree, by means of:

− **initial state** IS
− the rules describing how to generate successors to each current state using **construction/move operators** op from a certain set OP, that realize transitions between states, and
− the **termination criteria** that (implicitly or explicitly) define a **set of goal states** GS, each representing a complete solution.

A particular **constructive search** searches (a part of the) state space, by starting from the initial state IS, and (selectively and/or in a certain order) applying the construction operators to the initial state or to the previously constructed states (in this way constructing the successive states), until some goal states from GS are reached.

The **basic beam-search** is a variation of breadth-first search, where only a limited number of the most promising alternatives are explored in parallel. It requires two **basic data structures**:

− **present states** PS (that contains the set of states which have been constructed earlier and are considered to be extended presently), and
− **candidate states** CS (that contains the set of states which are being created as a direct extension of the present states).

A third supplementary data structure, **final states** FS, contains the states that cannot be further extended.

The **basic beam-search** applies all possible to apply construction operators op from OP to each present state ps from PS, constructing this way a new generation of candidate states CS, and subsequently, applies its only selection mechanism **Select States** to the candidate states CS in order to include the best of them into a new generation of present states PS.

The **double-beam search**, developed by me as an extension and elaboration of the basic beam-search, uses **two selection mechanisms**: **Select Moves** and **Select States**. Move operators are evaluated and selected in relation to a certain state ps (dynamically). Only a few of the most promising move operators are chosen for a certain present state ps by **Select Moves**, using problem-specific heuristic choice strategies and evaluation functions. By applying the selected move operators to each current state from PS a new generation of candidate states CS is created. The work of the second selection mechanism, **Select States**, is twofold: it scans the candidate states CS for states that cannot be further extended (represent complete solutions), in order to include them into the set of final states FS, and it examines the rest of the candidate states in order to include the best of them into a new generation of present states PS. The beam-search stops if the set PS becomes empty. Some of the constructed final states FS that satisfy some extra problem specific criteria are the goal states GS.

The selection mechanisms Select Moves and Select States must ensure that a solution that violates the hard constraints will not be constructed and they should try to robustly construct only some strictly optimal or near-optimal solutions by limited expenditure of computation time and memory space:

- Select Moves will select only those move operators that, applied to a given state, do not lead to the violation of hard constraints
- Select Moves and Select States will select only a limited number of the most promising operators or states, respectively, by using the estimations provided by some heuristic evaluation functions.

The selection mechanisms and evaluation functions determine together the search extent and quality of the results.

In a number of constructive double-beam algorithms that we developed and tested (Jóźwiak and Kolsteren 1991; Jóźwiak 1992a; Jóźwiak and Volf 1992; Jóźwiak 1992b; Jóźwiak and Volf 1995; Jóźwiak et al. 2001; Jóźwiak and Chojnacki 2001), we implemented certain heuristic elaborations of the following **general decision rule** for selection of the construction operators and partial solutions: *"at each stage of the search take a decision that has the highest chance (or certainty) of leading to the optimal solution according to the estimations given by the evaluation criteria; if there are more decisions of comparable certainty, try a number of them in parallel, preferring the decisions that bring more information for deciding on the successive issues"*.

According to the above rule, **Select Moves** will apply those move operators which maximize the choice certainty in a given present state and it will leave the operators which are open to doubts for future consideration. Since information contained in the partially constructed solutions and used by the evaluation functions grows with the progress of computations, the uncertainty related to both the partial solutions and construction operators decreases. In a particular computation stage, Select Moves will maximize the conditional probability that the application of a certain move operator to a certain partial solution leads to the optimal complete solution. Under this condition, it will maximize the growth of the information in the partial solution, which will be used to estimate the quality of choices in the successive computation steps, enhancing this way the choice certainty in the successive steps. The quality $Q(op)$ of a given move operator op is decided by these two factors. Select Moves is controlled by two parameters:

- $MAXMOVES$: the maximum number of the move operator alternatives explored in relation to each present state, and
- $OQFACTOR$: the operator quality factor.

In relation to a certain present state ps from PS, Select Moves selects no more than $MAXMOVES$ of the highest quality move operators op from OP, so that: $Q(op) \geq OQFACTOR * Qop_{max}$, where Qop_{max} is the quality of the best alternative operator that can be applied to a given state ps. Poor quality alternatives are not taken into account.

The task of **Select States**, in addition to selecting the final states, is to choose the most promising candidate states for a new generation of current states. Select States is controlled by two parameters:

- $MAXSTATES$: the maximum number of state alternatives explored at a certain computation stage, and
- $SQFACTOR$: the state quality factor.

Select States selects no more than $MAXSTATES$ of the highest quality alternative states cs from CS, for which: $Q(cs) \geq SQFACTOR * Q_{max}$, where $Q(cs)$ denotes the quality of an alternative cs from CS and Q_{max} denotes the quality of the best state alternative in CS. Poor quality alternatives are not taken into account. $Q(cs)$ can be computed by cumulating the qualities of the choices of move operators that took place during the construction of a certain cs and prediction of the quality of the best possible future operator choices on the way to the best complete solution possible to arrive at from this cs. Another possibility consists of predicting the quality of the best complete solution that can be achieved from a certain candidate state cs. The double-beam search scheme is graphically represented in Figure 2.

As signalled in Section 3, the selection of operators and partial solutions is generally performed with some uncertainty that decreases with the progress of computations, because both the "sure" information contained in partial solutions and the quality of prediction grow with this progress. Moreover, in the first phase of the search, the choices of operators can be made with much more certainty than the choices of partial solutions, because, in this phase, partial solutions almost do not exist and almost anything can happen to them on the way to achievable complete solutions. Therefore, in the first phase, the search should be performed almost exclusively based on the choices of operators and, with the progress of computations, more and more on the choices of partial solutions. In our double-beam algorithm, this is achieved by giving a relatively low value to $MAXMOVES$ compared to $MAXSTATES$ and a relatively high value to $OQFACTOR$ compared to SQFACTOR. Since the uncertainty of estimations decreases with the progress of computations, $MAXMOVES$ and $MAXSTATES$ can decrease and $OQFACTOR$ and $SQFACTOR$ can increase with the progress of computations, increasing

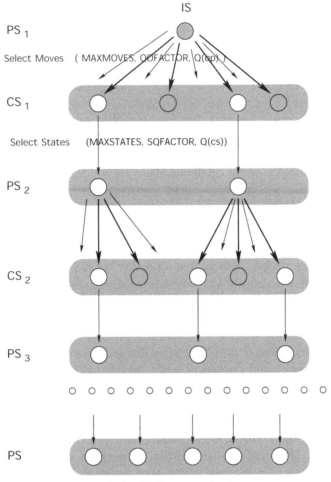

Figure 2. Double-beam search scheme.

this way the search efficiency. In the first phase, the double beam-search is divergent to high degree, i.e. a large number of the most promising directions in the search space are tried. In the second phase, when it is already possible to estimate the search directions and operators with a relatively high degree of certainty, the search becomes more and more convergent. The highly divergent character of the double-beam search in the first phase, composed with the continuous and growing competition between the partial solutions in the second phase, result in its global character, effectiveness and efficiency. In this way, the double beam-search allows for an effective and efficient decision-making under changing uncertainty.

The double-beam search scheme was implemented in numerous circuit synthesis methods and tools solving different kinds of synthesis problems, and when tested on benchmarks, it efficiently produced very good results for all considered applications (Jóźwiak and Kolsteren 1991; Jóźwiak 1992a; Jóźwiak and Volf 1992; Jóźwiak 1992b; Jóźwiak and Volf 1995; Jóźwiak et al. 2001; Jóźwiak and Chojnacki 2001).

4.2. *Relationships between the double-beam search and some other AI search schemes*

The double-beam search, being a sophisticated extension and elaboration of the basic beam search, is related to the family of the best-first search approaches known from the literature, and especially, the most sophisticated and informed algorithms of this family, as A^* or Z^* (Hart et al. 1968; Pearl 1984).

A^* is an enhanced version of the basic best-first search. It is a goal-directed search that uses a global state evaluation function. At each stage, the most promising state cs is selected for further expansion from the set of "open" (candidate) states CS. The global selection of the most promising candidate state $cs \in CS$ requires adequate measures of the **initial path cost g(cs)** from the initial state IS to cs, and of the predicted **lowest-cost-path cost h(cs)** from cs **to a goal state** gs. The **total cost measure** of a given state cs is thus: $\mathbf{f(cs)} = \mathbf{g(cs)} + \mathbf{h(cs)}$, where: $g(IS) = 0$, $g(cs') = g(cs) + c(cs, cs')$ and $c(cs, cs')$ is the cost of move from cs to cs', for every direct successor cs' of cs. $h(cs)$ is a heuristic estimate of a lower bound on the cost of getting from cs to a goal state gs that is based on the problem specific knowledge. Since a preferred solution is given by a path from IS to a $gs \in GS$ with the lowest possible cost, at each stage A^* selects a cs that is included in such a path, i.e. with the lowest value of $f(cs)$. If $cs = gs \in GS$, the cost of gs is $g(cs)$, and the path from IS to gs is an optimum complete solution path. If $CS = \emptyset$, no solution path exists. Otherwise A^* computes all direct successors cs' of cs, adds them to CS, and computes $g(cs') = g(cs) + c(cs, cs')$ and $h(cs')$. If $h(cs) \leq h^*(cs)$ for all cs, where $h^*(cs)$ is the actual cost of a preferred path from cs to a $gs \in GS$, then A^* is guaranteed to find an actual preferred solution path if one exists (Hart et al. 1968; Pearl 1984). Thus A^* can produce non-optimal result only if the estimated $h(cs)$ exceeds the actual $h^*(cs)$. More precise descriptions of A^* can be found in (Hart et al. 1968; Pearl 1984). The Z^* algorithm is essentially the same as A^*, with one extension: any arbitrary function of the form $f(cs') = F(f(cs), m(cs), h(cs'))$ can be used for calculation of $f(cs')$, where $m(cs)$ is a set of local parameters characterizing cs or move from cs to cs'.

The basic beam search can be seen as simplified version of Z^* or A^* where only a limited set of the most promising states is considered to be extended. This way the basic beam search explores only the most promising paths and saves a lot of memory and computation time comparing to Z^* and A^*, on the cost of not guaranteeing the strict optimality in the special cases where A^* guarantees it, but often producing optimal or close to optimal solutions.

The double-beam search is much more sophisticated than the basic beam search. Its decisions are based not only on estimation and selection of states as in the basic beam search, A^* or Z^*, but also (and even in the first phase mainly) on the estimation and selection of move operators. Its evaluation and selection mechanisms enable effective and efficient decision-making under changing uncertainty. What is probably even more important, is the fact that A^* and Z^* are heavily based on the high quality of estimation of a distance from a certain state cs (partially constructed solution) to a goal state $gs \in GS$ (completely constructed optimal solution). However, for the most practical problems of the digital circuit and system design automation, the optimum solutions (goal states) and their reasonable characteristics are completely unknown. In most cases, it is therefore impossible or extremely difficult to estimate the distance from a certain partially constructed solution to an optimum solution and guarantee a satisfactory quality of such estimation. In particular, it is virtually impossible to estimate it so precisely as to satisfy the A^* optimality condition: $h(cs) \leq h^*(cs)$ for all cs. The heuristic search algorithms for solving the complex problems of the digital circuit and system design automation cannot therefore assume having the knowledge that does not exists and is extremely difficult to obtain (i.e. its obtaining basically requires solving or almost solving the problem a given algorithm is intended to solve). In particular, the circuit or system synthesis algorithms cannot assume having a satisfactory knowledge of the optimal solutions or their features, before or after just starting to construct them. In most cases, they have to use less direct, less precise and smaller pieces of information related to the solution being constructed, but in a much more sophisticated way than A^* or similar algorithms, to guarantee a very high chance of remaining on some paths to the optimum or very close to optimum solutions. Therefore, in the double-beam search the solutions are gradually constructed by combining the most promising building blocks (the most promising construction operators). Moreover, the construction/move operators are estimated dynamically, in the context of a just created state to which they potentially can be applied. No prior estimation of moves, as e.g. $c(cs, cs')$ assumed in A^*, is possible with a reasonable precision for the complex synthesis problems to which double-beam is typically applied. Also, in the double-beam the selection of both the construction operators and partial solutions is based on the comparison of chances of remaining on some paths to the (near-)optimum solutions for the alternative construction operators or partial solutions, and not on the actual solution cost estimation as in A^* and similar algorithms. This is very important, because the chance estimates are not required to directly represent the actual solution costs, only in the simplest cases may do it. Moreover, the estimates can involve estimation offset that is automatically eliminated by comparison. Thus, to well perform, the double-beam search requires much less direct and less precise knowledge than A^* and similar algorithms. If however, in simple practical situations, the knowledge required by A^* would be given to the double-beam than the double-beam would perform at least equally well or better than the basic beam, thus comparable to A^*. Summing up, the double-beam search performs practically so well as A^* and similar algorithms for reasonably simple problems, where the direct very precise knowledge of the possible to achieve solutions is available, but it can be successfully applied for the much more complex problems, where much less direct and/or precise knowledge related to the achievable solutions is available than required by A^*.

In the above two sections, the double-beam search scheme has been introduced and discussed. In order to apply this search scheme for a particular problem, the problem specific features should be used for defining the precise decision-making procedures of the search process and the procedures to estimate the state and operator quality. In this way, the generic double beam-search scheme will be transformed into a problem specific double-beam algorithm. In the following section we demonstrate the application of the double-beam search to Boolean function and finite state machine (FSM) decomposition, being an important problem of modern circuit synthesis.

4.3. Decomposition of Boolean functions and FSMs

4.3.1. Introduction to decomposition

General structural decomposition refers to transforming a system into a structure of two or more cooperating sub-systems in such a way that the original system's behaviour is retained, specific (structural) constraints are satisfied and certain objectives optimized. It is the main system design activity that enables the designers to solve or reduce the design problems related to complex systems, implement the complex systems with less complex sub-systems, and account for the structural constraints of sub-systems and interconnection. With growing system complexity and (long) interconnections becoming to be the dominant factor in deciding the circuit speed and other parameters, the methods and CAD-tools for decomposing complex digital systems and circuits have attracted great interest recently. The structural decomposition can be performed much faster and better with a CAD-tool than manually, because intensive analysis of the system and precise synthesis are necessary for high quality results. A survey of decomposition types, methods, algorithms and results can be found in (Jóźwiak 1995; Jóźwiak et al. 2001; Jóźwiak and Chojnacki 2001).

In general, system decomposition involves a certain kind of partitioning. A network of collaborating sub-systems being the decomposition result is constructed based on the result of this partitioning. In solutions of complex problems various decomposition sorts are mixed together, and construction of the decomposition based on the partitioning results may still be a complex task. Nevertheless, the generic solution model of a specific Boolean function or sequential machine decomposition problem can be found by defining a set of some atomic computations and a certain way of constructing partitions on the set of the atomic computations. Then, each partition block is implemented as a separate sub-function/component machine and interconnections between the sub-functions/component machines are made, based on the interrelations between the corresponding partition blocks. With such a generic partitioning-based solution model its corresponding MOCOP model can be associated, by defining a set of some problem-specific objectives, constraints, and trade-off information. So modelled decomposition problems that involve hard constraints can be solved by special multi-dimensional and multi-objective packing approaches (Jóźwiak and Kolsteren 1991; Jóźwiak and Volf 1992; Jóźwiak and Volf 1995). Problems without hard constraints can be solved by special multi-objective clustering (Jóźwiak 1992; Jóźwiak et al. 2001). The partitioning process, which produces partitions on the set of atomic computations is preceded by analysis of characteristic features of an original function/machine related to the characteristics of building blocks. In particular, the input, output and state information, and their interrelations are analysed. This analysis enables us to distinguish elementary computations and to characterize them and their interrelations. Its results are used to guide the heuristic packing or clustering processes.

The partitioning problem is represented in terms of a space of states, where each state corresponds to a particular (partial) solution, being a (partially constructed) partition. Partitions in packing algorithms are constructed by putting unallocated atomic computations successively into the partition blocks (Jóźwiak and Kolsteren 1991; Jóźwiak and Volf 1992; Jóźwiak and Volf 1995). Clustering algorithms construct the successor partitions by merging some of the partition blocks (Jóźwiak 1992; Jóźwiak et al. 2001). The selection mechanisms of the constructive heuristic search approaches use estimations delivered by their estimation mechanisms to select the most promising partial solutions and to develop them further when applying only the best move operators.

For each particular decomposition problem, the problem specific features are used to distinguish and characterize the atomic computations, and to perform their partitioning processes effectively and efficiently. In this way, the generic packaging or clustering concepts and the generic search schemes are transformed into some problem specific algorithms.

4.3.2. *Model and general solution method for the output decomposition*

An assigned sequential machine can be described by the following set of Boolean equations:

$$y_i = f_{y_i}(x_1, \ldots, x_n, s_1, \ldots, s_k) \tag{1}$$
$$s_j = f_{s_j}(x_1, \ldots, x_n, s_1, \ldots, s_k) \tag{2}$$
$$i = 1, \ldots, m, \ j = 1, \ldots, k$$

where:

$F_y = \{f_{yi}\}$ – a set of incompletely specified Boolean functions for representing the output function λ of a sequential machine,

$F_s = \{f_{sj}\}$ – a set of incompletely specified Boolean functions for representing the next-state function δ of a sequential machine,

$X = \{x_l\}$ – a set of binary input variables,

$Y = \{y_i\}$ – a set of binary output variables,

$S = \{s_j\}$ – a set of binary state variables.

In general this model represents a sequential circuit having memory (internal state variables) and involving feedback for computation of new values of the state variables. A special case of the above, when no internal state variables or feedback will occur, i.e. $y_i (i = 1 \ldots m)$ do not depend upon $s_1, \ldots, s_k$, represents a multiple output combinational circuit. Let $O = Y \cup S$ be a set of binary output variables of a multiple output Boolean function with or without feedback, let $I = X \cup S$ be a set of binary input variables for that function and let $F = F_y \cup F_s$ be a definition of that function. Let T be a set of the input product terms (patterns of the function's input variable values) of F.

Output decomposition consists of partitioning the set of outputs O into two or more disjoint subsets $O_i (\cup O_i = O)$. Such output decomposition separates particular functions f_i that compute values for single outputs o_i of a multiple output function F into two or more component multiple-output functions $F_i (\cup F_i = F)$, so that each component function F_i can be implemented with a separate logic building block.

The sequel to this section will be devoted to a brief discussion of an affective and efficient double-beam search algorithm for the output decomposition. More extensive discussion of this algorithm can be found in (Jóźwiak and Volf 1992; Jóźwiak and Volf 1995). For simplicity's sake, the decomposition method is introduced using an example of programmable logic array (PLA) building blocks; however, the method will work in a strictly analogous way for any other kind of building blocks because its computations are based on only the structural limitation characteristics of building blocks such as the total number of outputs, inputs, terms, internal flip-flops etc. In the case of PLA building blocks, the limitation characteristics are given by NO, NI and NT which denote the total number of output bits, input bits and terms of a PLA, respectively. Those three parameters define the **hard constraints** of the partitioning problem. The **objectives** consist of using a minimal number of such constrained building blocks when implementing a certain function F and, under that condition, a minimal number of connections between the blocks. A **connection** between blocks is defined as a primary input variable common to two or more blocks, or a state variable calculated in one block that acts as an input variable to another block. Each extra occurrence of a variable from I as an input variable to one of the building blocks is taken into account as an extra connection.

Each variable from O is a function of variables from I, but not necessarily it is a function of all the variables from I. A similar situation is valid with product terms. So, we can assume that we have two functions $\mathbf{AI(f_i)}$ and $\mathbf{AT(f_i)}$ which will determine the **active input variables** (both primary and feedback variables) and the **active terms**, respectively, for each single output Boolean function f_i from F, and similar functions $\mathbf{AI(F_j)}$ and $\mathbf{AT(F_j)}$ for determining the active input variables and terms for multiple output Boolean functions $F_j \subseteq F$.

For **example**, for the multiple-output Boolean function F defined in Figure 3 by a tabular PLA product-term representation, where "i1–i8" represent inputs and "o1–o10" outputs of F, and "a" denotes that a particular input term (pattern of the function's input variable values) is active for a given output o_k, $AT(f_1) = \{1, 4, 5, 7, 7, 8, 9, 10, 15, 16\}$ gives the numbers of the input terms (thus also the terms) that are active for $f_1(o_1)$, and $AI(f_1) = \{1, 2, 3, 4, 5, 6\}$ gives the numbers of the inputs that occur in the terms from $AT(f_1)$, and thus, are involved in computation of $f_1(o_1)$.

An output decomposition of F can be modelled by an NB-block partition π_o on O:

$$\pi_o = \{B_1, B_2, \ldots, B_{NB}\} \tag{3}$$

The **partitioning problem** can now be formulated as follows: *find the smallest value of NB and the corresponding NB-block partition π_o on O for which the following statements hold together*:

$$\forall_{i=1\ldots NB} \quad |AI(B_i)| \leq NI \tag{4}$$

$$\forall_{i=1\ldots NB} \quad |AT(B_i)| \leq NT \tag{5}$$

$$\forall_{i=1\ldots NB} \quad |B_i| \leq NO \tag{6}$$

$$\sum_{i=1}^{NB} |AI(B_i)| \text{ is minimal} \tag{7}$$

For a given NB and $|F|$, the number of partitions, $NP(|F|, NB)$, is given by Stirling numbers of the second kind:

$$NP(|F|, NB) = \frac{1}{NB!} \sum_{i=1}^{NB} (-1)^{NB-i} \binom{NB}{i} (i)^{|F|} \tag{8}$$

```
"FILE: tue5.def
"i i i i i i i i   o o o o o o o o o o
"1 2 3 4 5 6 7 8   1 2 3 4 5 6 7 8 9 10
 1 0 0 1 1 0 x x   a a a . . . . . . .
 x x 1 0 0 1 x x   . . . a a a . . . .
 x x 1 0 0 0 x x   . . a a . . . . . .
 1 0 1 1 0 1 x x   a . a . . a . . . .
 0 1 0 0 x x x x   a . a . . a . . . .
 1 0 0 1 0 1 x x   a a . a . . . . . .
 1 1 0 x x x x x   a . . . a . . . . .
 1 1 1 0 1 0 x x   a . . . . . . . . .
 0 0 1 x x x x x   a . . . a . . . . .
 0 0 0 x x x x x   a . . a . a . . . .
 0 1 1 x x x x x   . . a a . . . . . .
 x x 1 1 0 0 x x   . . . a a . . . . .
 x x 1 1 1 0 x x   . . . a a . . . . .
 x x 0 1 0 x x x   . . . a a . . . . .
 1 1 1 1 x x x x   a a . a . a a a a a
 1 0 1 0 x x x x   a a . a . a a a a a
 1 0 x x x x 1 1   . . . . . . a a . .
 0 1 x x x x 1 0   . . . . . . . . a a
 1 0 x x x x 0 1   . . . . . . a . . a
 0 1 x x x x 0 0   . . . . . . . a a .
 0 0 0 1 x x 1 1   . . . . . . a a a a
 1 1 1 0 x x 1 0   . . . . . . a . a .
 0 0 x x x x 0 1   . . . . . . . a . a
 1 1 x x x x 0 0   . . . . . . a . . a
```

Figure 3. Example Boolean function.

This number is astronomical, even for functions F with a moderate number $|F|$ of output variables and partitions with just a few blocks NB. However, we even do not know a value for NB in our problem. The problem is more complex than multi-dimensional packing and even one-dimensional packing is known as an NP-hard problem. Since the published packaging and clustering methods do not work properly with nominal data and multiple constraints in many structural dimensions, they cannot be applied without extensive modifications to solve the problem. We developed therefore a special partitioning method that uses the double-beam search scheme explained in Section 4.1.

To avoid the combinational explosion, our double-beam algorithm constructs only the most promising partitions using knowledge of the relationships between the internal structure of the multiple output function F that has to be decomposed and the structural limitations of building blocks. The following characteristics are extracted from function F and constraints imposed by the building blocks before the partitioning:

- active inputs/terms, AI and AT, respectively, for each single output function $f_i \in F$,
- the lower and upper bound of the number of blocks, LNB and HNB, respectively,
- the expected minimal number of blocks, ENB,
- the expected number of input bits, output bits and terms per partition (building) block, ENI, ENO and ENT respectively.

The expected values show how difficult it is to satisfy each of the constraints with a given number of blocks. That difficulty indicates the amount of attention that must be paid to each of the constraints during the partitioning process (trade-off information). All the extracted knowledge is used to guide the search during the partitioning process.

The **decomposition method** is composed of the following main steps:

1. Input of the function F to decompose and the building block limitations.
2. Analyses of F and building block limitations to compute the problem characteristics: AI, AT, ENI, ENO, ENT, LNB, HNB and ENB.
3. **Partitioning**: selection of the number of partition blocks NB and constructive double-beam search for the most promising NB-block partitions.
4. Evaluation of the constructed partitions using the minimal interconnection criterion (formula (7)) and selection of the best partitions.
5. Construction of the best decompositions for F based on the selected partitions.
6. Output of the best decompositions of F.

The **partitioning process**, being the heart of the decomposition method and being of our main interest, is composed of two cooperating sub-processes: a block number selection process and partition construction process. The **block number selection process** selects the number of building blocks NB, and the **partition construction process** performs the constructive double-beam search to construct the most promising NB-block partitions on the set O. The block number selection process starts with $NB = ENB$. Depending on whether the partition construction process can find NB-block partitions which satisfy the hard constraints or not, the selection algorithm will decrease or increase NB, respectively. These steps need to be repeated until some most promising partitions with the minimal number of blocks are produced. The outcome is a set of partitions that do not violate the hard constraints, use the minimal number of building blocks and take into account the minimal interconnection criterion (formula (7)). Below, the application of the double-beam search is discussed for construction of the most promising NB-block partitions.

4.4. *Application of the double-beam search to Boolean function and FSM decomposition*

The double beam-search is used in the above outlined decomposition method for construction of a small set of the most promising NB-block partitions on O. The partitions are constructed successively, by putting the unallocated elements of O into the partition blocks.

The partitioning problem is represented in the terms of a space of states, where each **state** corresponds to a pair $PS = (\pi_o, UO)$ where π_o is a (partially constructed) partition on the set of the output variables O and UO is a set of unallocated elements of O.

The **initial state** is defined as:

$$IS = (\pi_o^\emptyset, \emptyset) \tag{9}$$

where $\pi_o^\emptyset$ is an NB-block partition with all its blocks empty.

The **set of goal states** is defined as:

$$GS = \{(\pi_o^{best}, \emptyset)\} \tag{10}$$

where π_o^{best} is an NB-block partition on O that minimizes:

$$\sum_{i=1}^{NB} |AI\,(B_i(\pi_o))| \tag{11}$$

and $\emptyset$ is an empty set.

In this particular instantiation of the double beam-search scheme outlined in Section 4.1, the data structure **current states** contains the set of partitions that have earlier been partially constructed and are considered to be extended presently, **candidate states** contains the set of more advanced partitions that are being created as a direct extension of the partially constructed partitions from current states, and **final states** contains the states with completed partitions $\{(\pi_o, \emptyset)\}$.

The **construction/move operator**, that can be applied to a given current state (π_o, UO), is defined as a pair $(o, B_i(\pi_o))$ where $o \in UO$. It selects an unallocated element o from UO and the block $B_i(\pi_o)$ of a partition π_o in which this element will be placed. The solutions will be constructed by putting in each step one or more unallocated elements of O into one or more partition blocks in result of application of some most promising construction operators.

The method constructs the partition blocks simultaneously, and not sequentially as in the methods for solving similar simpler problems known from literature. The simultaneous construction of blocks makes it possible to use for decision making more precise information about distribution of the elements from O among the different partition blocks and UO at earlier stages of the search process. This results in better decisions, and consequently, in superiority of our method (Jóźwiak and Volf 1992).

The two **selection mechanisms**: Select Moves and Select States must ensure that a solution that violates the hard constraints will not be constructed and they will try to satisfy the objective of finding the NB-block partitions with least interconnections by limited expenditure of computation time and memory space. In order to fulfil the first task, Select Moves will select only those move operators that, when applied to a given state, do not lead to the violation of hard constraints. In order to fulfil the second task, Select Moves and Select States will select a number of the most promising move operators or states, respectively, by using the estimations provided by some heuristic evaluation functions. The selection mechanisms and evaluation functions together determine the extent of the search and quality of the results. As explained in Section 4.1, the selection mechanisms use heuristic elaborations of one coherent decision rule: "at each stage of the search, take a decision which has the greatest chance of leading to the optimal solution, i.e. a decision which is the most certain according to the estimations given by the heuristic evaluation functions". If there are more decisions of the same or comparable quality, a number of them will be tried in parallel (beam-search).

According to the above rule, **Select Moves** will apply those move operators which maximize the choice certainty in a given current state and it will leave the operators which are open to doubts for future consideration. Since information contained in the partial solutions and used by the evaluation functions grows with the progress of computations, the uncertainty related to partial solutions and operators decreases. In a particular computation stage, Select Moves will maximize the conditional probability that application of a certain move operator (o_k, B_i) to a certain solution state (π_o, UO) leads to the optimal complete solution and, under this condition, it will maximize the growth of the information in the partial solution which will be used then in the successive computations steps in order to estimate the quality of choices. As explained in Section 4.1, Select Moves is controlled by two parameters: $MAXMOVES$ (the maximal number of the move operator alternatives explored) and $OQFACTOR$ (the quality factor for operators).

In the initial state, Select Moves first checks if there are mutually exclusive sets (MES) of elements from O, i.e. such sets that any two elements of a certain MES can not be placed together in one block of π_o without violation of constraints. It is evident that the elements from each MES must be placed in separate blocks, and Select Moves does it. For a certain current state (π_o, UO) not being the initial state, Select Moves selects in the first place operators $(o, B_i(\pi_o))$ for elements $o \in UO$, with exactly one possible placement in a block of π_o without violation of constraints. In each other situation, the selection is based on similarity between the particular elements $o \in UO$, and sets of elements contained in particular blocks $B_i(\pi_o)$ of π_o.

The **similarity measure** is defined for two sets of elements O_1 and O_2, $O_1, O_2 \subseteq O$ in the following manner:

$$SIM(O_1, O_2) = ICI * ISIM(O_1, O_2) + ICT * TSIM(O_1, O_2) \quad (12)$$

where $ISIM(O_1, O_2)$ is the input similarity measure between O_1 and O_2 and $TSIM(O_1, O_2)$ is the term similarity measure between O_1 and O_2. ICI and ICT are the importance coefficients for inputs and terms, respectively. The initial values for ICI and ICT are computed taking into account values of the structural parameters (ENI and ENT) and block limitations (NI, NT and NB). ICI and ICT are changed adaptively during computations dependent of the progress in satisfying the constraints. The input similarity measure $ISIM$ is defined as:

$$ISIM(O_1, O_2) = \sum_{i \in AI(O_1) \cap AI(O_2)} IC(i) \quad (13)$$

where $IC(i)$ is an importance coefficient for a certain input variable i that describes the expected maximal savings in the number of occurrences of a certain input variable i with functions from different blocks of π_o, and is computed taking into account the structural parameters and block limitations. $TSIM$ is defined analogously to $ISIM$ (Jóźwiak and Volf 1992).

Select_Moves uses values of the following three different estimations for the similarity measure $SIM(o, B_i(\pi_o))$, where $o \in UO$ and $B_i(\pi_o)$ is a constructed or empty block, in order to select operators:

- $MINSIM(o, Bi(\pi_o))$, the lower bound for SIM which is computed using only the completely certain information contained in the partial solution (without any prediction),
- $MAXSIM(o, B_i(\pi_o))$, the upper bound for SIM computed using the certain information and predictions,
- $AVESIM(o, B_i(\pi_o))$, the average value for SIM computed using the certain information and predictions.

Select_Moves applies the following four selection criteria which differ in the selection strength and the extend of predictions (operator $(o, B_i(\pi_o))$ will be selected):

$$1. MINSIM(o, B_i(\pi_o)) \geq MAXSIM(o, B_j(\pi_o)) \tag{14}$$

$$2. MINSIM(o, B_i(\pi_o)) \geq AVESIM(o, B_j(\pi_o)) \tag{15}$$

$$3. AVESIM(o, B_i(\pi_o)) \geq MAXSIM(o, B_j(\pi_o)) \tag{16}$$

$$4. AVESIM(o, B_i(\pi_o)) \geq AVESIM(o, B_j(\pi_o)) \tag{17}$$

The selection strength and degree of certainty decrease from Criterion 1 to Criterion 4. The criteria are used in the order given above. If a certain criterion results in a too small set of candidate move operators, the algorithm will switch to the next weaker criterion. If there are no elements $o \in UO$, that satisfy any of the above criteria for all $B_j : B_j \neq B_i$, the elements will be selected which satisfy the strongest possible condition for the highest number of $B_j : B_j \neq B_i$.

Select States works precisely as explained in Section 4.1. The only extra comment is that $Q(ps)$ is computed by cumulating the qualities of the choices of move operators that took place during the construction of the particular ps and prediction of the quality of the possible future choices (Jóźwiak and Volf 1992). However, the cumulated measure for the quality of choices corresponds to the degree of certainty of being on the way to some optimal or close to optimal solutions, and not to the distance from the initial state as by A^*. The prediction of the quality of the possible future choices is also expressed as certainty of remaining on a path to a (near-)optimal solution, and not as a lower bound on the distance to a (completely unknown) optimum solution (goal state).

For **example**, let us consider the multiple-output Boolean function F given in Figure 3. This function has to be implemented using a minimal number of constrained PLA building blocks and with minimal connections between the blocks. The constraints imposed by the PLA are as follows: $NI = 8$,

$NO = 6$, $NT = 16$. The function F is characterized by the following para-meters: $|I| = 8$, $|O| = 10$, $|T| = 24$. Since $|O| > NO$ and $|T| > NT$, an implementation with at least two PLA blocks is necessary. After analyzing the function F and PLA limitations, our decomposition program performed the search for the "sub-optimal" two-block partition with $MAXMOVES = 1$, $MAXSTATES = 1$.

Initial state: $\pi_o^{\emptyset} = \{\emptyset, \emptyset\}$, $UO = O - O_1$ was selected as the output with most terms and inputs, and the highest value of the similarity measure with other outputs, thus giving most information for the successive decisions.

State 1: $\pi_o^1 = \{\{O_1\}, \emptyset\}$; $UO = O - \{O_1\} - O_2$ was selected according to the strongest selection criterion, i.e. Criterion 1 (formula 14).

State 2: $\pi_o^2 = \{\{O_1, O_2\}, \emptyset\}$; $UO = O - \{O_1, O_2\} - O_7$ was selected according to Criterion 1.

State 3: $\pi_o^3 = \{\{O_1, O_2\}, \{O_7\}\}$; $UO = O - \{O_1, O_2, O_7\} - O_6$ was selected according to Criterion 1.

State 4: $\pi_o^4 = \{\{O_1, O_2, O_6\}, \{O_7\}\}$; $UO = O - \{O_1, O_2, O_6, O_7\} -$ Elements O_8, O_9, O_{10} were placed in the second block of π_o^3, as they had exactly one possible placement in view of the constraints.

State 5: $\pi_o^5 = \{\{O_1, O_2, O_6\}, \{O_7, O_8, O_9, O_{10}\}\}$; $UO = \{O_3, O_4, O_5\} - O3$ was selected according to Criterion 1.

State 6: $\pi_o^6 = \{\{O_1, O_2, O_3, O_6\}, \{O_7, O_8, O_9, O_{10}\}\}$; $UO = \{O_4, O_5\} - O_4$ was selected according to Criterion 1.

State 7: $\pi_o^7 = \{\{O_1, O_2, O_3, O_4, O_6\}, \{O_7, O_8, O_9, O_{10}\}\}$; $UO = \{O_5\} - O_5$ was selected according to Criterion 1.

Final state: $\pi_o = \{\{O_1, O_2, O_3, O_4, O_5, O_6\}, \{O_7, O_8, O_9, O_{10}\}\}$; $UO = \emptyset$ Subfunctions F_1 and F_2 that were constructed according to the partition π_o are given in Figures 4 and 5. The two subfunctions have four common inputs: I_1, I_2, I_3 and I_4.

By performing an exhaustive search, we found other six two-block decompositions; however, each of them with six common inputs, thus more interconnections. So, our method constructed the unique strictly optimal decomposition.

Table 1 and 2 present some **experimental results** that were obtained when using the software implementation of the output decomposition method presented above. In Table 1 the results from our method are compared with the results computed from an exhaustive search. Our double-beam search could find strictly optimal decompositions in all the cases. In many cases unique decompositions were found. Table 2 shows some decomposition results for large functions. The results from our double beam-search are compared to the results computed from the time consuming exhaustive search (for misex3) or to the best results from the random search performed ten times

```
"Subfunction 1 of 2
" Number of inputs: 6
" Number of outputs: 6
" Number of terms: 16}

"i i i i i i | o o o o o o
"1 2 3 4 5 6 | 1 2 3 4 5 6
"-----------+--------------
 1 0 0 1 1 0 | A A A . . .
 x x 1 0 0 1 | . . . A A A
 x x 1 0 0 0 | . . A A . .
 1 0 1 1 0 1 | A . A . . A
 0 1 0 0 x x | A . A . . A
 1 0 0 1 0 1 | A A . A . .
 1 1 0 x x x | A . . . . A
 1 1 1 0 1 0 | A . . . . .
 0 0 1 x x x | A . . . . A
 0 0 0 x x x | A . . A . A
 0 1 1 x x x | . . A A . .
 x x 1 1 0 0 | . . . . A A
 x x 1 1 1 0 | . . . . A A
 x x 0 1 0 x | . . . A A .
 1 1 1 1 x x | A A . A . A
 1 0 1 0 x x | A A . A . A
```

Figure 4. Subfunction 1.

```
"Subfunction 2 of 2
" Number of inputs: 6
" Number of outputs: 4
" Number of terms: 10}

"i i i i i i | o o o o
"1 2 3 4 7 8 | 7 8 9 10
"-----------+------------
 1 1 1 1 x x | A A A A
 1 0 1 0 x x | A A A A
 1 0 x x 1 1 | A A . .
 0 1 x x 1 0 | . . A A
 1 0 x x 0 1 | A . . A
 0 1 x x 0 0 | . A A .
 0 0 0 1 1 1 | A A A A
 1 1 1 0 1 0 | A . A .
 0 0 x x 0 1 | . A . A
 1 1 x x 0 0 | A . . A
```

Figure 5. Subfunction 2.

Table 1. Experimental results from decomposition with double-beam search

Benchmark name	Function			Constraint			Decomposition			
							Method		Exhaustive	
	I	O	T	NI	NO	NT	NB	NC	NC	ND
Tue1	7	8	14	4	4	8	2	1	1	1
Tue2	10	8	38	8	6	16	3	12	12	2
Tue3	12	10	26	8	6	16	3	6	6	1
Tue4	12	10	24	8	6	16	2	0	0	1
Tue5	8	10	24	8	6	16	2	4	4	1
Tue6	17	11	18	8	8	8	4	15	15	12
Tbk	11	8	276	12	6	136	5	44	44	70
Mark1	9	18	19	12	8	12	3	15	15	343
Ex6	8	11	28	8	8	24	2	8	8	73
Cse.1	17	17	73	16	8	48	3	31	31	5836
Ex1.1	14	24	57	12	8	32	3	22	22	3495

NB, NC are the number of blocks and extra connections, respectively; ND is the total number of decompositions with a given number of blocks NB and minimum extra connections NC that satisfy the constraints.

Table 2. Experimental results from decomposition with double-beam search for large functions

Name	Function			Constraint			Decomposition		Remarks
	I	O	T	NI	NO	NT	NB	NC	
Apex1	45	43	428	40	14	128	5	147	(1)
Gpio	131	91	253	32	24	48	7	60	(1)
Misex3	14	14	1000	14	8	256	5	56	(2)
Zeegers	18	18	590	18	6	128	5	72	(3)

(1) Random search did not find any solution satisfying the constraints
(2) Strictly optimal solution
(3) Random search: NB = 6, NC = 89.

longer than our method used for producing the results (for the remaining large benchmarks the exhaustive search would take an excessive time). In all the cases our double beam-search was able to find the best solution. All results from our method were obtained from a very limited search in a very short time (below one second).

For a given function F, the method is able to efficiently discover the best implementation structure of this function in relation to the actual constraints imposed by the logic building blocks and optimisation objectives. How

Table 3. Distribution of large functions on PLAs

Benchmark name	PLA1	PLA2	PLA3	PLA4	PLA5	PLA6	PLA7
				$|I|$			
Apex1	40	40	39	34	39		
Gpio	32	26	32	32	32	16	21
Misex3	14	14	14	14	14		
Zeegers	18	18	18	18	18		
				$|T|$			
Apex1	122	108	128	128	125		
Gpio	38	47	44	39	47	45	16
Misex3	241	227	242	237	243		
Zeegers	126	126	120	124	117		
				$|O|$			
Apex1	11	5	12	4	11		
Gpio	19	11	20	17	12	6	6
Misex3	2	2	2	3	5		
Zeegers	4	3	4	4	3		

versatile are the decompositions constructed by our double beam-search can be concluded from the results for the large functions. Unfortunately, it was virtually impossible to present the tables of these functions and their sub-functions in this paper, because the tables are large and would use too many pages of paper. Fortunately, it is possible to present the decomposition results in a compressed form, by showing the distribution of these large functions on PLAs (see Table 3). It is easy to observe, that despite the fact that there are strong interrelations between the inputs, terms and outputs of each of these functions, the inputs and terms are very regularly distributed among the PLAs. Just the distribution, that is as regular as possible and remains within the constraints imposed by the PLAs used to implement the functions, enables to implement the functions with the minimum number of PLAs. Our double-beam search enables to construct such extremely regular distributions by a very limited search. The experimental results demonstrate that our double-beam search scheme, adequately instantiated in relation to the particular problem, enables a very effective and efficient solution of circuit decomposition problems.

Moreover, the double-beam search scheme is quite general. After an adequate modelling and representation in terms of a space of states, virtually any discrete optimisation problem can be successfully tackled with it. We applied with success the double-beam search scheme in numerous circuit synthesis methods and tools solving different kinds of synthesis problems (Jóźwiak and Kolsteren 1991; Jóźwiak 1992a; Jóźwiak and Volf 1992; Jóźwiak 1992b; Jóźwiak and Volf 1995; Jóźwiak 1995; Jóźwiak et al. 2001; Jóźwiak and Chojnacki 2001). The experimental results and other arguments presented above and in (Jóźwiak and Kolsteren 1991; Jóźwiak 1992a; Jóźwiak and Volf 1992; Jóźwiak 1992b; Jóźwiak and Volf 1995; Jóźwiak 1995; Jóźwiak et al. 2001; Jóźwiak and Chojnacki 2001) demonstrate that the double-beam search scheme is suitable for solving the circuit synthesis problems and its application results in very effective and efficient synthesis algorithms.

5. *QuickScan* Algorithm

5.1. *Input support minimisation*

Reduction of the number of inputs to information processing systems is one of the most important design aspects in many fields of information technology. In circuit synthesis, it enables satisfaction of the interconnection constraints, input constraints of building blocks, and minimization of both the circuit active and interconnection area (Jóźwiak and Konieczny 1996). Although our research of the minimal input support problem (MISP) is primarily motivated by applications in digital circuit synthesis (Jasinski et al. 1989; Jóźwiak and Kolsteren 1991; Jóźwiak 1992a; Jóźwiak and Volf 1992; Jóźwiak 1992b; Jóźwiak and Volf 1995; Jóźwiak 1995; Jóźwiak and Ong 1996; Jóźwiak and Konieczny 1996; Jóźwiak et al. 1998; Jóźwiak et al. 2001; Jóźwiak and Chojnacki 2001), solutions to this problem can also be applied in many other fields, such as decision systems, database design, pattern recognition, machine learning and data mining (for minimizing the number of attributes in relation tables) (Lin and Cercone 1997; Luba and Rybnik 1992; Luba and Rybnik 1992). *QuickScan* is a simple, but quite effective and extremely efficient problem-specific search algorithm for input support minimisation of discrete relations, functions and sequential machines.

In general, the original specification of a (binary or multi-valued) discrete function or relation defines an incompatibility relation on the function's input symbols (terms/patterns of the function's input variable values). The input terms for which the function/relation has different output values are incompatible. In each (hardware or software) implementation of the original function/relation, the incompatibility relation must be preserved in order to preserve the originally specified behaviour. Input support minimization consists of finding such a minimal sub-set of input variables for a given function or relation that all incompatible input terms are distinguished by the input variables from this sub-set, making it possible to construct a function (relation) with variables from this sub-set that realizes the original function.

MISP can be formulated as follows. Given:

− a set of (multi-valued) inputs I, $I = \{x_1, \ldots, x_n\}$,

− a set of terms (vectors) T, $T \subseteq \{(t_1, \ldots, t_n) \mid t_i \in Val(x_i) \cup \text{"−"}\}$ (where $Val(x_i)$) denotes the set of possible values for x_i and "−" denotes "don't care"), and

− an incompatibility relation $\nsim$, $\nsim \subset T \times T$, imposed by a certain function (relation)

find a minimal input support U, $U \subseteq I$ such that:

$$\forall_{t',t'' \in T} \left[t' \nsim t'' \Rightarrow \exists_{i:x_i \in U} \left(t'_i \neq t''_i \wedge t'_i \neq \text{"−"} \wedge t''_i \neq \text{"−"} \right) \right] \qquad (18)$$

MISP may be easily expressed as a covering problem. For any pair of incompatible input patterns, at least one input variable must distinguish these patterns (we may say that the input variable "covers" the given pair). If we define the a 0-1 matrix M, for which: (i) each row m^i corresponds to a different incompatibility pair, (ii) each column m_j represents one input variable, (iii) the element m^i_j is equal to 1 if and only if the input symbol j allows us to distinguish the incompatibility pair related to the row m^i, then we obtain a matrix that represents a minimum cover problem. To solve this problem is to find a minimal set of columns such that for every row there is a column in the set that has 1 in the given row. The row cover problem is NP-hard (Garey and Johnson 1979; Jóźwiak and Konieczny 1996) and so is MISP.

Since MISP is an important problem, much research has been carried out in the past in relation to it. In particular, a number of exact solution algorithms have been proposed. We have tested the three most sophisticated exact algorithms that were available: the algorithm proposed by Luba and Rybnik (1992a, 1992b), the cover algorithm proposed by Rudell (1989) and used in *Espresso* for minimizing the number of terms, and a similar algorithm with the bound heuristics improved by Coudert (Coudert O. 1994) and used in a new two-level minimizer *Scherzo*. The test results clearly demonstrated that exact algorithms become inefficient for larger and difficult problem instances (the computations took hours for functions with more than 25 inputs). However, this was expected, because any exact algorithm collapses for some large enough and difficult instances of an NP-hard problem. Later we implemented a BDD-based MISP algorithm similar to this described in (Lin 1993), and Mishchenko at al. developed and implemented an exact MISP algorithm based on binary-encoded multi-valued decision diagrams (BEMDDs) (Mishchenko A. at al. 2000). Experimental results from Mishchenko's and our algorithms confirmed the above observation. The decision diagram based exact algorithms are also inefficient for larger difficult problem instances (e.g. for benchmarks "Flag" and "Mashroom" having 28 and 22 inputs, correspondingly (Mishchenko et al. 2000)).

5.2. QuickScan

Inefficiency of exact algorithms for larger instances of MISP motivates the development of faster heuristic algorithms. Although the heuristic algorithms cannot guarantee the optimal solution, they often find an optimal or near-optimal solution in a reasonable amount of time. In particular, we developed a quite effective and very efficient *QuickScan* algorithm that performs a heuristic "quick scan" through the search space (Jóźwiak and Konieczny 1996).

In the *QuickScan* we applied the following encoding of the problem:
- every input support is represented by a binary pattern;
- the binary value of a position i of the pattern is the value of a Boolean variable denoting if a certain input variable x_i is (value 1) or is not (value 0) included in the support;
- the quality of a support is equal to the number of zeros in its binary pattern.

The search space with this encoding is represented in Figure 6. It forms a lattice with the greatest lower bound (GLB) being the logical product of patterns and the least upper bound (LUB) being the logical sum of patterns.

QuickScan performs a quick traverse through several local minima (non-redundant supports). It works in two modes: the down-mode, when the algorithm seeks a leftmost (according to the lexicographic order) local minimum and the up-mode, when the algorithm escapes from the local minimum in the rightmost direction (see Figure 6). In the up-mode, the algorithm constantly tries to discover a new local minimum, before it makes the next step upwards. In result, it walks on the surface consisting of patterns representing valid supports, located in the search space just above the patterns that do not represent valid supports (thus on the surface representing the non-redundant valid supports). In order to prevent revisiting a previous local minimum, the algorithm considers only those supports that are on the right of the previously visited supports. The best local minima visited constitute the result of the algorithm. Although the global minimum is not guaranteed, the deeper a local minimum is, the more possible ways lead to it. Therefore, the chance is little that the algorithm will skip a solution that is much better than the best visited so far. The pseudo-code of the algorithm is given in Figure 7.

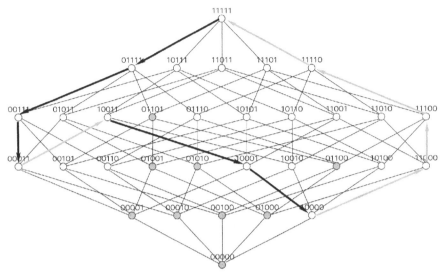

(grey nodes represent infeasible solutions, black arrows – the downward moves to the leftmost direction, grey arrows – the upward moves to the rightmost direction)

Figure 6. The search space and the scan path of the QuickScan algorithm.

QuickScan has been compared with the heuristic version of the cover algorithm implemented in *Espresso*, which is a best-first search equipped in pre-processing and quite sophisticated selection heuristics (Rudell 1989). Table 4 shows results from these two algorithms. The test benchmarks do not contain either essential or dominated inputs. This makes the benchmarks "difficult", because they cannot be solved or reduced by pre-processing techniques (i.e. recursively removing the essential columns together with the rows covered by them, and the dominated columns and rows). Although these benchmarks have from 10 to 100 inputs, they correspond to the hard cores of much larger practical benchmarks obtained from the large practical benchmarks after their pre-processing. Table 4 contains the file name of a benchmark, the number of inputs (#*inputs*), the number of product terms (#*terms*), the number of minimal supports (#*min*), the number of inputs in a minimal support ($Q(min)$), quality of the best support found by a particular algorithm (Q) and the time (T) consumed by the algorithms tested. Time is given in the format hours:minutes:seconds (SGi PowerChallenge R8000).

It should be noted, that *QuickScan* solves even the hardest problem instances (intractable with exact algorithms) within a few seconds. It is able to process very quickly even very large (100 inputs) and difficult problem instances. *QuickScan* is quite robust, but its effectiveness anyway goes somewhat down with increase of the problem dimensions: for the small (but

```
procedure QuickScan
{
  current support = full support;
  best support = none;
  while (current support exists)
  {
    if (last move was down or first move)
    {
      next support = leftmost down feasible from
                       current support;
      if (next support exists)
      {
        current support = next support;
      }
      else
      { // local minimum
        if (current support < best support)
        {
          best support = current support;
        }
        current support = rightmost up from current support;
      }
    }
    else // last move was up
    {
      next support = leftmost down feasible but to the
                       right from previous support;
      if (next support exists)
      {
        current support = next support;
      }
      else
      {
        current support = rightmost up from current support;
      }
    }
  }
  return best support;
}
```

Figure 7. Pseudo-code of QuickScan.

Table 4. The test results of heuristic algorithms for MISP

File	#inputs	#terms	#min	Q min scan	Q quick scan	T quick heuristic	Q espresso heuristic	T espresso
i10.pla	10	180	1	7	7	<1	7	<1
i20s04a.pla	20	20	31	4	4	<1	4	<1
i20s05a.pla	20	30	66	5	5	<1	5	<1
i20s06a.pla	20	40	72	6	6	<1	6	<1
kaz.pla	21	31	35	5	6	<1	5	<1
i25.pla	25	100	196	10	11	4	11	23:26
i30s08a.pla	30	120	7	8	9	2	9	29:44
i30s08b.pla	30	120	17	8	9	2	9	26:10
i30s08c.pla	30	120	13	8	9	2	9	25:30
i30s10a.pla	30	240	3	10	11	6	?	(f)
i40s09a.pla	40	160	1632	9	10	5	?	(f)
i40s09b.pla	40	160	1632	9	10	5	10	4:35:25
i40s10a.pla	40	240	716	10	11	9	?	(f)
i40.pla	40	240	716	10	11	9	?	(f)
i50.pla	50	100	1	7	9	2	8	19:29
i100.pla	100	200	–	9	11	30	?	(f)

(f) The program could not be completed because of enormous memory claims.

difficult) problems with up to 20 inputs, it finds almost always an exact minimum within an extremely short time, for the problems with 20 to 50 inputs, it almost always finds a solution being on distance one from the optimum, and for the large and difficult problems (with more than 50 inputs), it almost always finds a solution being on distance 2 from the optimum. *QuickScan* is a simple sequential algorithm. Its extremely high efficiency follows from its simplicity, but its reasonably high effectiveness (robustness) follows from its problem-specific character. *QuickScan* directly exploits the knowledge of the input support minimisation problem constraints (expressed by the incompatibility relation) and objectives (finding the minimum support, being the smallest of the non-redundant valid supports), and in result traverses only the surface of the valid (satisfying relation) and non-redundant supports, and thus does not loose time for visiting any other supports. Since *QuickScan* is extremely efficient, but also very effective for small enough problem instances, we exploit this algorithm with success inside a number of electronic design automation (EDA) tools for combinational and sequential circuit synthesis, especially in the places where repetitive support minimization must performed, i.e. the extremely high efficiency of this algorithm is indispensable (Józwiak 1992: Józwiak and Volf 1995; Józwiak and Konieczny 1996; Józwiak et al. 2001; Józwiak and Chojnacki 2001). This is a very good illustration for the fact that, although the parallel constructive search approaches are in general more suitable for solving the circuit synthesis problems, even a very simple (but problem-specific, robust and extremely fast) sequential search can be very useful in particular situations, especially, when the computation speed is of primary importance. In the synthesis tools described in (Józwiak et al. 2001) and (Józwiak and Chojnacki 2001) that serve for sequential and combinational circuit synthesis targeted to LUT-based FPGAs, several double-beam and *QuickScan* searches are used in combination. We performed an extensive experimental research of these tools using thousands of different sort circuits, including hundreds of contemporary industrial circuits, in which we compared results from these tools to the available results from several academic tools and to the results from several state-of-the-art industrial tools. The tools were applied to problems of practical size and a large part of the experiments was performed in a practical industrial setting. The results of this research are partially reported in (Józwiak et al. 2001) and (Józwiak and Chojnacki 2001). Our tools constructed in most cases much better circuits, regarding both the circuit speed and area, than all other tools used for the experiments, and they robustly constructed such high-quality circuits within the time comparable to the run-time of the commercial tools used for the experiments. Results of this research give much experimental evidence that the double-beam search, *QuickScan* and their software implementations are capable to effectively and efficiently solve the contemporary industrial problems of practical size.

6. Genetic Engineering Algorithm (GEA)

6.1. *The concept of a genetic engineering algorithm*

Genetic algorithms (GAs) originate from Fraser's work on the simulation of genetic systems (Fraser 1962) and Holland's study on adaptation in natural and artificial systems (Holland 1992). The basic GA imitates some evolutionary characteristics of natural populations. It works with some collections of the search-space elements called *populations*. The members of a population are evaluated using a *fitness function*. Elements that are better than others in the same population according to the fitness function have a higher chance of generating offspring (survival of the fittest principle). Two operators are used for offspring generation: crossover and mutation. The *crossover* constructs two *children* by combining the *chromosomes* of their parents. The *mutation* generates one child from one parent by changing the parent's *gene* values. Children of the selected members of the current population form a new population. Depending on a particular stop criterion, GA decides for each population whether a new population has to be created or the algorithm has to terminate. More precise description of the basic GA can be found in (Jóźwiak and Postuła 2002). Although some research on the application of genetic algorithms in hardware design and electronic design automation (EDA) is reported in the literature (e.g. Almaini et al. 1995; Benten and Sait 1994; Chattopadhyay and Chaurhuri 1997; Dill and Perkowski 1998; Dill and Perkowski 2001), many problems remained unsolved, and this motivated us to carry out the research reported below.

As explained in Section 3, a search algorithm adequate for circuit synthesis should robustly and efficiently produce the optimal or near-optimal results. This can be achieved by appropriately combining the exploration of the search space in many promising directions with the fast convergence to some optimal or near-optimal solutions (Davis 1991; Goldberg 1975; Holland 1995; Jóźwiak 1995; Jóźwiak and Postuła 2002). GAs have several important features that predestine them to solve synthesis problems. An advantage of a GA is its ability to generate globally good results if the exploration and exploitation are combined appropriately and the algorithm is given a long enough run-time. Another advantage is its generality. After appropriate problem encoding, virtually any synthesis problem can be solved with a GA. Furthermore, GAs work with some collections of partial or full solutions, and therefore, they enable interplay among the solutions by their combination, comparison and selection. Moreover, when appropriately used, GAs enable constructive parallel searches in which the most promising construction operators have a higher chance of being applied to the most promising (partial or complete) solutions. The main disadvantage of a GA, especially when applied to large instances of complex synthesis problems, is its excessively long run-time that is needed to deliver some satisfactory results.

The main conclusion of our research on GAs is that the effective and efficient application of the GA concept to the synthesis problem solving requires substitution of the basic GA's "natural evolution" by "genetic engineering" (GE).

Below, we discuss the concept of a genetic engineering algorithm (GEA), show how to apply the GEA to solve synthesis problems, and demonstrate the high effectiveness and efficiency of the genetic engineering algorithm for solving the minimal input support problem (MISP) discussed in the previous section.

The basic GA imitates natural evolution. Nature has time: it creates its beautiful structures in hundreds, thousands or millions of years. Engineers have much less time: they have to create their complex system structures within days, months or years. The EDA tools, that help the designers, have to solve the complex circuit synthesis problems within seconds, minutes or hours.

In the basic GA both the genetic operators and the selection mechanisms are probabilistic. Such a purely probabilistic evolution is good for exploration of the solution space in many different directions, but for complex design problems, it results in an excessively long time before satisfactory solutions are delivered. Moreover, the fitness function's computation time is problem dependent and often excessively long. Application of the GA concept to the automatic circuit synthesis requires therefore a big increase of the basic GA's efficiency in parallel with increase of its effectiveness.

In brief, our **genetic engineering** concept consists of perturbing the natural evolution of the basic GA to some (often very high) degree, by implementing some specific goal-oriented deterministic or semi-deterministic decisions. In general, the degree of determinism introduced can range from no to full determinism, dependent on the availability of information for the deterministic decision making of acceptable quality and other characteristics of a particular problem. In addition, a particular genetic engineering algorithm (GEA) can be controlled in such a way that it can change its characteristics from completely probabilistic, through partly deterministic, to completely deterministic, even during its run-time. Moreover, if the fitness function for a particular problem has an excessively long computation time, it is replaced by a proxy function and/or an approximate fitness value is computed.

The concept of the GEA exploits the positive features of the basic GA and eliminates its weaknesses. This concept is implemented by supplementing the basic GA with the deterministic and pseudo-deterministic crossover and mutation operators, deterministic and pseudo-deterministic construction of the initial population, mixed and deterministic selection mechanisms, and adaptation of the application probabilities of operators with the progress of computations. Moreover, if the computation time of the original fitness function for a particular problem is excessively long, the original fitness function is replaced by a proxy fitness function being highly positively correlated with the original function, but much easier to compute, and/or the fitness function's computation process is simplified without too much sacrificing the result quality, by computing an approximate fitness value in a reasonable time instead of the precise value.

6.2. *The genetic engineering scheme*

A search algorithm can only be effective and efficient if each of its particular parts used repetitively (i.e. operators, fitness function computation, selection mechanisms, etc.) is effective and efficient, and all parts are appropriately composed together. The quality of the total algorithm is a composition of its particular part's quality, being rather a product and not a sum of the part's quality. A good algorithm should appropriately combine exploration of the search space in many promising directions with fast convergence to some optimal or near-optimal solutions (Davis 1991; Goldberg 1975; Holland 1995; Jóźwiak 1995; Jóźwiak and Postuła 2002).

In a GA, *selection mechanism* has an important role in maintaining an appropriate mixture of exploration and exploitation. Each of the two most frequently used selection mechanisms, being roulette wheel and tournament selection, has its advantages and disadvantages. An important advantage of the roulette wheel selection is that the selection probability is proportional to the fitness, and consequently, the population members with similar fitness values are expected to be selected an equal number of times. The disadvantage is the danger of the super-fit element dominance leading to premature convergence. The advantage of the tournament selection is its speed and the fact that a super-fit element will not dominate the next generation. The disadvantage of the tournament selection is that the population members with similar fitness are not expected to be selected an equal number of times (Jóźwiak and Postuła 2002). To overcome both the super-fit element dominance and the similar fitness problem, we combined both selection mechanisms. The mixed probabilistic selection can be further mixed with the deterministic selection. An appropriate part of the selection can be performed purely deterministically, based on the estimations with the same or different fitness function as used for the probabilistic selection.

To speed up the convergence of the algorithm towards some very good solutions, we developed two extra *deterministic operators*: *merge and repair*, next to the basic probabilistic genetic operators. Each of them implements a sort of very robust and fast local search. The underlying principle for the development of these two deterministic operators is imitation of some human design characteristics. When human designers have some prototype solutions they evolve them, but not randomly. They refine the prototypes by changing or combining some parts of the prototypes in such a way that the prototypes' good features are preserved and the bad ones are eliminated. The *merge operator* is a kind of deterministic crossover, but instead of combining two population members at random, it tries to combine two population members in such a way that their good features are combined and bad ones eliminated. The *repair operator* is a kind of complex deterministic mutation. However, it is applied exclusively to the population members that represent non-valid solutions. It transforms a pattern representing a non-valid solution into a pattern representing a valid solution. It tries to find the best valid solution in the surrounding of the original non-valid one.

The *mutation* operator of our GEA works differently for population members representing valid and non-valid solutions. For valid solutions, the design's hard constraints are satisfied, and therefore, the preferred direction of the gene's change aims at optimizing the design objectives. For non-valid solutions, the design hard constraints are violated, and therefore, the preferred direction of the gene's change aims at giving the non-valid solutions a high chance of being transformed into some valid solutions, without too much sacrificing the design objectives.

The *crossover* operator used in our GEA is a special modified two-point crossover. The basic two-point crossover (Beasley et al. 1993a, 1993b) treats the chromosomes as a ring structure, thus completely symmetrically, what is its advantage. Two cut points, the same for both parent chromosomes, are selected at random, and all positions within the range specified by these two points are exchanged between the chromosomes. A disadvantage of the two-point crossover is a high chance of producing children identical or very similar to their parents by converging populations (i.e. if the parent chromosomes are similar to each other). This may result in the premature convergence. Our modified crossover ensures that both children are different from their parents. Parents that are the same or different in only one position are not allowed to cross. Moreover, one of the cut points is in the section that contains the different positions, while the second one is just outside of this section. The experimental results presented in this paper were achieved with this modified two-point crossover.

Although the modified two-point crossover prevents the creation of children identical to their parents, it does not prevent the creation of children that are very similar to their parents. This can be achieved by avoiding crossover of the chromosomes that are too similar to each other. We proposed two particular solutions to this problem based on two different similarity (or distance) measures between the parent chromosomes.

The *first solution* of the premature convergence is based on the simple Hamming distance between the parent chromosomes, and it results in creation of *children different from their parents*. It realizes exploration of the solution space in *many different directions*. The basic concept of this solution consists of avoiding crossover of chromosomes with a too small Hamming distance between them. This basic concept can be of course realized in many various ways (Jóźwiak and Postuła 2002). Perhaps the best way that we proposed is based on the computation of the Hamming distance distribution for a given chromosome population, by randomly choosing a representative sample of chromosome pairs, followed be computation of the Hamming distance for each pair from the sample, and computation of the average distance and standard deviation. Based on the distance distribution parameter computed this way, an appropriate minimum Hamming distance value is selected that guarantees high enough number of chromosome pairs for crossover and high enough chromosome diversity at the same time. This method is thus both effective and efficient.

The *second solution* we proposed is based not on the simple Hamming distance, but rather on the weighted (Hamming type) distance, and therefore it results in *children different from their parents at "important" gene positions*. It realizes exploration of the solution space in *many different "most prom-ising" directions*. It is analogous to the above-proposed solution using the simple Hamming distance and computing the distance distribution to decide the minimum crossover distance. However, instead of computing the simple Hamming distance, it computes the weighted distance, with weights of partic-ular genes growing in relation to their importance as expressed by the chance of a particular gene's value appearing in the optimal solution. This weighted distance measure is computed by summing up the weights of the genes at which the two parent chromosomes differ.

The operators of our GEA are applied with various probabilities that evolve with the progress of computations. At the very beginning, the mutation and merge operators have a small chance of being applied, but crossover and repair are often used. Mutation is mainly meant to perform a local search around already found good solutions and maintain the diversity of the popula-tion. However, the search space has to be globally explored first (to find some good solutions) and the population is quite diverse at the beginning. Merge is aimed at combining good features and eliminating weaknesses of some already advanced prototype solutions that have to be constructed and selected during the GA execution. Crossover mainly aims at global exploration of the solution space in many directions. Repair is often used, because at the beginning the non-valid solutions have quite a high chance of being created. With the progress of the process, the merge operator becomes more important at the cost of the crossover and the chance of mutation grows.

The initial population is created not randomly, but semi-randomly. The probability of each gene's value appearing in a pattern that is a member of the initial population grows with the gene's value importance as expressed by a chance of this gene's value appearing in the optimal solution. In this way, the chance that high-quality solutions are included in the initial population is high while the diversity of the population is still maintained. In some cases, the initial population can even be created purely deterministically. This is achieved by constructing some (or all) combinations of gene values with the highest importance. The importance of a particular gene value may depend on the context of a certain partially constructed combination and it may dynamically change during the construction process.

By appropriately controlling the choice, mixture and extent of application of the above probabilistic and deterministic mechanisms of the GEA, one can obtain various particular algorithms: from purely probabilistic to purely deterministic. If the control is dynamic, i.e. realized during the run-time of the GEA algorithm, the algorithm can evolve during the run-time. For example, it can evolve from a purely probabilistic genetic algorithm in the begin of the search process (when a divergent search of the solution space in many directions is needed and there is a very limited knowledge about the solutions quality), to a purely deterministic algorithm by the end of this process (when there already is much knowledge about the quality of the solutions found).

In order to check how well the proposed GEA scheme works and at the same time solve an important practical synthesis problem, we developed a genetic engineering algorithm that solves the MISP problem discussed in the previous section. In case of MISP, the fitness function represents the number of inputs in a particular support and is very easy to compute. It is thus not necessary to replace it by any proxy function or to perform any approximate fitness computation. For some other problems however, too complex fitness computation can make the algorithm impractical, and therefore, the replacement of the original fitness function by a proxy function and/or computation of an approximate fitness value instead of the precise value may be necessary (Dill and Perkowski 2001; Jóźwiak 1992; Jóźwiak and Postuła 2002). Application of the proposed GEA scheme to MISP resulted in a very effective and efficient algorithm for this problem. Although all the new concepts proposed above influenced the algorithm's effectiveness and efficiency positively, the main gain in computation time and quality of the results is due to the deterministic crossover operator (merge).

6.3. *Application of the GE scheme to MISP*

Below, we only very briefly comment on our genetic engineering algorithm for MISP, which applies the GE scheme outlined in Sections 6.1 and 6.2. More detailed discussion can be found in (Jóźwiak and Konieczny 1996; Jóźwiak et al. 1998; Jóźwiak and Postuła 2002).

Our GEA for MISP uses the same encoding as *QuickScan* (see Section 5). The population in our GEA for MISP is a sub-set of the search space elements, each being a binary pattern representing a certain subset of input variables (support). The *initial population* is created semi-deterministically. For each variable, its probability of being included in a support that is a member of the initial population is proportional to its importance. If the considered pattern represents a valid support, then its *fitness* decreases with the number of ones in the pattern. In the case of a non-valid support, our GEA estimates how many extra variables are minimally required to "repair" it to a valid support, and the total of this estimation and of the number of variables in the non-valid support is used for the calculation of its fitness. The defined this way fitness function is used by the mixed selection mechanism as described in Section 6.2. Also, the mutation operator works differently for population members representing valid supports and non-valid supports. For valid supports, the chance of changing 1 into 0 is higher than the chance of changing 0 into 1, and for non-valid supports this is vice versa.

Two *deterministic operators*: merge (complex deterministic crossover) and repair (complex deterministic mutation) implement a sort of robust and very fast (local) search, which primarily involves some pre-processing steps and, if necessary, some best-first search steps. The *merge operator* tries to find the best local minimum of the search subspace being a sub-lattice defined by the LUB and GLB of the two population members, i.e. by the logical sum and logical product of their representation patterns. The *repair operator* is exclusively applied to the population members that represent non-valid supports. It transforms a pattern representing a non-valid support into a pattern representing a valid support, by trying to change a minimal number of the original pattern's zeros into ones. The operators of our GEA are applied with various probabilities, which evolve with the progress of computations as described in Section 6.2.

Table 5 presents the test results of the GEA compared to the results from the *QuickScan* computed at the HP9000/735 processor under HP-UX 9.05. The test instances are "difficult", because they cannot be solved or reduced by the preprocessing techniques (i.e. recursively removing the rows covered by essential columns, and the dominated columns and rows). The test files are described more precisely in (Jóźwiak and Konieczny 1996). In almost all cases, the results from the GEA are strictly optimal, and only once per a number of runs are for some circuits on distance one from optimum. They are never worse and often one or two support variables better than the results from the *QuickScan*, but the execution time for large problem instances is longer (but still acceptable) for the GA. The effectiveness of the GEA almost does not decrease with the growing problem dimensions, but its efficiency slowly decreases, while the efficiency of *QuickScan* almost does not decrease with the growing problem dimensions, but its effectiveness slowly decreases. It is thus evident that *QuickScan* should be used for reasonably small problem instances or when the computation speed is of primary importance, while GEA can be used even for the large instances when the quality of results is the main aim. The *GEA realizes a very good trade-off between the effectiveness and efficiency*. The deterministic *merge operator is very powerful*, and the

Table 5. Test results of the genetic engineering algorithm (GE) and QuickScan (QS)

File	Q (min)	Q (QS)	T (QS)	Q (GE−)[1]	T (GE−)[1]	Q (mg)	T (mg)	Q (GE)[2]	T (GE)[2]
i10	7	7	<1	7	<1	7	<1	7	<1
i20s04a	4	4	<1	4	1	4	<1	4	<1
i20s05a	5	5	<1	5	1	5	<1	5	2
i20s06a	6	6	<1	6.17	2	6	<1	6	2
Kaz	5	6	<1	5.17	2	5	<1	5	1
i30s08a	8	9	2	9.17	30	8	6	8	40
i30s08b	8	9	2	9.17	22	9	6	8	38
i30s08c	8	9	2	9.17	21	8	6	8.13	34
i30s10a	10	11	6	11.17	1:37	11	58	10.88	2:39
i40s09a	9	10	5	9.83	56	9	19	9	1:52
i40s09b	9	10	5	9.83	55	9	19	9	1:53
i40s10a	10	11	9	11.17	2:53	11	1:47	10	6:47
i40	10	11	9	11	3:09	11	1:47	10	6:32
i50	7	9	2	8.67	20	8	4	8	48
i100	9	11	30	10.83	3:31	9	1:02	9	9:58

Q – quality of the best support found, Q(min) – quality of the minimum support, T – computation time, QS – *QuickScan*, GE – GEA algorithm, GE- – GEA without merge, mg – merge performed on the full original support.
[1] Average over 6 runs.
[2] Average over 8 runs.

GEA with merge is much better than without. *This shows that supplementing the basic GA with deterministic operators is a very successful concept.* An appropriate *combination of the non-determinism* of the basic GA *with pseudo-determinism* by generation of the initial population *and determinism* of merge and repair *resulted in a very effective and efficient algorithm* for MISP.

The experimental results and other arguments presented above demonstrate that the proposed GEA scheme is suitable for solving the synthesis problems and its application results in very effective and efficient synthesis algorithms. The GA concept and scheme are general, but their application is problem-specific, in many cases not trivial, and sometimes difficult. Therefore, the quality of the final GEA is determined both by the GEA scheme presented above and by the quality of its instantiation for a particular problem.

The experimental results and other arguments presented above demonstrate that the proposed GEA scheme is suitable for solving the synthesis problems and its application results in very effective and efficient synthesis algorithms. The GA concept and scheme are general, but their application is

problem-specific, in many cases not trivial, and sometimes difficult. Therefore, the quality of the final GEA is determined both by the GEA scheme
presented above and by the quality of its instantiation for a particular problem.

7. Conclusion

The fast progress in microelectronic technology, and especially, the recent
shift from the multi-chip systems to systems-on-a-chip opens new possibilities, but also creates new difficulties in development and implementation
of the complex and heterogeneous microelectronic circuits and systems.
Development and application of a new generation of more suitable design
paradigms, methods and tools that will solve some of the main difficulties is
therefore extremely important for the modern system industry. This paper
showed the importance of the AI search techniques for the circuits and
systems design space exploration, explained what sorts of search techniques
are useful for this aim, and discussed the place, role and way of use of the AI
search techniques. In particular, the paper explained the importance and usage
of the heuristic search techniques for the automatic construction and selection
of the most promising solutions to the circuit synthesis problems, when using
as examples three effective and efficient search algorithms, being representative to three different algorithm classes, and showing their application to
important circuit synthesis problems.

The main conclusions of the paper are the following. In general, to
solve the complex, multi-aspect, large and "irregular" circuit synthesis
problems effectively and efficiently, it is necessary to apply the complete
apparatus of the solution invention methodologies. Therefore, the parallel
search approaches that work with populations of solutions (e.g. genetic type
algorithms or double-beam search), enable interplay among various solutions
(by combination of their parts or elements, their comparison and selection),
and make constructive search possible (in which the most promising construction operators are applied to the most promising partial solutions), have a
good chance of performing this task adequately. What makes the search
effective and efficient is the usage of a search scheme that closely matches the
kind of problem to be solved, and usage of the problem-specific knowledge
in the algorithm itself or in the information for the decision-making in the
scope of the algorithm. Moreover, the circuit or system synthesis algorithms
cannot assume having a satisfactory knowledge of the optimal solutions
(goal states) or their features, before or after just starting to construct them,
what many AI algorithms known from literature assume (including $A*$ and
similar algorithms). In most cases, the synthesis algorithms have to use less
direct, less precise and smaller pieces of information related to the solution

being constructed, but in a much more sophisticated way than A^* or similar algorithms, to guarantee a very high chance of remaining on some paths to the optimum or very close to optimum solutions.

The conclusions of the paper were illustrated with examples of three effective and efficient search algorithms: *Double-Beam Search*, *QuickScan* and *Genetic Engineering Algorithm*, and experimental results from their application to two important circuit synthesis problems: output decomposition and input support minimization. The experimental results and other arguments presented in the paper demonstrate that the quite generally applicable parallel search schemes (Double-Beam and Genetic Engineering), appropriately applied and provided with an adequate problem knowledge, result in very effective and efficient constructive synthesis algorithms. However, *QuickScan*, being a simple sequential, but problem specific search is also extremely efficient, and very effective for small enough problem instances. This is a very good illustration for the fact that, although the parallel (constructive) search approaches are in general more suitable for solving the circuit synthesis problems, even a very simple (but problem-specific, robust and extremely fast) sequential search can be very useful in particular situations, especially when search of a reasonably small solution space or an extremely quick search is required (e.g. for the improvement of the earlier found near-optimal solutions). To effectively and efficiently solve the complex practical synthesis problems several different search algorithms with characteristics closely matching the nature of the particular sub-problems should be used in combination. For instance, in the circuit synthesis tools described in (Jóźwiak et al. 2001) and (Jóźwiak and Chojnacki 2001) several double-beam and *QuickScan* searches are used in combination. An extensive experimental research of these tools performed using thousands of different sort circuits, including hundreds of contemporary industrial circuits, provides much experimental evidence that the double-beam search, *QuickScan* and their software implementations are capable to effectively and efficiently solve the contemporary circuit synthesis problems of practical size.

The knowledge presented in this paper mainly resulted from my own research and experience related to circuits, systems, electronic design automation and artificial intelligence, but it also resulted from various interactions and discussions with other researchers and practitioners. Also, a number of students helped me with implementation of numerous algorithms and experimental research through the years. I would like to thank them all, and also to express my high appreciation to all the people that join their effort in order to build, and not to destroy.

References

Almaini, A. E. A. et al. (1995). State Assignment of Finite State Machines Using a Genetic Algorithm. *IEE Proc. Comput. Digit. Tech.* **142**(4): 279–285.

Beasley, D. et al. (1993a). An Overview of Genetic Algorithms – Part 1: Fundamentals. *University Computing* **15**(2): 58–69.

Beasley, D. et al. (1993b). An Overview of Genetic Algorithms – Part 2: Research Topics. *University Computing* **15**(4): 170–181.

Benten, M. S. T. & Sait, S. M. (1994). GAP: A Genetic Algorithm to Optimize Two-bit Decoder PLAs. *Int. J. Electronics* **l76**(1): 99-106.

Chattopadhyay, S. & Chaurhuri, P.P. (1997). Genetic Algorithm Based Approach for Integrated State Assignment and Flip-Flop Selection in Finite State Machine Synthesis. In Proc. of the *Int. Conf. On VLSI Design*, 522–527.

Coudert, O. (1994). Two-Level Logic Minimization: An Overview. *INTEGRATION: The VLSI Journal* **17**: 97–140.

Davis, L. (1991). *Handbook of Genetic Algorithms*. New York: Van Nostrand Reinhold.

Dill, K. M. & Perkowski, M. A. (1998). Evolutionary Minimization of Generalized Reed-Muller Forms. In Proc. of *ICCIMA'98 Conference*, 727–733. Melbourne: World Scientific.

Dill, K. M. & Perkowski M. A. (2001). Baldwinian Learning Utilizing Genetic and Heuristic Algorithms for Logic Synthesis and Minimization of Incompletely Specified Data with Generalized Reed-Muller (AND-EXOR) Forms. *Journal of Systems Architecture* **47**(6): 477–489.

Fraser, A. S. (1962). Simulation of Genetic Systems. *Journal of Theoretical Biology*: 329–349.

Garey, M. R. & Johnson, D. S. (1979). *Computers and Intractability: A Guide to the Theory of NP-Completeness*. New York: W.H. Freeman and Company.

Goldberg, D. E. (1975). *Genetic Algorithms in Search, Optimization, and Machine Learning*. Reading, MA: Addison-Wesley.

Hart, P. E. et al. (1968). A Formal Basis for the Heuristic Determination of Minimum Cost Paths. *IEEE Trans. Syst. Sci. & Cybern.* **4**(2): 100–107.

Holland, J. H. (1992). *Adaptation in Natural and Artificial Systems: An Introductory Analysis with Applications to Biology, Control, and Artificial Intelligence*. Cambridge, MA: Cambridge MIT Press.

Holland, J. H. (1995). *Hidden Order: How Adaptation Builds Complexity*. Reading, MA: Addison-Wesley.

Jasinski, K. et al. (1989). Parallel Decomposition in Logic Synthesis. In Proc. of *15th European Solid-State Circuits Conference*, 113–116.

Jóźwiak, L. & Kolsteren, J. C. (1991). An Efficient Method for the Sequential General Decomposition of Sequential Machines. *Microprocessing and Microprogramming* **32**: 657–664.

Jóźwiak, L. (1992a). An Efficient Heuristic Method for State Assignment of Large Sequential Machines. *Journal of Circuits, Systems and Computers* **2**(1): 1–26.

Jóźwiak, L. & Volf, F. (1992). An Efficient Method for Decomposition of Multiple Output Boolean Functions and Assigned Sequential Machines. In Proc. of *EDAC – The European Conference on Design Automation*, 114–122. Brussels: IEEE Computer Society Press.

Jóźwiak, L. (1992b). *A Method for General Simultaneous Full-Decomposition of Sequential Machines: Algorithms and Implementation*, EUT-Report 92-E-267. The Netherlands: Eindhoven University of Technology.

Jóźwiak, L. & Volf, F. A. M. (1995). Efficient Decomposition of Assigned Sequential Machines and Boolean Functions for PLD Implementations. In *Proc. of International*

Conference on Electronic Technology Directions, 259–266. Adelaide: IEEE Computer Society Press.

Jóźwiak, L. (1995). General Decomposition and Its Use in Digital Circuit Synthesis. *VLSI Design* **3**(3–4): 225–248.

Jóźwiak, L. & Ong, A. (1996). Quality-Driven Decision Making Methodology for System-Level Design. In *Proc. of EUROMICRO'96 Conference*, 8–18. Prague: Computer Society Press.

Jóźwiak, L. & Konieczny, P. A. (1996). Input Support Minimization for Efficient PLD and FPGA Synthesis. In *Proc. of IWLAS'96*, 30–37. Grenoble: IFIP.

Jóźwiak, L. et al. (1998). Solving Synthesis Problems with Genetic Algorithms. In *Proc. of EUROMICRO'98 Conference*, 1–7. Vasteras: IEEE Computer Society Press.

Jóźwiak, L. et al. (2001). Fast and Compact Sequential Circuits Through the Information-Driven Circuit Synthesis In *Proc. of DSD'2001 – Euromicro Symposium on Digital System Design*, 46–53. Warsaw: IEEE Computer Society Press.

Jóźwiak, L. & Chojnacki, A. (2001). Effective and Efficient FPGA Synthesis through Functional Decomposition Based on Information Relationship Measures. In *Proc. of DSD'2001 – Euromicro Symposium on Digital System Design*, 30–37. Warsaw: IEEE Computer Society Press.

Jóźwiak, L. (2001). Quality-Driven Design in the System-on-a-Chip Era: Why and How? *Journal of Systems Architecture* **47**(3–4): 201–224.

Jóźwiak, L. & Postuła, A. (2002). Genetic Engineering versus Natural Evolution: Genetic Algorithms with Deterministic Operator. *Journal of Systems Architecture* **48**(1–3): 99–112.

Lin, B. (1993). Efficient Symbolic Support Manipulation In *Proc. of ICCD'93*, 513–516. IEEE Computer Society Press.

Lin, T. Y. & Cercone, N. (eds.) (1997). *Rough Sets and Data Mining – Analysis of Imprecise Data*. Boston/Dordrecht/London: Kluwer Academic Publishers.

Louis, S. J. & Rawlins, G. J. E. (1992). Syntactic Analysis of Convergence in Genetic Algorithms. In Rawlins, G. J. E. (ed.) *Foundation of Genetic Algorithms 2*, 141–151. Morgan Kaufman Publishers Inc.

Luba, T. & Rybnik, J. (1992a). Algorithm of Elimination of Attributes and Arguments Based on Unate Complement Concept. *Bulletin of the Polish Academy of Sciences* **40**(3): 313–322.

Luba, T. & Rybnik, J. (1992b). Rough Sets and Some Aspects of Logic Synthesis. In Slowinski, R. (ed.) *Intelligent Decision Support*. Boston/Dordrecht/London: Kluwer Academic Publishers.

Mishchenko, A. et al. (2000). Implicit Algorithms for Multi-Valued Input Support Manipulation In *Proc. of 4th Intl. Workshop on Boolean Problems*. Freiberg: University of Freiberg.

Pearl, J. (1984). *Heuristics*. Reading, MA: Addison-Wesley.

Rudell, R. (1989). *Logic Synthesis for VLSI Design*. Ph.D. diss., University of California, Berkeley, CA.

Saaty, T. L. (1980). *The Analytic Hierarchy Process*. New York: McGraw-Hill.

Sakawa, M. (1993). *Fuzzy Sets and Interactive Multiobjective Optimization*. New York: Plenum Press.

Vincke, P. et al. (1992). *Multicriteria Decision-Aid*. Chichester: John Wiley and Sons.

Wierzbicki, A. P. (1982). A Mathematical Basis for Satisficing Decision Making *Mathematical Modeling* **3**: 391–405.

Wierzbicki, A. P. (1986). On the Completness and Constructiveness of Parametric Characterizations to Vector Optimization Problems. *OR Spektrum* **8**: 73–87.
Yu, P. L. (1985). *Multiple-Criteria Decision Making*. New York/London: Plenum Press.

Exploiting Functional Properties of Boolean Functions for Optimal Multi-Level Design by Bi-Decomposition

BERND STEINBACH[1] and CHRISTIAN LANG[2]
[1]*TU Bergakademie Freiberg, Institute of Computer Science, 09596 Freiberg, Germany (E-mail: steinb@informatik.tu-freiberg.de);* [2]*IMMS gGmbH Erfurt, 99099 Erfurt, Germany (E-mail: christian.lang@imms.de)*

Abstract. This paper introduces the theory of bi-decomposition of Boolean functions. This approach optimally exploits functional properties of a Boolean function in order to find an associated multilevel circuit representation having a very short delay by using simple two input gates. The machine learning process is based on the Boolean Differential Calculus and is focused on the aim of detecting the profitable functional properties available for the Boolean function.

For clear understanding the bi-decomposition of completely specified Boolean functions is introduced first. Significantly better chance of success are given for bi-decomposition of incompletely specified Boolean functions, discussed secondly. The inclusion of the weak bi-decomposition allows to prove the the completeness of the suggested decomposition method. The basic task for machine learning consists of determining the decomposition type and dedicated sets of variables. Lean on this knowledge a complete recursive design algorithm is suggested.

Experimental results over MCNC benchmarks show that the bi-decomposition out-performs SIS and other BDD-based decomposition methods in terms of area and delay of the resulting circuits with comparable CPU time.

By switching from the ON-set/OFF-set model of Boolean function lattices to their upper- and lower-bound model a new view to the bi-decomposition arises. This new form of the bi-decomposition theory makes a comprehensible generalization of the bi-decomposition to multivalued function possible.

Keywords: area, bi-decomposition, Boolean differential calculus, Boolean function, delay, derivative operation, lattice, multi-level circuit

Abbreviations: BDC – Boolean Differential Calculus; BDD – Binary Decision Diagram; ISF – Incompletely Specified Function

1. Introduction

Boolean functions are widely used in control systems, computers and communication system as well. Such systems need a representation of the Boolean functions by simple Boolean operations directly mappable to logic gates.

There are in general two procedures to transform the behavioral description of a Boolean function into a structural description. First, the covering methods try to cover all function values 1 by a small number of cubes. This is the most studied topic of the switching theory, starting by George Boole in the 19th century. Second, the decomposition methods try to decompose a given Boolean function into two or more simpler ones. This task is much more difficult and the first paper about this topic was published by Ashenhurst (1957). This first suggested method methods was not practical, yet. Several years later Curtis (1962) published a more general decomposition method together with an practicable algorithm. Similar to Ashenhurst (1957), Curtis (1962) cuts in his decomposition the Boolean function into two large parts, connected in a chain. The difference was that Ashenhurst used one connection wire between the decomposed blocks and Curtis several wires.

A completely different approach is the bi-decomposition, first suggested by Bochmann (1991) to his PhD student (Böhlau 1987). The new idea based on the observation, that the final output gate in a circuit structure creates the Boolean function. If this final gate is a simple two input gate and for each Boolean function exist two subfunctions, each of them depending on fewer Boolean variables than the given Boolean function, a recursive application of this decomposition terminates in trivial Boolean functions of a single Boolean variable. Originally this type of decomposition was called *grouping*, because three groups of variables influence the decomposition, the first group of variables is part of the support of the first subfunction, the second group of variables is part of the support of the second subfunction, and the third group of variables controls both subfunctions. In the last years the term *bi-decomposition* is preferred for this type of decomposition, because the given Boolean function is split into exactly two subfunctions on the same circuit level.

Approximately two decades ago, a group of researchers first at the University of Chemnitz and later on at the University of Freiberg extended the knowledge about the theory and application of the bi-decomposition. Collaborations with researcher groups from the Protland State University (Oregon, USA) (Mischchenko et al. 2001) and Belarussian Academy of Science intensified the improvements (Steinbach and Zabrevski 1998). Also a group in Japan (Sassao and Butler 1997) studied this type of decomposition, but thier special method prefers disjoint bi-decomposition or restrict to small common sets. Many diploma thesises, PhD thesises (Dresig 1992; Le 1989), papers in international conferences and chapters in books (Bochmann and Steinbach 1991; Steinbach et al. 1993; Steinbch and Le 1990) give answers to selected questions around the bi-decomposition. This paper gives a compre-

a	b	c	f
1	-	-	1
0	1	0	1
0	0	-	0
0	1	1	0

c					f
0	0	1	1	1	
1	0	0	1	1	
	0	1	1	0	b
	0	0	1	1	a

(a) Table   (b) Karnaugh-Map

Figure 1. Representation of the Boolean function $f(a, b, c) = a \vee b \wedge \bar{c}$.

hensive and comprehensible introduction into the most important results of the bi-decomposition.

The structure of this paper is organized as follows. Basic terms and notations are introduced in section 2. In order to study the dynamic properties of Boolean functions the derivative operations of the *Boolean Differential Calculus (BDC)* are needed. The BDC is summarized in section 3. In the main section 4 the bi-decomposition will be represented starting with the simplest case and finishing with so far most powerful algorithm for bi-decomposition. Experimental results are discussed in section 5. A novel view to the bi-decomposition in section 6 eases the generalization of the bi-decomposition from Boolean function to the multi-valued case. Finally, section 7 concludes the paper.

2. Preliminaries

The one dimensional *Boolean space* **B** is defined by the set $\{0, 1\}$. A *Boolean variable* x can assume any value from this set. All Boolean vectors with n components form the n-dimensional *Boolean space* $\mathbf{B}^n$.

$$\mathbf{B}^n = \{X | X = (x_1, x_2, \ldots, x_n), x_i \in \mathbf{B}, \ \forall i = 1, \ldots, n\} \qquad (2.1)$$

DEFINITION 1. *Let* $X = (x_1, x_2, \ldots, x_n)$ *be a set of* n *Boolean variables. A unique mapping from* $\mathbf{B}^n$ *into* **B** *is a Boolean function* $f(X)$.

A Boolean function can be represented by a table or a Karnaugh-Map, see Figure 1. If two rows of the table differ only in one input variable than these rows can be merged. The new row contains a '$-$' for the variable that was different.

The third widely used representation of Boolean functions are expressions in which Boolean variables are combined by *Boolean Operators*. Figure 2 shows how a single Boolean variable defines a Boolean function and which

a	f_1
0	0
1	1

a	f_2
0	1
1	0

a	b	f_3
0	0	0
0	1	0
1	0	0
1	1	1

a	b	f_4
0	0	0
0	1	1
1	0	1
1	1	1

a	b	f_5
0	0	0
0	1	1
1	0	1
1	1	0

(a)   (b)   (c)   (d)   (e)

Figure 2. Definition of Boolean functions by Boolean variables and Boolean operators: (a) Boolean variable: $f_1(a) = a$, (b) NOT-operator: $f_2(a) = \bar{a}$, (c) AND-operator: $f_3(a, b) = a \wedge b = a \cdot b = ab$, (d) OR-operator: $f_4(a, b) = a \vee b$, (e) EXOR-operator: $f_5(a, b) = a \oplus b$.

Boolean functions are created by the Boolean operator NOT, AND, OR and EXOR. The order of calculation of the Boolean operation to evaluate an expression is controlled by the priority, ordered from the highest value for NOT over AND and OR to the lowest value for EXOR.

Sometimes a Boolean function is not *completely* specified.

DEFINITION 2. *Let $\mathbf{C} \subseteq \mathbf{B}^n$ and $\mathbf{D} = \mathbf{B}^n \setminus \mathbf{C}$. An incompletely specified function (ISF) $F(A)$ is a mapping $\mathbf{C} \rightarrow \mathbf{B}$. The set $\mathbf{C}$ is called care set and the set $\mathbf{D}$ don't care set (DC-set), respectively. The care set $\mathbf{C}$ is partitioned into the ON-set $\mathbf{A}_1 = \{A_i | F(A_i) = 1\}$, and the OFF-set $\mathbf{A}_0 = \{A_i | F(A_i) = 0\}$.*

The function value of the DC-set is labeled in the table by '$-$' and in the Karnaugh-Map by 'Φ'. The three sets ON-set, OFF-set and DC-set do not overlap each other and cover the Boolean space $\mathbf{B}^n$ completely. Thus, only two of these sets are necessary to specify an ISF. Each of these three sets separates the Boolean space $\mathbf{B}^n$ into two subsets, so that three associated characteristic Boolean functions are defined.

DEFINITION 3. *An ISF $F(A)$ defines the following three characteristic mark functions:*

$$ON - function: f_q(A_i) = \begin{cases} 1 & : A_i \in \mathbf{A}_1, \\ 0 & : otherwise. \end{cases}$$

$$OFF - function: f_r(A_i) = \begin{cases} 1 & : A_i \in \mathbf{A}_0, \\ 0 & : otherwise. \end{cases} \quad (2.2)$$

$$DC - function: f_\varphi(A_i) = \begin{cases} 1 & : A_i \in \mathbf{D}, \\ 0 & : otherwise. \end{cases}$$

Figure 3 shows an example of a simple ISF $F(a, b)$ together with all characteristic mark functions. Based on the three possible pairs of

a	b	F
0	0	1
0	1	-
1	0	-
1	1	0

(b) F: b rows / a columns — row $b=0$: [1, Φ]; row $b=1$: [Φ, 0]

(c) f_q: row $b=0$: [1, 0]; row $b=1$: [0, 0]

(d) f_r: row $b=0$: [0, 0]; row $b=1$: [0, 1]

(e) f_φ: row $b=0$: [0, 1]; row $b=1$: [1, 0]

(a)   (b)   (c)   (d)   (e)

Figure 3. Incompletely specified Boolean functions $F(a, b)$: (a) Table of $F(a, b)$, (b) Karnaugh-Map of $F(a, b)$, (c) ON-function: $f_q(a, b)$, (d) OFF-function: $f_r(a, b)$, (e) DC-function: $f_\varphi(a, b)$.

mark functions the ISF $F(a, b)$ can be expressed by $F\langle f_q(a, b), f_r(a, b)\rangle$, $F\langle f_q(a, b), f_\varphi(a, b)\rangle$ and $F\langle f_r(a, b), f_\varphi(a, b)\rangle$, respectively.

Because both function values of **B** are allowed for the DC-set of $F(A)$, $2^{|\mathbf{D}|}$ functions $f_i(A)$ match with $F(A)$ in **C**. All these Boolean functions form a lattice defined as follows.

DEFINITION 4. *The characteristic function set $F(A)$ of an ISF $F(A)$ with the care set **C** is a set of all Boolean functions $f(A)$ with $f(A_i) = F(A_i)$ for all $A_i \in \mathbf{C}$. The smallest Boolean function of this set is called lower bound function $f_l(A)$ and the largest one upper bound function $f_u(A)$, respectively.*

From Definitions 3 and 4 follows that:

$$f_l(A) = f_q(A) = \overline{f_r(A) \vee f_\varphi(A)}, \tag{2.3}$$
$$f_u(A) = f_q(A) \vee f_\varphi(A) = \overline{f_r(A)}. \tag{2.4}$$

Figure 4 shows all Boolean functions of the characteristic function set $F(A)$ of the ISF $F(A)$ from Figure 3. The lattice property is easy to check. The conjunction and the disjunction of each pair of these four functions create a function of the same set. Using the lower and upper bound function, the characteristic function set can be represented as an interval $F\left[f_l(A), f_u(A)\right]$.

Finally the terms Boolean equation and Boolean inequality are introduced.

$$f(A) = g(A) \tag{2.5}$$

The solution of a Boolean equation (2.5) is the set $\mathbf{L} \subseteq \mathbf{B}^n$ defined by formula (2.6).

$$\mathbf{L} = \{A_i | (f(A_i) = 0 \text{ and } g(A_i) = 0)$$
$$\text{or } (f(A_i) = 1 \text{ and } g(A_i) = 1)\} \tag{2.6}$$

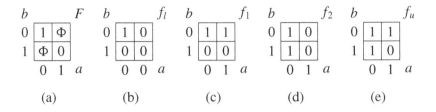

Figure 4. Characteristic function set $F(a, b)$ of the ISF $F(a, b)$: (a) Characteristic function set $F(a, b)$, (b) lower bound function $f_l(a, b)$, (c) function: $f_1(a, b)$, (d) function: $f_2(a, b)$, (e) upper bound function : $f_u(a, b)$.

If the right hand side is constant 0 or 1 the Boolean equation (2.5) is a homogeneous one. The special case of $f(A) = 1$ is called characteristic equation, and $f(A) = 0$ restrictive equation, respectively. A restrictive equation can be read as: "It is forbidden that $f(A)$ is valid".

$$f(A) \neq g(A) \tag{2.7}$$

The solution of a Boolean inequality (2.7) is the set $\mathbf{L}' \subseteq \mathbf{B}^n$ defined by formula (2.8).

$$\begin{aligned} \mathbf{L}' &= \{A_i | (f(A_i) = 0 \; and \; g(A_i) = 1) \\ &or \; (f(A_i) = 1 \\ &and \; g(A_i) = 0)\} \end{aligned} \tag{2.8}$$

The solution sets $\mathbf{L}$ and $\mathbf{L}'$ of the Boolean equation (2.5) and the Boolean inequality (2.7) are complementary to each other.

$$f(A) \leq g(A) \tag{2.9}$$

The solution of a Boolean inequality (2.9) is the same set of minters like the Boolean equation (2.10).

$$f(A) \wedge \overline{g(A)} = 0 \tag{2.10}$$

3. Boolean Differential Calculus

The Boolean Differential Calculus (BDC) was introduced by Bochmann and Posthoff in 1981 and is very helpful to study all problems around the dynamic, the testing and the influence of inputs to outputs of Boolean functions. The BDC supports the representation of the bi-decomposition in a comprehensible manner. For a better understanding the necessary definitions are recalled and important interpretations are explained as well.

DEFINITION 5. *Let be* $f(A, b = 0)$ *the negative cofactor and* $f(A, b = 1)$ *the positive cofactor of the Boolean function* $f(A, b)$ *then the simple derivative operations over a single variable b are defined by the following formulas.*

— *The simple derivative:*

$$\frac{\partial f(A, b)}{\partial b} = f(A, b = 0) \oplus f(A, b = 1) \tag{3.1}$$

— *The simple minimum:*

$$\min_{b} f(A, b) = f(A, b = 0) \wedge f(A, b = 1) \tag{3.2}$$

— *The simple maximum:*

$$\max_{b} f(A, b) = f(A, b = 0) \vee f(A, b = 1) \tag{3.3}$$

Note, the results of these operations are Boolean functions too. All of them do not depend on the variable b. The simple derivative of f is equal to 1 for such minterms A where a change of the value of the variable b causes a change in the value of the given function f. The simple minimum is equal to 1 for such minterms A where a change of the value of the variable b does not change the value 1 of the given function f. Finally, the simple maximum is equal to 1 for such minterms A where a change of the value of the variable b leads at least once to the value 1 of the given function f.

The derivative operations introduced above can be executed iteratively for several variables.

DEFINITION 6. *Let be* $B = (b_1, b_2, \ldots, b_k)$ *then the k-times derivative operations of the Boolean function* $f(A, B)$ *over the set of variables B are defined by the following formulas.*

— *The k-times derivative:*

$$\frac{\partial^k f(A, B)}{\partial b_1 \partial b_2 \ldots \partial b_k} = \frac{\partial}{\partial b_1} (\frac{\partial}{\partial b_2} (\ldots \frac{\partial f(A, B)}{\partial b_k} \ldots)) \tag{3.4}$$

— *The k-times minimum:*

$$\min_{B}^{k} f(A, B) = \min_{b_1}(\min_{b_2}(\ldots \min_{b_k} f(A, B) \ldots)) \tag{3.5}$$

— *The k-times maximum:*

$$\max_{B}^{k} f(A, B) = \max_{b_1}(\max_{b_2}(\ldots \max_{b_k} f(A, B) \ldots)) \tag{3.6}$$

Table 1. Semantics of k-times derivative operation

k-times derivative operation	Condition for a function value 1 in the result
k-times derivative	odd number of 1's in the subspace $A = const.$
k-times minimum	only 1's in the subspace $A = const.$
k-times maximum	at least one 1 in the subspace $A = const.$
Δ-operation	not constant in the subspace $A = const.$

– *The Δ-operation:*

$$\Delta_B f(A, B) = \min_B^k f(A, B) \oplus \max_B^k f(A, B) \qquad (3.7)$$

The Δ-operation was already used by Akers (1959). Table 1 shows the conditions for which a function value 1 in an k-times derivative operation occurs. The results of all k-times derivative operations depend only on the set of variables A, and thus they are independent on the variable set B. The function values of the derivatives are defined by all function values of the original function in the subspaces $A = const.$

4. Bi-Decomposition of Boolean Function

4.1. *Completely specified Boolean function*

The bi-decomposition is based on the facts, that
1. each of the two-input gates AND, OR or EXOR can create a more complex Boolean function at the output from the two simpler Boolean functions at the inputs (in special case the inputs may be Boolean variables),
2. each Boolean function can be created by a final two-input gate in a multilevel circuit.

The key idea of the suggested iterative design method is to decompose a given Boolean function into two simpler ones. The connection of these two simpler functions by an AND-, OR- or EXOR-gate creates the given function. It is not trivial to measure the complexity of the selected Boolean functions, but in general a Boolean function depending on fewer Boolean variables is simpler than a Boolean function depending on more Boolean variables. An iterative procedure of decomposition terminates if the number of independent Boolean variables of the function is reduced at least by one in each step.

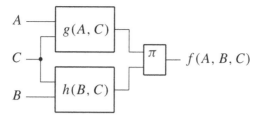

Figure 5. General structure of the π-bi-decomposition of $f(A, B, C)$.

DEFINITION 7. *A bi-decomposition of the Boolean function $f(A, B, C)$ with respect to the decomposition operator $\pi(x, y)$ and the dedicated sets A and B is a pair of functions $\langle g(A, C), h(B, C) \rangle$ with*

$$\boxed{f(A, B, C) = \pi\ (g(A, C), h(B, C)).}$$ (4.1)

The sets A and B must not be empty. The set C is called common set. If the set C is empty, the bi-decomposition is called disjoint. The Boolean functions $g(A, C)$ and $h(B, C)$ are called decomposition functions.

In the following, the term π-*bi-decomposition* abbreviates the term "bi-decomposition with respect to the decomposition operator $\pi(x, y)$". The associated circuit structure of the π-bi-decomposition is shown in Figure 5.

Let us study the the simplest case of a disjoint OR-bi-decomposition

$$f(A, B) = g(A) \vee h(B),$$ (4.2)

first. An example for this decomposition is shown in Figure 6. The sets $A = (a, b)$ and $B = (c, d)$ are selected such that the minterms of A define the columns of the Karnaugh-Map of $f(a, b, c, d)$ and the minterms of B the rows, respectively. In order to calculate the OR-operation between $g(a, b)$ and $h(c, d)$ both functions must be transformed in the common Boolean space $\mathbf{B}^4$ which expands each value 1 of $g(a, b)$ into a column of four values 1 and each value 1 of $h(c, d)$ into a row of four values 1. The result of the OR-operation $g(a, b) \vee h(c, d)$ is displayed in the Karnaugh-Map of $f(a, b, c, d)$ in Figure 6. There are two types of columns in this Karnaugh-Map, one of them includes values 0 and the other one contains only values 1. Because the value 1 dominates in the OR-operation, the function $g(a, b)$ must be 0 for such minterms (a, b) where in the associated column of $f(a, b, c, d)$ exists a value 0. If there is no value 0 in a column of $f(a, b, c, d)$, the function $g(a, b)$ may be 1 for the associated minterms (a, b). Corresponding conditions are valid for the function $h(c, d)$ and the rows of $f(a, b, c, d)$.

c	d	$h(c,d)$
0	0	0
0	1	1
1	1	0
1	0	1

$f(a,b,c,d)$

c	d				
0	0	0	1	1	0
0	1	1	1	1	1
1	1	0	1	1	0
1	0	1	1	1	1
		0	1	1	0 b
		0	0	1	1 a

$g(a,b)$

0	1	1	0
0	1	1	0 b
0	0	1	1 a

Figure 6. Disjoint OR-bi-decomposition of $f(a,b,c,d)$ into $g(a,b)$ and $h(c,d)$.

An OR-bi-decomposition for the function $f(a,b,c,d)$ in Figure 6 exists, because all values 1 are covered by the values 1 of the functions $g(a,b)$ and $h(c,d)$.

Assume, the function $f(a,b,c,d)$ is changed into $f'(a,b,c,d)$ by switching the value $f(0,0,0,0) = 0$ into the value $f'(0,0,0,0) = 1$. The Karnaugh-Map of $f'(a,b,c,d)$ contains the value 0 in the first column at $f'(0,0,1,1) = 0$ and in the first row at $f'(1,0,0,0) = 0$. Due to these values 0, the only possible OR-bi-decomposition functions must be partially defined by $g'(0,0) = 0$ and $h'(0,0) = 0$. Consequently the result of the OR-operation is $f'(0,0,0,0) = g'(0,0) \lor h'(0,0) = 0$, which is a contradiction to the assumption $f'(0,0,0,0) = 1$. This example proves, that not every Boolean function is disjointly OR-bi-decomposable for given dedicated sets A and B.

The chance to find an OR-bi-decomposition is enlarged for the non-disjoint case. By moving of variables from the sets A or B into the common set C, the function $f(A, B, C)$ can be divided into $2^{|C|}$ subfunctions $f(A, B, C = const.)$. Each of these subfunctions can be checked separately for the OR-bi-decomposition because both decomposition functions $g(A, C)$ and $h(B, C)$ depend on the common set C.

The above introduced function $f'(a,b,c,d)$ was not disjoint OR-bi-decomposable for dedicated sets $A = (a, b)$ and $B = (c, d)$. If the variable a is moved into the set $C = (a)$, the dedicated sets for a possible OR-bi-decomposition are $A = (b)$ and $B = (c, d)$. Figure 7 shows that this non-disjoint OR-bi-decomposition exist. The analysis for the non-disjoint OR-bi-decomposition is done in Figure 7 on the left hand side in the Boolean subspace defined by $a = 0$ and on the right hand side for $a = 1$, respectively.

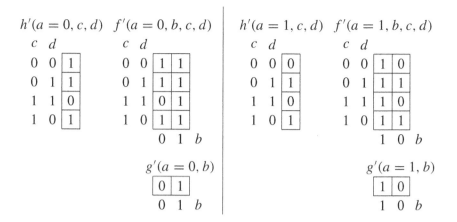

Figure 7. Non-Disjoint OR-bi-decomposition of the Boolean function $f(a, b, c, d)$ into $g(a, b)$ and $h(a, c, d)$.

Note, in contrast to Figure 6, the function $h'(a = 0, b, c)$ is different from $h'(a = 1, b, c)$ so that the decomposition functions $g'(a, b)$ and $h'(a, b, c)$ are created.

The number of variables in the set C is not restricted. The chances that an OR-bi-decomposition exists increase with the number of variables int the common set C. Theorem 1 gives the formal mathematical condition that a function $f(A, B, C)$ is OR-bi-decomposable. This is the case, if the projections in both directions A and B of the values 0 expressed by $\overline{f(A, B, C)}$ do not overlap with values 1 expressed by $f(A, B, C)$.

THEOREM 1. *The Boolean function $f(A, B, C)$ is OR-bi-decomposable with respect to the dedicated sets A and B iff*

$$f(A, B, C) \wedge \max_{A}{}^{k} \overline{f(A, B, C)} \wedge \max_{B}{}^{k} \overline{f(A, B, C)} = 0. \qquad (4.3)$$

There are several possibilities to calculate the decomposition functions $g(A, C)$ and $h(B, C)$ of an OR-bi-decomposition. One of them is based on the condition, that a value 0 in a subspace defined by $A = const.$ and $C = const.$ requires a value 0 for the selected minterm of $g(A, C)$. This describes directly the property of the k-times minimum. An equivalent condition is valid for decomposition functions $h(B, C)$ with respect to the subspace defined by $B = const.$ and $C = const.$ and therefore, the decomposition functions of the OR-bi-decomposition can be calculated by (4.4) and (4.5).

$$g(A, C) = \min_{B}{}^{k} f(A, B, C) \qquad (4.4)$$

$$h(B, C) = \min_{A}{}^{k} f(A, B, C) \qquad (4.5)$$

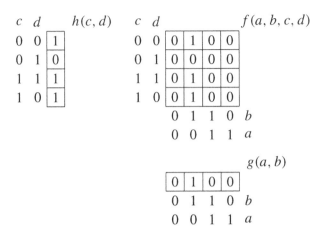

Figure 8. Disjoint AND-bi-decomposition of $f(a, b, c, d)$ into $g(a, b)$ and $h(c, d)$.

The condition for a disjoint AND-bi-decomposition

$$f(A, B) = g(A) \wedge h(B) \qquad (4.6)$$

are inverse to the disjoint OR-bi-decomposition. Figure 8 shows an example of the AND-bi-decomposition of $f(a, b, c, d)$ with respect to the dedicated sets $A = (a, b)$ and $B = (c, d)$. The AND-operation of the AND-bi-decomposition require values 1 in both functions $g(a, b)$ and $h(c, d)$ in order to create a value 1 in the function $f(a, b, c, d)$. Thus, an AND-bi-decomposition does not exist, if the projections of the values 1 expressed by $f(A, B)$ in the directions of A and B overlap with values 0 expressed by $\overline{f(A, B)}$. The generalization to the non-disjoint case is comparable to the OR-bi-decomposition and expressed in Theorem 2.

THEOREM 2. *The Boolean function $f(A, B, C)$ is AND-bi-decomposable with respect to dedicated sets A and B iff*

$$\overline{f(A, B, C)} \wedge \max_{A}^{k} f(A, B, C) \wedge \max_{B}^{k} f(A, B, C) = 0. \qquad (4.7)$$

There are several possibilities to calculate the decomposition functions $g(A, C)$ and $h(B, C)$ of an AND-bi-decomposition, too. Here, a value 1 in a suitable subspace requires a value 1 for the selected minterm of the decomposition functions. This describes directly the property of the k-times maximum and the decomposition functions of the AND-bi-decomposition can be calculated by (4.8) and (4.9).

$$g(A, C) = \max_{B}^{k} f(A, B, C) \qquad (4.8)$$

$$h(B, C) = \max_{A}^{k} f(A, B, C) \qquad (4.9)$$

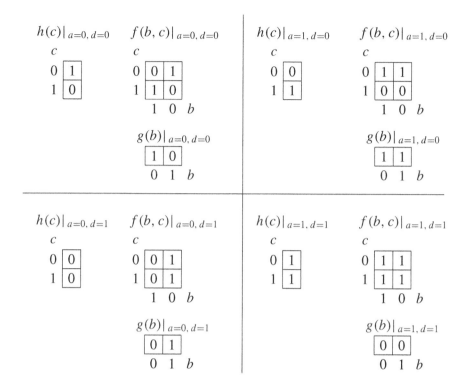

Figure 9. Non-disjoint EXOR-bi-decomposition of $f(a, b, c, d)$ into $g(a, b, d)$ and $h(a, c, d)$ with respect to the dedicated sets $A = (b)$, $B = (c)$ and the common set $C = (a, d)$.

Quite different properties exist for the EXOR-bi-decomposition. It is easier to start with such an non-disjoint EXOR-bi-decomposition where each of the dedicated sets A and B includes only one single variable. Figure 9 shows an example of an non-disjoint EXOR-bi-decomposition of $f(a, b, c, d)$ with respect to the dedicated sets $A = (b)$ and $B = (c)$.

The Karnaugh-Maps of $f(b, c)$ in Figure 9 show four of the eight Boolean functions which can decomposed by an EXOR into the decomposition functions $g(b)$ and $h(c)$. These eight functions are the result of an EXOR-operation between all pairs of the four Boolean functions depending on one variable; partially this can be seen in Figure 9. The enumeration of these EXOR-bi-decomposable functions of two variables is: $f(b, c) = 0$, $f(b, c) = 1$, $f(b, c) = b$, $f(b, c) = \bar{b}$, $f(b, c) = c$, $f(b, c) = \bar{c}$, $f(b, c) = a \oplus b$, and $f(b, c) = a \oplus \bar{b}$. Exactly these 8 of all 16 Boolean functions of two variables have an even number of function values 1, a property indicated by the complement of the 2-times derivative as stated in the Theorem 3.

THEOREM 3. *The Boolean function $f(a, b, C)$ is EXOR-bi-decomposable with respect to the single variables a and b iff*

$$\frac{\partial^2 f(a, b, C)}{\partial a \partial b} = 0. \tag{4.10}$$

As can be seen in the Karnaugh-Maps of $f(b, c)$ in Figure 9, the rows include either the function g or $\bar{g}$. Because each row covers only two values of the function, the derivative with respect to the variable b must be constant in each subspace defined by $a = const$ and $d = const$. Based on this observation the Theorem 3 can be generalized to the Theorem 4 for a single variable a and set of variables B.

THEOREM 4. *The Boolean function $f(a, B, C)$ is EXOR-bi-decomposable with respect to the single variable a the and set of variables B iff*

$$\Delta_B \left(\frac{\partial f(a, B, C)}{\partial a} \right) = 0. \tag{4.11}$$

There exists no closed formula to decide whether a Boolean function $f(A, B, C)$ is EXOR-bi-decomposable with respect to the dedicated sets A and B. By means of an indirect method it is easy to answer this question. As discussed above, in each subspace $C = const.$ only two functions $g(A, C = const.) = f(A, B = const., C = const.)$ and $\overline{g(A, C = const.)}$ are allowed. One function must always be the complement of the other one. Theorem 5 takes advantages of this property and delivers, if possible, an allowed pair of decomposition functions for the EXOR-bi-decomposition.

THEOREM 5. *Let (4.12) be the potential EXOR-decomposition function $g(A, C)$ of the Boolean function $f(A, B, C)$*

$$g(A, C) = f(A, B = (0, 0, \ldots, 0), C) \tag{4.12}$$

then $f(A, B, C)$ is EXOR-bi-decomposable with respect to the dedicated sets A and B iff

$$\Delta_A \left(f(A, B, C) \oplus f(A, B = (0, 0, \ldots, 0), C) \right) = 0. \tag{4.13}$$

If (4.13) is valid, then (4.12) and (4.14)

$$h(B, C) = \max_A^k (f(A, B, C) \oplus f(A, B = (0, 0, \ldots, 0), C)) \tag{4.14}$$

are allowed decomposition functions of the EXOR-bi-decomposition of $f(A, B, C)$ with respect to the dedicated sets A and B.

4.2. *Incompletely specified Boolean function*

Two results follow from the previous section for the question, whether a bi-decomposition exists for a completely specified Boolean function.

1. There are Boolean functions possessing the property of a selected π-bi-decomposition with respect to dedicated sets A and B, but other Boolean function can not be decomposed in this manner.

2. The larger the set of common variables C the more bi-decomposable Boolean functions exist.

In order to design a small and fast circuit, as few as possible variables should be in the common set C. Is there a chance to break this contradiction? Yes, the probability to find a Boolean function that is π-bi-decomposable for a small set of common variables C increases, if not only one completely specified Boolean function, but the whole set of Boolean functions, defined by an incompletely specified Boolean function, is take into account.

DEFINITION 8. *A bi-decomposition of the Boolean ISF $F(A, B, C)$, associated with the characteristic function set $\mathbf{F}(A, B, C)$, with respect to the decomposition operator $\pi(x, y)$ and the dedicated sets A and B is a pair of functions $\langle g(A, C) \in \mathbf{G}(A, C), h(B, C) \in \mathbf{H}(B, C) \rangle$ with*

$$\boxed{f(A, B, C) = \pi (g(A, C), h(B, C)).} \qquad (4.15)$$

where $f(A, B, C) \in \mathbf{F}(A, B, C)$. The sets A and B must not be empty. The set C is called common set. If the set C is empty, the π-bi-decomposition is called disjoint. The Boolean functions $g(A, C)$ and $h(B, C)$ are called decomposition functions.

The generalized decomposition problem for an ISF $F(A, B, C)$ can be solved if each completely specified Boolean function of the associated characteristic function set $\mathbf{F}(A, B, C)$ is checked separately by the condition from section 4.1. This is a very time consuming procedure for a large DC-set **D**, because $2^{|\mathbf{D}|}$ functions must be checked. Much better it is to answer the question: "Is there at least one Boolean function in the characteristic function set $\mathbf{F}(A, B, C)$ for that a π-bi-decomposition with respect to dedicated sets A and B exists?". In the following this question will be answered using the characteristic ON-function $f_q(A, B, C)$ and the characteristic OFF-function $f_r(A, B, C)$ of the ISF $F(A, B, C)$.

An example of a disjoint OR-bi-decomposition of the ISF $F(a, b, c, d)$ into the ISFs $G(a, b)$ and $H(c, d)$ with respect to the dedicated sets $A = (a, b)$ and $B = (c, d)$ is shown in Figure 10. The number of minterms in the

	c	d	H(c, d)
	0	0	0
	0	1	1
	1	1	0
	1	0	Φ

c d $\qquad$ $F(a, b, c, d)$

c	d				
0	0	0	1	Φ	Φ
0	1	1	1	Φ	1
1	1	0	1	Φ	0
1	0	Φ	Φ	Φ	Φ
		0	1	1	0	b
		0	0	1	1	a

$G(a, b)$

0	1	Φ	0
0	1	1	0	b
0	0	1	1	a

Figure 10. Disjoint OR-bi-decomposition of the incompletely specified Boolean function $F(a, b, c, d)$ into the ISFs $G(a, b)$ and $H(c, d)$.

DC-set of $F(a, b, c, d)$ is $|\mathbf{D}| = 8$ so that $2^8 = 256$ Boolean functions of the associated characteristic function set $\mathbf{F}(a, b, c, d)$ have to be checked for the OR-bi-decomposition. All of them have the same function values in the care set that influence the decomposition result significantly. Due to the OR-operation of the decomposition, a function value 0 in $F(a, b, c, d)$ requires associated function values 0 in the decomposition functions. As can be seen in Figure 10, the 0 values from $F(a, b, c, d)$ are projected to the bottom into $G(a, b)$ and to the left into $H(c, d)$.

Assume, $F(1, 0, 0, 0)$ in Figure 10 is not Φ but 1, an OR-bi-decomposition with respect to the dedicated sets $A = (a, b)$ and $B = (c, d)$ is not possible due to the necessary 0 values of $G(a = 1, b = 0)$ and $H(c = 0, d = 0)$. This shows that first, a larger DC-set improves the chance to find an OR-bi-decomposition, and second, the DC-set does not restrict the property of OR-bi-decomposition of the ISF $F(a, b, c, d)$. Thus, by means of the mark functions it is possible to decide whether the ISF $F(a, b, c, d) = F\langle f_q(a, b, c, d), f_r(a, b, c, d)\rangle$ is OR-bi-decomposable.

Analogous to the previous section, the disjoint OR-bi-decomposition can be generalized to the non-disjoint OR-bi-decomposition of the ISF $F(A, B, C)$.

The general condition for an non-disjoint OR-bi-decomposition of the ISF $F(A, B, C)$ is given in the Theorem 6 that describes the disjoint case for an empty set C, too. An OR-bi-decomposition exists for the ISF $F\langle f_q(A, B, C), f_r(A, B, C)\rangle$ if the projections in both directions A and B of the values 0 expressed by the OFF-function $f_r(A, B, C)$ do not overlap with values 1 expressed by the ON-function $f_q(A, B, C)$.

THEOREM 6. *The characteristic function set $F(A, B, C)$ which is associated with the ISF $F(A, B, C) = F \langle f_q(A, B, C), f_r(A, B, C) \rangle$ includes at least one Boolean function $f(A, B, C)$ that is OR-bi-decomposable with respect to the dedicated sets A and B iff*

$$f_q(A, B, C) \wedge \max_A^k f_r(A, B, C) \wedge \max_B^k f_r(A, B, C) = 0. \tag{4.16}$$

As can be seen in Figure 10, the result of the decomposition can be incompletely specified functions too. Formulas (4.17) and (4.18) specify necessary and sufficient conditions to calculate the DC-functions $g_\varphi(A, C)$ and $h_\varphi(B, C)$ independently of each other for all types of bi-decomposition of the ISF $F \langle f_q(A, B, C), f_\varphi(A, B, C) \rangle$.

$$g_\varphi(A, C) = \min_B^k f_\varphi(A, B, C) \tag{4.17}$$

$$h_\varphi(B, C) = \min_A^k f_\varphi(A, B, C) \tag{4.18}$$

There are larger DC-functions, but these depend from each other. Assume for an OR-bi-decomposition in the subspace $C = C_i$ the partial ISF $F(A, B, C = C_i) = 1$, if the ON-function $g_q(A, C = C_i) = 1$ then the DC-function may be $h_\varphi(B, C = C_i) = 1$ due to the OR-operation and alternatively if the ON-function $h_q(B, C = C_i) = 1$ then the DC-function may be $g_\varphi(A, C = C_i) = 1$. This condition can be generalized. For an OR-bi-decomposition only such values 1 of $F(A, B, C)$ must be covered by the ON-function $g_q(A, C)$ that do not overlap with $\max_A^k f_r(A, B, C)$ because these values 1 can not be covered by $h_q(B, C)$. Due to the OR-operation, all values 0 must be mapped from $F(A, B, C)$ into $G(A, C)$. The mark functions for $G(A, C)$ of an OR-bi-decomposition of $F(A, B, C)$ can be calculated by (4.19) and (4.20).

$$g_q(A, C) = \max_B^k \left(f_q(A, B, C) \wedge \max_A^k f_r(A, B, C) \right) \tag{4.19}$$

$$g_r(A, C) = \max_B^k f_r(A, B, C) \tag{4.20}$$

Formulas (4.19) and (4.20) describe the ISF $G(A, C)$ associated with the largest allowed characteristic function set $G(A, C)$ of the OR-bi-decomposition. Since the DC-functions $g_\varphi(A, C)$ and $h_\varphi(B, C)$ are not independent from each other, the selected function $g(A, C)$ from $G(A, C)$ must be used to calculate the mark functions for $H(B, C)$ of the OR-bi-decomposition of $F(A, B, C)$ by (4.21) and (4.22). The term $f_q(A, B, C) \wedge \overline{g(A, C)}$ of (4.21)

$H(a = 0, c, d)$

c	d	
0	0	Φ
0	1	1
1	1	0
1	0	1

$F(a = 0, b, c, d)$

c	d		
0	0	Φ	Φ
0	1	0	1
1	1	Φ	0
1	0	0	1
		0	1 $\;b$

$G(a = 0, b)$

0	1
0	1 $\;b$

$H(a = 1, c, d)$

c	d	
0	0	0
0	1	1
1	1	0
1	0	Φ

$F(a = 1, b, c, d)$

c	d		
0	0	Φ	1
0	1	Φ	0
1	1	Φ	1
1	0	Φ	Φ
		1	0 $\;b$

$G(a = 1, b)$

Φ	1
1	0 $\;b$

Figure 11. Non-disjoint AND-bi-decomposition of the ISF $F(a, b, c, d)$ into the ISFs $G(a, b)$ and $H(a, c, d)$ with common set $C = (a)$.

characterizes all values 1 of $F(A, B, C)$ not covered by $g(A, C)$ yet and hence must be covered by $h_q(B, C)$.

$$h_q(B, C) = \max_A^k \left(f_q(A, B, C) \wedge \overline{g(A, C)} \right) \qquad (4.21)$$

$$h_r(B, C) = \max_A^k f_r(A, B, C) \qquad (4.22)$$

Next, the non-disjoint AND-bi-decomposition is considered. As can be seen in Figure 11, the non-disjoint AND-bi-decomposition of the ISF $F(a, b, c, d)$ with respect to the dedicated sets $A = (b)$, $B = (c, d)$ and the common set $C = (a)$ is carried out separately for the incompletely specified subfunctions $F(a = 0, b, c, d)$ and $F(a = 1, b, c, d)$, respectively. The condition for an non-disjoint AND-bi-decomposition of the ISF $F(A, B, C)$ is given in Theorem 7. An AND-bi-decomposition exists for the ISF $F \langle f_q(A, B, C), f_r(A, B, C) \rangle$ if the projections in both directions A and B of the values 1 expressed by the ON-function $f_q(A, B, C)$ do not overlap with values 0 expressed by the OFF-function $f_r(A, B, C)$.

THEOREM 7. *The characteristic function set $F(A, B, C)$ which is associated with the ISF $F(A, B, C) = F \langle f_q(A, B, C), f_r(A, B, C) \rangle$ includes at least one Boolean function $f(A, B, C)$ that is AND-bi-decomposable with respect to the dedicated sets A and B iff*

$$f_r(A, B, C) \wedge \max_A^k f_q(A, B, C) \wedge \max_B^k f_q(A, B, C) = 0. \qquad (4.23)$$

Assume the IFS of Figure 11 is changed for one minterm such that $f_q(0, 0, 0, 0) = 1$. This new value 1 requires $g_q(0, 0) = 1$ and the values

0 of the IFS $F(a, b, c, d)$ for the minterms $(0, 0, 0, 1)$ and $(0, 0, 1, 0)$ can not be a result of an AND-bi-decomposition. Thus, larger DC-functions decrease possible contradictions to the condition of the AND-bi-decomposition (4.23). Only $2 * 2 = 4$ of the $2^8 = 256$ functions expressed by the ISF $F(a, b, c, d)$ in Figure 11 are AND-bi-decomposable with respect to the dedicated sets $A = (b)$, $B = (c, d)$.

The mark functions for $G(A, C)$ of an AND-bi-decomposition of $F(A, B, C)$ can be calculated by (4.24) and (4.25). In this case the values 1 of $F(A, B, C)$ must be covered by $G(A, C)$ shown in (4.24) and only such values 0 of $F(A, B, C)$ must be covered by $G(A, C)$ that can not be covered by $H(B, C)$, see (4.25).

$$g_q(A, C) = \max_{B}{}^k f_q(A, B, C) \tag{4.24}$$

$$g_r(A, C) = \max_{B}{}^k \left(f_r(A, B, C) \wedge \max_{A}{}^k f_q(A, B, C) \right) \tag{4.25}$$

The characteristic function set $G(A, C)$, specified by formulas (4.24) and (4.25), describes all possible Boolean function $g(A, C)$ of the AND-bi-decomposition. For each Boolean function $g(A, C) \in G(A, C)$ there exists an associated ISF $H(B, C) = H \langle h_q(B, C), h_r(B, C) \rangle$ for the AND-bi-decomposition specified by formulas (4.26) and (4.27).

$$h_q(A, C) = \max_{A}{}^k f_q(A, B, C) \tag{4.26}$$

$$h_r(A, C) = \max_{A}{}^k (f_r(A, B, C) \wedge g(A, C)) \tag{4.27}$$

The EXOR-bi-decomposition of an ISF $F(A, B, C)$ is difficult because in general the sets of decomposition functions do not fit to the characteristic function set of an ISF and there exists no closed formula to check the EXOR-bi-decomposition if both dedicated sets A and B include more than one variable. For that reason in Figure 12 the EXOR-bi-decomposition of an ISF $F(a, b, C)$ is considered where each of the dedicated sets includes only one variable.

The condition of the EXOR-bi-decomposition with respect to two single variables as dedicated sets is that a subfunction must cover an even number of values 1. Formula (4.10) can be generalized for the case of an ISF $F(a, b, C)$. As can be seen in Figure 12, if at least one minterm of the subspace $C = const.$ is a member of the DC-set a subfunction with an even number of values 1 can be constructed in this subspace. Theorem 8 describes this condition in a formal way.

THEOREM 8. *The characteristic function set $F(a, b, C)$ associated with the ISF $F(a, b, C) = F \langle f_q(a, b, C), f_\varphi(a, b, C) \rangle$ includes at least one Boolean*

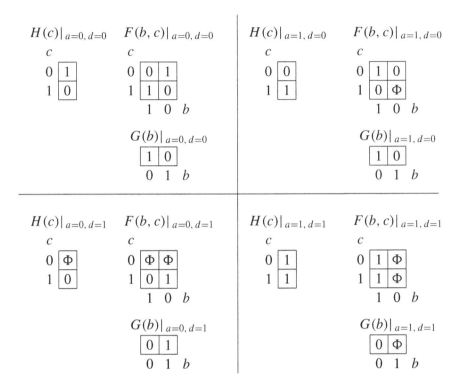

Figure 12. Non-disjoint EXOR-bi-decomposition of the ISF $F(a, b, c, d)$ into the ISFs $G(a, b, d)$ and $H(a, c, d)$ with respect to the dedicated sets $A = (b)$, $B = (c)$ and the common set $C = (a, d)$.

function $f(a, b, C)$ which is EXOR-bi-decomposable with respect to the dedicated sets $A = (a)$ and $B = (b)$ iff

$$\frac{\partial^2 f_q(a, b, C)}{\partial a \partial b} \wedge \overline{\max_{(a,b)}{}^2 f_\varphi(a, b, C)} = 0. \qquad (4.28)$$

Comparable to Theorem 4 all Boolean functions of a characteristic function set $F(a, B, C)$ associated with the ISF $F(a, B, C)$ can be checked together for the EXOR-bi-decomposition with respect to the single variable a and the dedicated set B. The derivative of all Boolean functions of $F(a, B, C)$ with respect to the variable a creates a new ISF $F^a(B, C) = F^a \langle f_q^a(B, C), f_r^a(B, C) \rangle$. The EXOR-bi-decomposition with respect to the the single variable a and the dedicated set B is possible if the characteristic function set $F^a(B, C)$ associated with $F^a(B, C)$ includes a constant function in each subspace $C = const$. Theorem 9 expresses this condition formally.

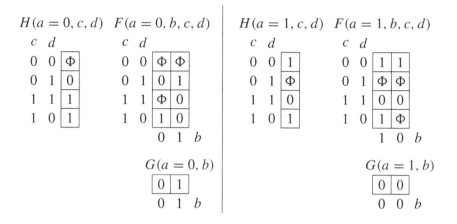

Figure 13. Non-disjoint EXOR-bi-decomposition of the ISF $F(a, b, c, d)$ into the ISFs $G(a, b)$ and $H(a, c, d)$ with respect to the dedicated sets $A = (b)$, $B = (c, d)$ and the common set $C = (a)$.

THEOREM 9. *The characteristic function set $F(a, B, C)$ which is associated with the ISF $F(a, B, C) = F\langle f_q(a, B, C), f_r(a, B, C)\rangle$ includes at least one Boolean function $f(a, B, C)$ that is EXOR-bi-decomposable with respect to the dedicated sets $A = (a)$ and B iff*

$$\max_{B}^{k} f_q^a(B, C) \wedge f_r^a(B, C) = 0 \qquad (4.29)$$

where

$$f_q^a(B, C) = \max_a f_q(a, B, C) \wedge \max_a f_r(a, B, C) \qquad (4.30)$$

$$f_r^a(B, C) = \min_a f_q(a, B, C) \vee \min_a f_r(a, B, C). \qquad (4.31)$$

Theorem 9 is illustrated for the dedicated sets $A = (b)$ and $B = (c, d)$ in Figure 13. The single variable of the dedicated set is $A = (b)$ in this example. Adjusted to this example, the results of (4.30) and (4.31) are $f_q^b(a, c, d) = \bar{a}(c \oplus d)$ and $f_r^b(a, c, d) = a(\bar{c} \oplus d)$, respectively. Condition (4.29) holds, because $\max_{(c,d)}^{k} f_q^b(b, c, d) = \bar{a}$, and consequently an EXOR-bi-decomposition exist.

The last open question is how the incompletely specified decomposition functions of such an EXOR-bi-decomposition can be found. There are only 4 different decomposition functions $g(a)$ in the subspace $C = const.$ If $f_q^a(B, C = const.) \neq 0$ from (4.30) then $g(a)$ cannot be a constant function so that there is a free choice between $g(a) = a$ or $g(a) = \bar{a}$. Using the notations from Theorem 9, the formulas (4.32), (4.33) and

(4.34), compute one combination of allowed decomposition functions of an EXOR-bi-decomposable ISF $F(a, B, C) = F\langle f_q(a, B, C), f_r(a, B, C)\rangle$.

$$g(a, C) \;=\; a \wedge f_q^a(B, C) \tag{4.32}$$

$$h_q(B, C) \;=\; \max_a \left((\overline{g(a, C)} \wedge f_q(a, B, C)) \vee (g(a, C) \wedge f_r(a, B, C)) \right) \tag{4.33}$$

$$h_r(B, C) \;=\; \max_a \left((\overline{g(a, C)} \wedge f_r(a, B, C)) \vee (g(a, C) \wedge f_q(a, B, C)) \right) \tag{4.34}$$

Using (4.33) and (4.34) the EXOR-operation between $g(a, C)$ and all functions of the characteristic function set $F(a, B, C)$ associated with the ISF $F(a, B, C)$ is computed first. Then the projection of these functions into the characteristic function set $H(B, C)$ associated with the ISF $H(B, C)$ is calculated. Figure 13 shows the decomposition functions of the EXOR-bi-decomposition calculated by (4.32), (4.33) and (4.34).

As mentioned above, there is no closed formula to check the EXOR-bi-decomposition of $F(A, B, C)$ if both dedicated sets A and B include more than one variable. This task can be solved by an algorithm which iteratively calculates possible decomposition functions based on the mutual dependencies. Because the decomposition ISFs $G(A, C)$ and $H(B, C)$ are alternately considered free or bound, the decomposition ISFs $F^{free}\langle f_q^{free}, f_r^{free}\rangle$ and $F^{bound}\langle f_q^{bound}, f_r^{bound}\rangle$ are introduced, where the dedicated sets are renamed into X^{free} and X^{bound} so that $F^{free}(X^{free}, C)$ and $F^{bound}(X^{bound}, C)$ are used mutually for $G(A, C)$ and $H(B, C)$. Formulas (4.35) and (4.36) show, how the bound decomposition ISF must be restricted.

$$f_q^{bound}(X^{bound}, C) \;=\; f_q^{bound} \vee \max_{X^{free}}{}^k \left((f_q^{free} \wedge f_r) \vee (f_r^{free} \wedge f_q) \right) \tag{4.35}$$

$$f_r^{bound}(X^{bound}, C) \;=\; f_r^{bound} \vee \max_{X^{free}}{}^k \left((f_r^{free} \wedge f_r) \vee (f_q^{free} \wedge f_q) \right) \tag{4.36}$$

After this restriction the IFS F to be EXOR-bi-decomposed can be expanded by the formulas (4.37) and (4.38).

$$f_q(X^{free}, X^{bound}, C) \;=\; f_q \wedge \overline{(f_q^{free} \vee f_r^{free})} \tag{4.37}$$

$$f_r(X^{free}, X^{bound}, C) \;=\; f_r \wedge \overline{(f_q^{free} \vee f_r^{free})} \tag{4.38}$$

The algorithm DECOMPTESTEXORMM is shown in Figure 14 and checks whether the characteristic function set F associated with the ISF $F(X^{free}, X^{bound}, C)$ includes a least one function that is EXOR-bi-decomposable with respect to the dedicated sets X^{free} and X^{bound} and calculates implicitly both decomposition ISFs. The algorithm SELECTMINTERM selects exactly one minterm from the given function. This algorithm is applied if the free decomposition ISF does not overlap with the remaining

$decomposable$ DECOMPTESTEXORMM (ISF F , DS X^{free}, DS X^{bound})
{ $\left(f_q^{free}, f_r^{free} \right) \leftarrow \langle 0, 0 \rangle$
$\left(f_q^{bound}, f_r^{bound} \right) \leftarrow \langle 0, 0 \rangle$
$decomposable \leftarrow true$
while$(decomposable \wedge (\langle f_q, f_r \rangle \neq \langle 0, 0 \rangle))$
   if $((f_q \vee f_r) \wedge (f_q^{free} \vee f_r^{free})) = 0$
     then $f_q^{free} \leftarrow f_q^{free} \vee$ SELECTMINTERM $(\max_{X^{bound}}{}^k (f_q \vee f_r))$
     $\left(f_q^{bound}, f_r^{bound} \right) \leftarrow$ RESTRICT $(f_q, f_r, f_q^{free}, f_r^{free}, X^{free})$
     $\left(f_q, f_r \right) \leftarrow$ EXPAND $(f_q, f_r, f_q^{free}, f_r^{free})$
   if $(f_q^{bound} \wedge f_r^{bound}) \neq 0$ **then** $decomposable \leftarrow false$
   SWAP $\left(\left\langle f_q^{free}, f_r^{free}, X^{free} \right\rangle, \left\langle f_q^{bound}, f_r^{bound}, X^{bound} \right\rangle \right)$
return $decomposable$ }

Figure 14. Algorithm DECOMPTESTEXORMM, which returns the Boolean value true, if the ISF $F(X)$ is EXOR-bi-decomposable with respect to the dedicated sets X^{free} and X^{bound}.

IFS F. The algorithm RESTRICT implements formulas (4.35) and (4.36) in order to restrict the bound decomposition IFS with respect to the free decomposition IFS and the given IFS F. Such minterms of F that can not cause further contradictions are removed by the algorithm EXPAND, which implements the formulas (4.37) and (4.38). If the ON-function and the OFF-function of the restricted bound decomposition ISF overlap then the ISF $F(X^{free}, X^{bound}, C)$ is not EXOR-bi-decomposable with respect to the dedicated sets X^{free} and X^{bound} so that the loop and the whole algorithm terminates. The algorithm SWAP swaps the mark functions and the associated dedicated sets of the free and bound decomposition ISFs. This is done because the new specified values in the bound decomposition ISF may create new conditions for the previously free decomposition ISF. An example of a former version of this algorithm is published in Steinbach and Wereszczynski (1995).

4.3. *Completeness and weak bi-decomposition*

The π-bi-decomposition, introduced in section 4.1, is a powerful design method. Only a simple two input gate is necessary to split a complex Boolean function into two simpler ones. If the dedicated sets A and B are enlarged as much as possible, the simplest decomposition functions can be calculated. The recursive application of the bi-decomposition to the decomposition functions may be the core of a complete design method for multilevel circuits. A necessary condition for such a design method is that a bi-decomposition exists for each Boolean function.

Unfortunately, already Böhlau (1987) found that not all Boolean function are bi-decomposable. An simple example of a not bi-decomposable Boolean

b	c	$h(b,c)$
0	0	0
0	1	0
1	1	1
1	0	0

b	c	$f(a,b,c)$	
0	0	0	0
0	1	0	1
1	1	1	1
1	0	0	1
		0	1 a

b	c	$G(a,b,c)$	
0	0	0	0
0	1	0	1
1	1	Φ	Φ
1	0	0	1
		0	1 a

Figure 15. Weak OR-bi-decomposition of the function $f(a, b, c) = ab \vee ac \vee bc$ with respect to the dedicated set $A = (a)$ into $G(A, C)$ and $h(C)$.

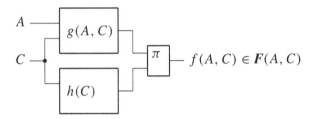

Figure 16. General structure of the weak π-bi-decomposition of $f(A, C)$.

function is $f(a, b, c) = ab \vee ac \vee bc$, visualized in the center of Figure 15. This is a symmetric function. Each selection of a pair of variables for the dedicated sets A and B divides the Boolean space $\mathbf{B}^3$ into two subspace $\mathbf{B}^2$, characterized by the value of the remaining variable. The values 1 of $f(a, b, c)$ are distributed between these subspaces such that one subspace includes one values 1 and the other one three values 1, respectively. Thus, one subfunction of $f(a, b, c)$ allows only an AND-bi-decomposition and the other one only an OR-bi-decomposition. There is no common π-bi-decomposition.

In order to overcome this gap, Le (1989) suggested the weak bi-decomposition. As can be seen in Figure 16, the formal difference between the weak π-bi-decomposition and the π-bi-decomposition is that there is no dedicated set B in the weak π-bi-decomposition of $f(A, C)$.

DEFINITION 9. *A weak bi-decomposition of the Boolean ISF $F(A, C)$, associated with characteristic function set $\mathbf{F}(A, C)$, with respect to the decomposition operator $\pi(x, y)$ and the not empty dedicated set A is a pair of functions $\langle g(A, C) \in \mathbf{G}(A, C), h(C) \in \mathbf{H}(C) \rangle$ with*

$$f(A, C) = \pi(g(A, C), h(C)) . \tag{4.39}$$

where $f(A, C) \in \boldsymbol{F}(A, C)$ *and*

$$\boldsymbol{G}(A, C) \supset \boldsymbol{F}(A, C). \tag{4.40}$$

The set C is called common set. The Boolean functions $g(A, C)$ and $h(C)$ are called decomposition functions.

Definition 9 describes the weak π-bi-decomposition for the general case of an ISF $F(A, C)$ and covers the special case of the weak π-bi-decomposition of a completely specified function $f(A, C)$, where $\boldsymbol{F}(A, C) = \{f(A, C)\}$. In contrast to the π-bi-decomposition, the decomposition function $g(A, C)$ of the weak π-bi-decomposition depends on the same number of variables as the function $f(A, C)$, which should be decomposed into simpler functions. Both decomposition functions $g(A, C)$ and $h(C)$ of a weak π-bi-decomposition may be easier than $f(A, C)$. Due to the formula (4.40) the function $g(A, C)$ can be selected from a larger characteristic function set $\boldsymbol{G}(A, C)$ than $\boldsymbol{F}(A, C)$ and the function $h(C)$ depends on a smaller number of variables. There is no Boolean function $h(C)$ for that an EXOR-operation satisfies the condition (4.40) and consequently only weak OR-bi-decompositions and AND-bi-decomposition exist. The reason for this is that the EXOR-operation exchanges only certain values of the care sets between the ON-function and OFF-function of the ISFs $F(A, C)$ and $G(A, C)$ by $g_q(A, C) = \overline{h(C)} \wedge f_q(A, C) \vee h(C) \wedge f_r(A, C)$ and $g_r(A, C) = \overline{h(C)} \wedge f_r(A, C) \vee h(C) \wedge f_q(A, C)$.

An example of a weak OR-bi-decomposition with respect to the dedicated set $A = (a)$ and the common set $C = (b, c)$ is shown in Figure 15. By means of the OR-operation condition (4.40) may be evaluated. The value 1 for the minterm $(b, c) = (1, 1)$ on the OR-gate input of $h(1, 1)$ causes the values 1 on the output $f(a, b = 1, c = 1)$ without any condition on the function $g(a, b = 1, c = 1)$. Consequently, $g_\varphi = b \wedge c$ and the relation (4.40) comes true. As specified in Theorem 10, the condition for a weak OR-bi-decomposition is that at least on value 1 expressed by $f_q(A, C)$ occurs in a subspace $C = const.$, which does not include any value 0 expressed by $\overline{\max_A^k f_r(A, C)}$.

THEOREM 10. *The characteristic function set $\boldsymbol{F}(A, C)$ associated with the ISF $F(A, C) = F\langle f_q(A, C), f_r(A, C)\rangle$ includes at least one Boolean function $f(A, C)$ that is weak OR-bi-decomposable with respect to the dedicated set A iff*

$$f_q(A, C) \wedge \overline{\max_A^k f_r(A, C)} \neq 0. \tag{4.41}$$

b c	$h(b,c)$
0 0	0
0 1	1
1 1	1
1 0	1

b c	$F(a,b,c)$	
0 0	Φ	0
0 1	0	1
1 1	1	1
1 0	0	1
	0	1 a

b c	$G(a,b,c)$	
0 0	Φ	Φ
0 1	0	1
1 1	1	1
1 0	0	1
	0	1 a

Figure 17. Weak AND-bi-decomposition of the function $F(a,b,c)$ with respect to the dedicated set $A = (a)$ into $G(a,b,c)$ and $h(b,c)$.

The mark functions for $G(A,C)$ of a weak OR-bi-decomposition of $F(A,C)$ can be calculated by the formulas (4.42) and (4.43).

$$g_q(A,C) = f_q(A,C) \wedge \max_A^k f_r(A,C) \qquad (4.42)$$

$$g_r(A,C) = f_r(A,C) \qquad (4.43)$$

The ON-function $g_q(A,C)$ in (4.42) includes such values 1 from the ON-function $f_q(A,C)$ which occur in a subspace $C = const.$ together with values 0 from the OFF-function $f_r(A,C)$. Because $g_q(A,C)$ in (4.42) is smaller than $f_q(A,C)$, (4.40) becomes true. The gain of the weak OR-bi-decomposition is visualized in Figure 15 and will be recognized in the next level of the π-bi-decomposition. The characteristic function set $G(A,C)$ includes the Boolean function $g(a,b,c) = a \wedge (b \vee c)$ for which the disjoint AND-bi-decomposition into $g'(a) = a$ and $h'(b,c) = b \vee c$ exists. Influenced by the selected function $g(A,C)$, the mark functions for $H(C)$ of a weak OR-bi-decomposition of $F(A,C)$ can be calculated by the formulas (4.44) and (4.45).

$$h_q(C) = \max_A^k \left(f_q(A,C) \wedge \overline{g(A,C)} \right) \qquad (4.44)$$

$$h_r(C) = \max_A^k f_r(A,C) \qquad (4.45)$$

An example of a weak AND-bi-decomposition of the ISF $F(A,C)$ with respect to the dedicated set $A = (a)$ and the common set $C = (b,c)$ is shown in Figure 17. Note, the characteristic function set $F(A,C)$ includes the weak OR-bi-decomposable function $f(A,C)$ from Figure 15. No Boolean function of $F(A,C)$ is π-bi-decomposable.

Similar to the OR-operation, the AND-operation of the weak bi-decomposition can also fulfill condition (4.40). The value 0 for the minterm $(b,c) = (0,0)$ on the AND-gate input of $h(0,0)$ specifies the values 0 on the output $f(a, b = 0, c = 0)$ without any condition on the function $g(a, b = 0, c = 0)$.

Consequently, $g_\varphi = \bar{b} \wedge \bar{c}$ and the relation (4.40) becomes true. As specified in Theorem 11, the condition for an weak AND-bi-decomposition is that at least on value 0 expressed by $f_r(A, C)$ occurs in a subspace $C = const.$ which does not include any value 1 expressed by $\overline{\max_A{}^k f_q(A, C)}$.

THEOREM 11. *The characteristic function set $F(A, C)$ associated with the ISF $F(A, C) = F\langle f_q(A, C), f_r(A, C)\rangle$ includes at least one Boolean function $f(A, C)$ which is weak AND-bi-decomposable with respect to the dedicated set A iff*

$$f_r(A, C) \wedge \overline{\max_A{}^k f_q(A, C)} \neq 0. \qquad (4.46)$$

The mark functions of $G(A, C)$ for a weak AND-bi-decomposition of $F(A, C)$ can be calculated by the formulas (4.47) and (4.48).

$$g_q(A, C) = f_q(A, C) \qquad (4.47)$$
$$g_r(A, C) = f_r(A, C) \wedge \max_A{}^k f_q(A, C) \qquad (4.48)$$

The OFF-function $g_r(A, C)$ (4.48) includes such values 0 from the OFF-function $f_r(A, C)$ which occur in a subspace $C = const.$ together with values 1 from the ON-function $f_q(A, C)$. Because $g_r(A, C)$ in (4.48) is smaller than $f_r(A, C)$, (4.40) becomes true. The gain of the weak OR-bi-decomposition is visualized in Figure 17 and will be recognized in the next level of the π-bi-decomposition. The characteristic function set $G(A, C)$ includes the Boolean function $g(a, b, c) = a \vee (b \wedge c)$ for which the disjoint OR-bi-decomposition into $g'(a) = a$ and $h'(b, c) = b \wedge c$ exists. Influenced by the selected function $g(A, C)$, the mark functions for $H(C)$ of a weak AND-bi-decomposition of $F(A, C)$ can be calculated by the formulas (4.49) and (4.50).

$$h_q(C) = \max_A{}^k f_q(A, C) \qquad (4.49)$$
$$h_r(C) = \max_A{}^k (f_r(A, C) \wedge g(A, C)) \qquad (4.50)$$

A complete multilevel design of each Boolean function is possible by the recursive application of all types of the π-bi-decomposition together with the weak OR-bi-decomposition and the weak AND-bi-decomposition. The formal proof of the corresponding Theorem 12 for completeness is more comprehensible than the explanation by an example so that the representation form is changed here.

THEOREM 12. *If the characteristic function set $F(A, C)$ associated with the ISF $F(A, C) = F\langle f_q(A, C), f_r(A, C)\rangle$ includes only Boolean function*

$f(A, C)$ which are neither weakly OR-bi-decomposable nor weakly AND-bi-decomposable with respect to a single variable in the dedicated set $A = (a)$ then $\mathbf{F}(A, C)$ includes at least one Boolean function $f(A, C)$ which is disjointly EXOR-bi-decomposable with respect to the single variable in the dedicated set $A = (a)$ and the dedicated set $B = C$.

Proof (of the Theorem 12 – Completeness). If $F(A, C)$ is not weakly OR-bi-decomposable (4.51) follows from (4.41), which is equivalent to the equations (4.52) and (4.53).

$$f_q(A, C) \wedge \overline{\max_A^k f_r(A, C)} = 0 \tag{4.51}$$

$$f_q(A, C) \wedge \min_A^k \overline{f_r(A, C)} = 0 \tag{4.52}$$

$$f_q(A, C) \wedge \min_A^k \left(f_q(A, C) \vee f_\varphi(A, C) \right) = 0 \tag{4.53}$$

By removing $f_\varphi(A, C)$ from the left side of (4.53) the valid equation (4.54) is built, from which the equivalent equation (4.55) follows.

$$f_q(A, C) \wedge \min_A^k f_q(A, C) = 0 \tag{4.54}$$

$$\min_A^k f_q(A, C) = 0 \tag{4.55}$$

If $F(A, C)$ is not weakly AND-bi-decomposable (4.56) follows from (4.46), which is equivalent to the equations (4.57) and (4.58).

$$f_r(A, C) \wedge \overline{\max_A^k f_q(A, C)} = 0 \tag{4.56}$$

$$f_r(A, C) \wedge \min_A^k \overline{f_q(A, C)} = 0 \tag{4.57}$$

$$f_r(A, C) \wedge \min_A^k \left(f_r(A, C) \vee f_\varphi(A, C) \right) = 0 \tag{4.58}$$

By removing $f_\varphi(A, C)$ from the left side of (4.58) the valid equation (4.59) is built, from which the equivalent equation (4.60) follows.

$$f_r(A, C) \wedge \min_A^k f_r(A, C) = 0 \tag{4.59}$$

$$\min_A^k f_r(A, C) = 0 \tag{4.60}$$

The substitution from (4.55) and (4.60) into (4.31) with $A = (a)$ results in (4.61).

$$f_r^a(C) = \min_a f_q(a, C) \vee \min_a f_r(a, C) = 0 \vee 0 = 0 \tag{4.61}$$

The final substitution of (4.61) into the condition (4.29) shows that for any dedicated subset $B \subseteq C$ and $C_0 = C \setminus B$ the specialized condition (4.62) of the EXOR-bi-decomposition holds.

$$\max_B^k f_q^a(B, C_0) \wedge f_r^a(C) = \max_B^k f_q^a(B, C_0) \wedge 0 = 0 \qquad (4.62)$$

4.4. Selection of decomposition type and sets of variables

The selection of the decomposition type and dedicated sets of variables influences significantly the the result of the designed circuit in terms of structure, area, delay, and power consumption. There is a direct relation between the total delay of a circuit and the longest path, measured in levels of gates. The larger the number of variables in the dedicated sets A and B of all selected π-bi-decompositions, the fewer levels of gates are necessary. Because the number of variables of the decomposition function $g(A, C)$ is not reduced in the case of a weak π-bi-decomposition, this decomposition type is used only for completeness if no strong π-bi-decomposition exists.

There is an extremely huge number of different selections of the dedicated sets A and B. The trivial method to check all of them for each type of the π-bi-decomposition is not practical for expenditure reasons. By means of the Theorem 13, the number of such checks can be drastically reduced.

THEOREM 13. *If the ISF $F(A, B, C)$ is π-bi-decomposable with respect to any dedicated sets A and B, then it is also π-bi-decomposable with respect to any dedicated sets $A' \subseteq A$ and $B' \subseteq B$.*

By changing the dedicated set A into A' or B into B', the excluded k variables from the dedicated sets become part of a common set $C' \supseteq C$. Theorem 13 follows from the fact, that the decomposition functions may be mapped without any restrictions from $\mathbf{B}^n \to \mathbf{B}$ to $\mathbf{B}^{n+k} \to \mathbf{B}$.

In the following, different partitions of the variables $X = (A, B, C) = (A', B', C')$ are considered, where $A = A' \cup C_A$, $B = B' \cup C_B$, and $C' = C \cup C_A \cup C_B$. The trivial conclusion from the Theorem 13 is, that no check for a π-bi-decomposition of the ISF $F(A', B', C')$ is necessary, if it is known that the ISF $F(A, B, C)$ is π-bi-decomposable. Vice versa, if it is known that the ISF $F(A', B', C')$ is π-bi-decomposable, then a π-bi-decomposition of the ISF $F(A, B, C)$ may exist. A simple algorithm of the complexity of $O(n^2)$ can be defined, that checks whether or not there is an initial π-bi-decomposition with respect to single variables in the dedicated sets A' and B'. The pseudo code of the algorithm FINDINITIALVARIABLES is shown in Figure 18 and uses two other algorithms. The algorithm SUPPORT returns

$\langle A, B \rangle$ FINDINITIALVARIABLES (DT π, ISF F)
{ $X \leftarrow$ SUPPORT(F)
    $Y \leftarrow X$
    for all $a \in X$ **do**
        $A \leftarrow \{a\}$
        $Y \leftarrow Y \setminus \{a\}$
        for all $b \in Y$ **do**
        $B \leftarrow \{b\}$
            if DECOMPOSITIONTESTOO(π, F, A, B)
                then return $\langle A, B \rangle$
    return $\langle \emptyset, \emptyset \rangle$ }

Figure 18. Algorithm FINDINITIALVARIABLES, which calculates the dedicated sets $\langle A = (a), B = (b) \rangle$ of a π-bi-decomposition of the given ISF $F(X)$ with respect to the decomposition type DT π or returns $\langle \emptyset, \emptyset \rangle$ if no π-bi-decomposition exists.

the support of the given ISF and the algorithm DECOMPOSITIONTESTOO implements the condition of Theorems 6, 7, 8 and returns the value *true* if the π-bi-decomposition the respect to the given single variables in the dedicated sets exists.

From Theorem 13 is known, that an initial π-bi-decomposition with respect to single variables in the dedicated sets may be extended to the π-bi-decomposition of the same type with respect to several variables in the dedicated sets. This can be done by an algorithm that tries for each variable of the common set C' whether the inclusion of this variable into the dedicated sets removes the property of the known π-bi-decomposition or not. Note, such a try and error algorithm has only a linear complexity of $O(n)$, because the extension of a dedicated set of a not π-bi-decomposable ISF $F(A', B', C')$ cannot create a π-bi-decomposable ISF $F(A, B, C)$, as stated in Theorem 14. This Theorem follows directly from the conditions of all π-bi-decompositions. If neither the dedicated set A nor the the dedicated set B of a π-bi-decomposition of an ISF $F(A, B, C)$ can be extended, the π-bi-decomposition is called *compact*.

THEOREM 14. *If the ISF $F(A', B', C')$ is not π-bi-decomposable with respect to any dedicated sets A' and B', then there does not exist any π-bi-decomposable ISF $F(A, B, C)$ with respect to any dedicated sets $A \supseteq A'$ and $B \supseteq B'$.*

As shown in Steinbach et al. (1993), the distribution of the variables to the dedicated sets should be done in such a way that the smallest of the dedicated sets includes as many variables as possible. The algorithm FIND-VARIABLESETS, shown in Figure 19, solves this condition and leads to

```
⟨A, B⟩ FINDVARIABLESETS ( DT π, ISF F, DS ⟨A, B⟩)
{ Z ← SUPPORT(F) \ (A ∪ B)
    for all c ∈ Z do
        if DECOMPOSITIONTESTMM(π, F, A ∪ {c}, B)
            then A ← A ∪ {c}
            else if DECOMPOSITIONTESTMM(π, F, A, B ∪ {c})
                then B ← B ∪ {c}
        if |A| > |B|
            then ⟨A, B⟩ ← SWAP(A, B)
    return ⟨A, B⟩ }
```

Figure 19. Algorithm FINDVARIABLESETS, which calculates the dedicated sets $\langle A, B \rangle$ of a compact π-bi-decomposition of the given ISF $F(X)$ with respect to the decomposition type DT π and the dedicated sets $\langle A = (a), B = (b) \rangle$ of an initial π-bi-decomposition.

```
⟨A⟩ FINDWEAKVARIABLE( DT π, ISF F )
{ X ← SUPPORT(F)
    for all a ∈ X do
        if DECOMPOSITIONTESTW(π, F, {a})
            then return ⟨{a}⟩
    return ⟨Ø⟩ }
```

Figure 20. Algorithm FINDWEAKVARIABLE, which calculates the dedicated set $\langle A = (a) \rangle$ of a weak π-bi-decomposition of the given ISF $F(X)$ with respect to the decomposition type DT π or returns $\langle \emptyset \rangle$ if no weak π-bi-decomposition exists.

several advantages of the designed circuit. First, the maximal number of variables in the dedicated sets A and B minimizes the decomposition functions, so that the switching activities and consequently the power consumption is minimized. Second, the nearly equal number of variables in the dedicated sets A and B leads to a low number of gate levels and a very short total delay. Third, the area of the designed circuit will be minimized implicitly by minimizing the decomposition function.

The pseudo code of the algorithm FINDVARIABLESETS is shown in Figure 19 and uses two other algorithms, not discussed yet. The dedicated sets A and B are swapped by the algorithm SWAP. The algorithm DECOMPOSITIONTESTMM checks the π-bi-decomposition of an ISF $F(X)$ with respect to the dedicated sets A and B where each of the dedicated sets can cover many variables. This algorithm is based on Theorems 6 and 7 in the case of the OR- or AND-bi-decomposition. The algorithm DECOMPOSITIONTESTMM uses Theorem 9 for the EXOR-bi-decomposition of an ISF $F(X)$ if only one dedicated set includes several variables, and otherwise if both dedicated sets

$\langle A, B, \pi \rangle$ SELECTBESTDECOMPOSITION (ISF F)
{ $X \leftarrow$ SUPPORT(F)
  $\langle best\_A, best\_B \rangle \leftarrow \langle \emptyset, \emptyset \rangle$
  $best\_cost \leftarrow 1$
  for all $\pi \in \{$'OR', 'AND', 'EXOR'$\}$ **do**
    $\langle A, B \rangle \leftarrow$ FINDINITIALVARIABLES (π, F)
    if $\langle A, B \rangle \neq \langle \emptyset, \emptyset \rangle$
      then $\langle A, B \rangle \leftarrow$ FINDVARIABLESETS (π, F, A, B)
        $cost \leftarrow (|X| * \min(|A|, |B|) + \max(|A|, |B|))^{-1}$
        if $cost < best\_cost$)
          $\langle best\_cost, best\_A, best\_B, best\_\pi \rangle \leftarrow \langle cost, A, B, \pi \rangle$
  if $best\_A = \emptyset$
    then $\pi \leftarrow$ 'Weak AND'
      $\langle best\_A, best\_\pi \rangle \leftarrow \langle$ FINDWEAKVARIABLE(F), π $\rangle$
  if $best\_A = \emptyset$
    then $\pi \leftarrow$ 'Weak OR'
      $\langle best\_A, best\_\pi \rangle \leftarrow \langle$ FINDWEAKVARIABLE(F), π $\rangle$
  return $\langle best\_A, best\_B, best\_\pi \rangle$ }

Figure 21. Algorithm SELECTBESTDECOMPOSITION, which returns the best possible decomposition type π and the dedicated sets $\langle A, B, \pi \rangle$ of a π-bi-decomposition of the given ISF $F(X)$, where π may indicate a weak decomposition too.

include several variables, the algorithm DECOMPTESTEXORMM, shown in Figure 14, is applied.

If the ISF $F(X)$ is not π-bi-decomposable then a weak π-bi-decomposition exists. The dedicated set A of the weak π-bi-decomposition can cover one or several variables. The larger the number of variables in this dedicated set A the smaller is the DC-set of the ISF $G(A, C)$ and the smaller is the probability that $G(A, C)$ is π-bi-decomposable. For this reason a weak π-bi-decomposition with respect to a single variable is preferred. The appropriate algorithm FINDWEAKVARIABLE is shown in Figure 20 and uses the algorithm DECOMPOSITIONTESTW that implements Theorems 10 and 11, respectively.

Figure 21 shows the pseudo code of the entire algorithm to select the locally best decomposition. This algorithm uses the previously introduced algorithms.

4.5. *Algorithm for recursive bi-decomposition*

Based on the introduced theory, an algorithm for recursive bi-decomposition is shown in Figure 22. The recursion of the algorithm DECOMPOSISF determines if the support of the ISF F that is to be decomposed includes less

---

f DECOMPOSISF (ISF F)
{ **if** | SUPPORT(F) | < 2
    then $f \leftarrow$ SIMPLE(F)
    else $\langle A, B, \pi \rangle \leftarrow$ SELECTBESTDECOMPOSITION (F)
        $G \leftarrow$ COMPUTEFREESET (π, F, A, B)
        $g \leftarrow$ DECOMPOSISF (G)
        $H_g \leftarrow$ COMPUTEBOUNDSET (π, F, A, g)
        $h \leftarrow$ DECOMPOSISF (H_g)
        $f \leftarrow$ ADDGATETODECOMPOSITIONTREE(g, h, π)
  return f }

Figure 22. Algorithm DECOMPOSISF, which calculates recursively the entire decomposition structure of the returned Boolean function $f(X) \in F(X)$ of the given ISF $F(X)$.

than 2 variables. In this case, the algorithm SIMPLE selects one of the Boolean functions '0', '1', 'x' or '$\overline{x}$', specified by $F(x)$. If the ISF F depends on two or more variables, the best π-bi-decomposition is selected by the algorithm SELECTBESTDECOMPOSITION. Remember, the decomposition function g and h cannot be chosen independently from each other. Hence, first the ISF $G(A, C)$ is calculated by the algorithm COMPUTEFREESET using the formulas form sections 4.2 and 4.3.

An important property of this ISF $G(A, C)$ is that the maximal possible DC-set is associated. For each function $g(A, C) \in G(A, C)$ there exists a decomposition function h, but not each combination of g and h is allowed. Therefore, the decomposition structure of the ISF $G(A, C)$ is designed completely by the recursive calculation of the algorithm DECOMPOSISF. During this process the Boolean function $g(A, C)$ that is represented in the chosen circuit structure is calculated and returned. Using the selected function $g(A, C)$, the ISF $H(A, C)$ is calculated by the algorithm COMPUTE-BOUNDSET using the formulas from sections 4.2 and 4.3 such, that a maximal possible DC-set occurs. Without any restriction, the decomposition structure of the ISFs $H(B, C)$ or $H(C)$ is designed completely by the second recursive calculation of the algorithm DECOMPOSISF. Now, both decomposition functions g and h and the decomposition operator π for the corresponding gate of the circuit are known. The algorithm ADDGATETO-DECOMPOSITIONTREE stores this information into a net list and returns the output function of the gate calculated by $f = \pi(g, h)$.

The suggested algorithm DECOMPOSISF uses the algorithm SELECT-BESTDECOMPOSITION so that fast circuits of low power consumption are designed. The calculation of compact π-bi-decompositions leads implicitly to a minimization of the needed area. Further reductions of the area are possible by including a method of π-bi-decomposition with respect to a certain Boolean function, published in Dresig (1992), or the additional check for reusing of gates (Mishchenko et al. 2001).

One more excellent property of the algorithm DECOMPOSISF is, that the designed circuits are completely testable for all single stuck-at-0 and stuck-at-1 fault on the inputs and outputs of the circuit and on all gate inputs and gate outputs as well. This nice property based on the theory of Le, published in Le (1989). The necessary requirements for the complete testability are compact π-bi-decomposition and maximal DC-sets. The strong theory of Le (1989) was applied in Steinbach and Stöckert (1994) to the practical method of the π-bi-decomposition presented in this paper. Additional to the proof the testability, Steinbach and Stöckert (1994) offers a simple extension to the algorithm DECOMPOSISF that calculates all test pattern in parallel to the design process. In order to solve the NP-complete task of test pattern generation, intermediate results of the π-bi-decomposition are reused, so that only an overhead of about 10% arises.

5. Experimental Results

The algorithms introduced above have been implemented in the program BI-DECOMP written in the object oriented language C++ using the BDD package CUDD (Somenzi 2001). The experimantal results of this program were completely published in Mishchenko et al. (2001) and are partially repeated here.

The program BI-DECOMP does not need extensive hardware resources. All benchmarks were calculatd by the program BI-DECOMP on a 300 Mhz Pentium II PC with 64 MByte RAM. The correctness of the resulting networks has been checked using a BDD-based verifier. In two separate experiments both delay and area of the circuits designed by the metheod of π-bi-decomposition implemented in the program BI-DECOMP was analysed. The used benchmarks are labeled by "name" in Tables 2 and 3 where the columns "in" and "out" specify the correponding numbers of inputs and outputs.

The first experiment aims at the delay of designed circuits of MCNC benchmarks into two-input NAND/NOR/EXOR/NEXOR gates. Table 2 shows the results of the well known program SIS (Sentovich 1992) and the method of π-bi-decomposition implemented in the program BI-DECOMP. For a delay-minimized design, the benchmarks have been preprocessed in SIS using script.rugged with speed_up followed by a delay-oriented mapping into a subset of mcnc.genlib containing the above listed two-input logic gates. Columns "gate", "lev", "area" and "del" give the number of gates, the number of logic levels, the needed area and the delay of the circuit. The

Table 2. Delay-oriented comparison between SIS and BI-DECOMP

Benchmark			SIS				π-bi-decomposition			
Name	In	Out	Gate	Lev	Area	Del	Gate	Lev	Area	Del
9sym	9	1	235	18	486	28.5	83	11	226	17.1
alu4.pla	14	8	211	26	490	27.8	279	12	619	20.4
cps	24	109	1188	21	2629	40.5	1733	13	3679	22.8
duke2	22	29	467	27	1280	39.0	684	12	1503	20.0
e64	65	65	253	125	798	32.3	1558	7	2999	11.2
misex3	14	14	717	30	1528	42.2	1017	15	2414	27.8
pdc	16	40	415	20	955	29.8	328	8	712	14.5
spla	16	46	658	21	1369	29.9	786	14	1691	23.3
vg2	25	8	107	14	234	13.3	259	11	601	18.3
Average			472	33.5	1085	31.4	747	11.4	1605	19.5
Ratio %			100	100	100	100	158	34	148	62

runtime for SIS was one minute on average and for BI-DECOMP less than 10 seconds.

Table 2 shows that BI-DECOMP designs are significantly faster circuits than SIS. In average the delay of the circuits designed by BI-DECOMP is only 62% of the delay-minimized SIS results. The reasons for that are the implicite selection of the functions having the best decomposition properties, the consequent utilization of these properties and the selection of the dedicated sets for a well-balanced π-bi-decomposition. The program BI-DECOMP works in contrast to SIS without any pre- or post-processing, so that a 48% larger area is needed.

The second experiment aims at the minimization of the the required area. Especially circuits that need many EXOR gates are analyzed. In this experiment the program BI-DECOMP is compared with the desing tool SIS (Sentovich 1992) (script.rugged + area-oriented mapping into mcnc.genlib) and BDS, a BDD-based logic synthesis system (C. Yang 2000). The results are shown in Table 3.

The output of BDS is a netlist of 2/3/4-input (N)AND-gates and (N)OR-gates, and 2-input (N)EXOR-gates. To make a fair comparison, the two-input-gates from the output of BI-DECOMP were translated into a set of gates comparable to the output of BDS. This is the reason why the gate count for 9sym in the Table 3 differs from that of the Table 2. The number of such more complex gates after the design is shown in column "gate" and includes also the number of EXOR-gates which is addionally listed in the column

Table 3. Area-oriented comparison between SIS, BDS, and BI-DECOMP for EXOR-intensive circuits

benchmark			SIS		BDS			π-bi-decomposition		
Name	In	Out	Gate	Area	Gate	Exor	Area	Gate	Exor	Area
5xp1	7	10	81	195	67	16	172	62	21	160
9sym	9	1	152	396	42	4	109	50	27	155
alu2	10	6	217	524	230	53	632	198	61	519
alu4	14	8	409	996	582	124	1655	508	126	1264
cordic	23	2	34	94	47	16	126	33	15	121
f51m	8	8	58	139	56	11	174	40	15	110
rd53	5	3	22	47	25	6	72	21	8	67
rd73	7	3	106	258	45	8	133	39	19	122
rd84	8	4	192	468	62	12	189	54	25	166
t481	16	1	407	1023	15	5	45	17	6	65
z4ml	7	4	20	59	20	6	53	30	6	61
Average			154.5	435	108.2	23.7	305	95.6	29.9	256
Ratio %			100	100	70	100	70	62	126	59

"exor". The column "area" shows the results of area-oriented mapping into mcnc.genlib.

Table 3 shows that BI-DECOMP produces good results in terms of area too. In comparison to SIS, BDS needs only 70% of the area, but BI-DECOMP outperforms the other two tools requiring only 59% of the circuit area desinged by SIS. One reasons for this result is that BI-DECOMP applies only weak π-bi-decomposition if no strong π-bi-decomposition exist. A further reasons why BI-DECOMP outperforms BDS is in the use of the DC-set. BDS uses the DC-set locally, to optimize the size of the BDD representation of one ISF. BI-DECOMP uses the DC-set both locally to optimize the decomposition *and* globally by passing the remaining DC-set together with the new generated DC-set to the next decomposition step.

6. Upper- and Lower-Bound Description for Bi-Decomposition

The π-bi-decomposition was introduced by means of mark functions of an ISF F. Mostly the ON-function f_q and the OFF-function f_r were used to describe the ISF F. Sometimes the OFF-function f_r was substituted by the

DC-function f_φ. The reason for that consist in the understandable view of the associated sets.

In a quite different view the ISF F is regarded as an interval characterized by the lower bound f_l and the upper bound f_u. On the one hand there are simple relation (2.3), (2.4) between these different mark functions but on the other hand the comprehension to the π-bi-decomposition of Boolean ISFs is changed into such a form which can easily be extended to intervals of multivalued functions.

First, the OR-bi-decomposition of the ISF $F\left[f_l, f_u\right]$ is considered. The substitution of (2.3) and (2.4) into the condition of the OR-bi-decomposition of the ISF $F\langle f_q, f_r\rangle$ (4.16) leads to the equivalent equation (6.1).

$$f_l(A, B, C) \wedge \max_A{}^k \overline{f_u(A, B, C)} \wedge \max_B{}^k \overline{f_u(A, B, C)} = 0. \tag{6.1}$$

Using the laws of the Boolean Differential Calculus, the laws of the Boolean Algebra and (2.9), (2.10), the equation (6.1) can be transformed into the inequality (6.4) of the alternative Theorem 15 of the OR-bi-decomposition by the following transformations.

$$f_l(A, B, C) \wedge \overline{\min_A{}^k f_u(A, B, C)} \wedge \overline{\min_B{}^k f_u(A, B, C)} = 0. \tag{6.2}$$

$$f_l(A, B, C) \wedge \overline{\min_A{}^k f_u(A, B, C) \vee \min_B{}^k f_u(A, B, C)} = 0. \tag{6.3}$$

THEOREM 15. *The characteristic function set $F(A, B, C)$ which is associated with the ISF $F(A, B, C) = F\left[f_l(A, B, C), f_u(A, B, C)\right]$ includes at least one Boolean function $f(A, B, C)$ that is OR-bi-decomposable with respect to the dedicated sets A and B iff*

$$f_l(A, B, C) \leq \min_A{}^k f_u(A, B, C) \vee \min_B{}^k f_u(A, B, C). \tag{6.4}$$

Condition (6.4) may be explained as follows. If the lower bound function $f_l(A, B, C) = 0$ for a certain minterm $(A, B, C) = const.$, inequality (6.4) holds without any additional condition. But, if the lower bound function $f_l(A, B, C) = 1$ for a certain minterm $(A, B, C) = const.$ then at least in one of the subspaces, defined by $B = const.$ or $A = const.$ the upper bound function $f_u(A, B = const., C = const.)$ or $f_u(A = const., B, C = const.)$ must be constant 1. This describes with other words exactly the previously introduced condition of an OR-bi-decomposition.

The formulas to calculate the ISFs $G\left[g_l, g_u\right]$ and $H\left[h_l, h_u\right]$ of the OR-bi-decomposition in the interval representation are created by similar transformation of (4.19) into (6.5), of (4.20) into (6.6), of (4.21) into (6.7), and of

(4.22) into (6.8). Note, equations (6.6) and (6.8) of the OR-bi-decomposition upper bound functions generalize directly the equation of the corresponding decomposition function of a completely specified function (4.4) and (4.5).

$$g_l(A, C) = \max_B^k \left(f_l(A, B, C) \wedge \overline{\min_A^k f_u(A, B, C)} \right) \qquad (6.5)$$

$$g_u(A, C) = \min_B^k f_u(A, B, C) \qquad (6.6)$$

$$h_l(B, C) = \max_A^k \left(f_l(A, B, C) \wedge \overline{g(A, C)} \right) \qquad (6.7)$$

$$h_u(B, C) = \min_A^k f_u(A, B, C) \qquad (6.8)$$

As shown in Theorem 16, the AND-bi-decomposition can be transformed in the interval representation, too.

THEOREM 16. *The characteristic function set $F(A, B, C)$ which is associated with the ISF $F(A, B, C) = F\left[f_l(A, B, C), f_u(A, B, C) \right]$ includes at least one Boolean function $f(A, B, C)$ that is AND-bi-decomposable with respect to the dedicated sets A and B iff*

$$f_u(A, B, C) \geq \max_A^k f_l(A, B, C) \wedge \max_B^k f_l(A, B, C). \qquad (6.9)$$

Condition (6.9) holds without any additional condition if the upper bound function $f_u(A, B, C) = 1$ for a certain minterm $(A, B, C) = const.$ Otherwise, if the upper bound function $f_u(A, B, C) = 0$ for a certain minterm $(A = const., B = const., C = const.)$ then in at least one of the subspaces, defined by $B = const.$ or $A = const.$ the lower bound function $f_l(A, B = const., C = const.)$ or $f_l(A = const., B, C = const.)$ must be 0. The results of the transformation to calculate the ISFs $G\left[g_l, g_u \right]$ and $H\left[h_l, h_u \right]$ of the AND-bi-decomposition in the interval representation are given in the formulas (6.10), (6.11), (6.12) and (6.13), where the lower bound functions $g_l(A, C)$ and $h_l(B, C)$ generalize the corresponding completely specified decomposition functions (4.8) and (4.9).

$$g_l(A, C) = \max_B^k f_l(A, B, C) \qquad (6.10)$$

$$g_u(A, C) = \min_B^k \left(f_u(A, B, C) \vee \overline{\max_A^k f_l(A, B, C)} \right) \qquad (6.11)$$

$$h_l(B, C) = \max_A^k f_l(A, B, C) \qquad (6.12)$$

$$h_u(B, C) = \min_A^k \left(f_u(A, B, C) \vee \overline{g(A, C)} \right) \qquad (6.13)$$

The one-to-many EXOR-bi-decomposition can be checked in the interval representation of an ISF using Theorem 17.

THEOREM 17. *The characteristic function set $F(a, B, C)$ which is associated with the ISF $F(a, B, C) = F\left[f_l(a, B, C), f_u(a, B, C)\right]$ includes at least one Boolean function $f(a, B, C)$ that is EXOR-bi-decomposable with respect to the dedicated sets $A = (a)$ and B iff*

$$\max_B^k f_l^a(B, C) \wedge \overline{f_u^a(B, C)} = 0 \qquad (6.14)$$

where

$$f_l^a(B, C) = \max_a f_l(a, B, C) \wedge \overline{\min_a f_u(a, B, C)} \qquad (6.15)$$

$$\overline{f_u^a(B, C)} = \min_a f_l(a, B, C) \vee \overline{\max_a f_u(a, B, C)}. \qquad (6.16)$$

To complete the theory of the π-bi-decomposition in the interval representation the weak OR- and weak AND-bi-decomposition are considered.

THEOREM 18. *The characteristic function set $F(A, C)$ associated with the ISF $F(A, C) = F\left[f_l(A, C), f_u(A, C)\right]$ includes at least one Boolean function $f(A, C)$ that is weakly OR-bi-decomposable with respect to the dedicated set A iff*

$$f_l(A, C) \wedge \min_A^k f_u(A, C) \neq 0. \qquad (6.17)$$

Condition (6.17) for a weak OR-bi-decomposition is that at least on value 1 expressed by the lower bound function $f_l(A, C)$ occurs in a subspace $C = const.$ which is completely covered by the upper bound function $f_u(A, C)$. The mark functions for $G(A, C)$ and $H(C)$ of a weak OR-bi-decomposition of $F(A, C)$ can be calculated by the formulas (6.18), (6.19), (6.20) and (6.21).

$$g_l(A, C) = f_l(A, C) \wedge \overline{\min_A^k f_u(A, C)} \qquad (6.18)$$

$$g_u(A, C) = f_u(A, C) \qquad (6.19)$$

$$h_l(C) = \max_A^k \left(f_l(A, C) \wedge \overline{g(A, C)} \right) \qquad (6.20)$$

$$h_u(C) = \min_A^k f_u(A, C) \qquad (6.21)$$

THEOREM 19. *The characteristic function set $F(A, C)$ associated with the ISF $F(A, C) = F\left[f_l(A, C), f_u(A, C)\right]$ includes at least one Boolean function $f(A, C)$ that is weakly AND-bi-decomposable with respect to the dedicated set A iff*

$$f_u(A, C) \vee \max_A^k f_l(A, C) \neq 1. \qquad (6.22)$$

Condition (6.22) for a weak AND-bi-decomposition is that at least one subspace $C = const.$ is not completely covered by the upper bound function $f_u(A, C)$ and does not include any value 1 expressed by the k-times maximum of the lower bound function $f_l(A, C)$. The mark functions for $G(A, C)$ and $H(C)$ of a weak AND-bi-decomposition of $F(A, C)$ can be calculated by the formulas (6.23), (6.24), (6.25) and (6.26).

$$g_l(A, C) = f_l(A, C) \tag{6.23}$$

$$g_u(A, C) = f_u(A, C) \vee \overline{\max_A{}^k f_l(A, C)} \tag{6.24}$$

$$h_l(C) = \max_A{}^k f_l(A, C) \tag{6.25}$$

$$h_u(C) = \min_A{}^k \left(f_u(A, C) \vee \overline{g(A, C)} \right) \tag{6.26}$$

7. Conclusion

The π-bi-decomposition is an excellent method to design a multilevel circuit. It can be used for completely specified Boolean functions and ISFs as well. During the design process ISFs for the decomposition functions are created even if a completely specified Boolean function is given. Due to the proved completeness no other design method is necessary.

The optimal utilization of functional properties for the π-bi-decomposition leads to circuit structures having a low number of levels and and thus a very short delay. The area of the circuits, designed by the method of π-bi-decomposition, is implicitly minimized especially if many EXOR-bi-decompositions occur.

Generally, there are several different pairs of decomposition function if a π-bi-decomposition with respect to the decomposition operator π and dedicated sets exists. Using the compact π-bi-decomposition, formulas and algorithms suggested in this paper the designed circuit is completely testable for single stuck-at faults.

This paper summarizes the most important results of the π-bi-decomposition from the last two decades and focuses on a well understandable presentation. With the exception of one formal proof all other theorems were explained by means of comprehensible examples. The proofs of these Theorems can be found in the original literature. For the first time this paper includes the theory of the π-bi-decomposition of incompletely specified Boolean function represented by mark functions especially $F \langle f_q(X), f_r(X) \rangle$ and the lower and upper bound function of the interval $F [f_l(X), f_u(X)]$.

Around the presented theory several extensions exist. Pre- and post processing steps allow the design of unlimited large circuits (Steinbach and

Hesse 1996). Special methods (Steinbach and Zhang 1997) guarantee the testability in this case, too. The area of the circuit may be further reduced by additional methods, published in Dresig (1992) or Mischchenko et al. (2001). The interval representation of the π-bi-decomposition is suited for the generalization to multi-valued function (Lang 2003).

References

Akers, S. B. (1959). On a Theory of Boolean Functions. *J. Soc. Ind. Appl. Math.* **7**(4).

Ashenhurst, R. (1957). The Decomposition of Switching Functions. In *International Symposium on the Theory of Switching Functions*, 74–116.

Bochmann, D., Dresig, F. & Steinbach, B. (1991). A New Decomposition Method for Multilevel Circuit Design. In *European Conference on Design Automation*, 374–377. Amsterdam, Holland.

Bochmann, D. & Posthoff, C. (1981). *Binäre dynamische Systeme*. München, Germany: Oldenbourg Verlag.

Bochmann, D. & Steinbach, B. (1991). *Logikentwurf mit XBOOLE*. Berlin, Germany: Verlag Technik.

Böhlau, P. (1987). Eine Dekompositionsstrategie für den Logikentwurf auf der Basis funktions-typischer Eigenschaften. Dissertation thesis, Technical University Karl-Marx-Stadt, Germany.

Curtis, H. (1962). *A New Approach to the Design of Switching Circuits*. Princeton, USA: Van Nostrand.

Dresig, F. (1992). *Gruppierung – Theorie und Anwendung in der Logiksynthese*. Düsseldorf, Germany: VDI-Verlag.

Lang, C. (2003). *Bi-Decomposition of Function Sets Using Multi-Valued Logic*. Dissertation thesis, Freiberg University of Mining and Technology, Germany.

Le, T. Q. (1989). Testbarkeit kombinatorischer Schaltungen – Theorie und Entwurf. Dissertation thesis, Technical University Karl-Marx-Stadt, Germany.

Mishchenko, A., Steinbach, B. & Perkowski, M. (2001). An Algorithm for Bi-Decomposition of Logic Functions. In *38th Design Automation Conference*, 18–22. Las Vegas, USA.

Sasao, T. & Butler, J. (1997). On Bi-Decompositions of Logic Functions. In *International Workshop on Logic Synthesis*, 18-21. Lake Tahoe, USA

Sentovich, e. a. (1992). *SIS: A System for Sequential Circuit Synthesis*. University of California, Berkeley, California, USA. Technical Report UCB/ERI, M92/41, ERL, Dept. of EECS.

Somenzi, F. (2001). Binary Decition Diagram (BDD) Package: CUDD v. 2.3.1. University of Colorado at Boulder. URL http://vlsi.colorado.edu/~fabio/CUDD/cuddIntro.html.

Steinbach, B. & Hesse, K. (1996). Design of Large Digital Circuits Utilizing Functional and Structural Properties. In *2nd Workshop on Boolean Problems*, 23–30. Freiberg, Germany.

Steinbach, B. & Le, T. Q. (1990). Entwurf testbarer Schaltnetzwerke. Wissenschaftliche Schriftenreihe 12/1990, Technical University Chemnitz, Germany.

Steinbach, B., Schuhmann, F. & Stöckert, M. (1993). Functional Decomposition of Speed Optimized Circuits. In Auvergne, D. & Hartenstein, R. (eds.) *Power and Timing Modelling for Performance of Integrated Circuits*, 65–77. Bruchsal, Germany: IT Press.

Steinbach, B. & Stöckert, M. (1994). Design of Fully Testable Circuits by Functional Decomposition and Implicit Test Pattern Generation. In *12th IEEE VLSI Test Symposium*, 22–27.

Steinbach, B. & Wereszczynski, A. (1995). Synthesis of Multi-Level Circuits Using EXOR-Gates. In *IFIP WG 10.5 Workshop on Applications of the Reed-Muller Expansion*, 161–168. Chiba, Japan.

Steinbach, B. & Zakrevski, A. (1998). Three Models and Some Theorems on Decomposition of Boolean Functions. In *3rd International Workshop on Boolean Problems*, 11–18. Freiberg, Germany.

Steinbach, B. & Zhang, Z. (1997). Synthesis for Full Testability of Partitioned Combinational Circuits Using Boolean Differential Calculus. In *6th International Workshop on Logic and Synthesis*, 1–4. Granlibakken, USA.

Yang, M. C. (2000). BDD-Based Logic Optimization System. Technical Report CSE-00-1.

Evolutionary Approach to Quantum and Reversible Circuits Synthesis

MARTIN LUKAC[1], MAREK PERKOWSKI[1], HILTON GOI[1], MIKHAIL PIVTORAIKO[2], CHUNG HYO YU[1], KYUSIK CHUNG[1], HYUNKOO JEE[1], BYUNG-GUK KIM[1] and YONG-DUK KIM[1]

[1]*Department of Electrical Engineering and Computer Science, Korea Advanced Institute of Science and Technology, 373-1 Guseong-dong, Yuseong-gu, Daejeon 305-701, Korea;*
[2]*Department of Electrical Engineering, Portland State University, Portland, Oregon, 97207-0751, USA (E-mail: lukacm@ece.pdx.edu; mperkows@ee.kaist.ac.kr)*

Abstract. The paper discusses the evolutionary computation approach to the problem of optimal synthesis of Quantum and Reversible Logic circuits. Our approach uses standard Genetic Algorithm (GA) and its relative power as compared to previous approaches comes from the encoding and the formulation of the cost and fitness functions for quantum circuits synthesis. We analyze new operators and their role in synthesis and optimization processes. Cost and fitness functions for Reversible Circuit synthesis are introduced as well as local optimizing transformations. It is also shown that our approach can be used alternatively for synthesis of either reversible or quantum circuits without a major change in the algorithm. Results are illustrated on synthesized Margolus, Toffoli, Fredkin and other gates and Entanglement Circuits. This is for the first time that several variants of these gates have been automatically synthesized from quantum primitives.

Keywords: genetic algorithm, minimizing transformation, Quantum CAD, Quantum Logic Synthesis

1. Introduction

Quantum computing is a flourishing and very attractive research area (William and Clearwater 1998; Hirvensalo 2001; Nielsen and Chuang 2000). Inheriting properties from Quantum Mechanics, it allows theoretically to build computers much more efficient than the existing ones. For instance, certain problems non solvable in polynomial time in classical domain can be solved in polynomial time in quantum domain. Similarly, the complexity of other problems can be reduced while transforming them into the Quantum domain (Nielsen and Chuang 2000). Moreover, Quantum Circuits (QC) have an advantage of being able to perform massively parallel computations in one time step (William and Clearwater 1998; Hirvensalo 2001; Nielsen and Chuang 2000). The motivation to develop automated CAD tools for quantum circuits becomes recently quite high because according to the results of

201

Cirac and Zoller 1995; Monroe et al. 1997, 1996, Di Vincenzo 1995; Ekert and Jozsa 1996 such computers can be physically build and the progress of these realizations is fast. It is already possible to perform quantum-mechanically simple operations with trapped ions or atoms. Simplified but complete quantum circuits were constructed using Nuclear Magnetic Resonance technology (Nielsen and Chuang 2000). The state of the art in year 2002 is a 7 qubit quantum computer (Van Der Sypen et al. 2001). For this size of computer the problem of quantum circuit synthesis and optimization is not trivial and cannot be solved by hand so some kind of design automation becomes necessary. While quantum mechanics and quantum computing are established research areas, systematic design methods of quantum computers, and especially logic circuit design for such computers still remain only at the beginning stages of exploring available possibilities. It can be compared to the state of the art of logic design in 1940s when the first standard binary computers were built and the minimization algorithms like those from Quine and Shestakov started to appear. Currently, the works in quantum logic design are on: designing new universal gates and investigating their properties, creating methods of composing gates to circuits, and building elementary quantum circuits for practical applications, for instance arithmetics. New algebraic models for quantum logic and circuit design are also investigated. The Computer Aided Design of Quantum Circuits is even less developed and there exist only few programs to design such circuits automatically or with a limited user intervention. This paper is related to new approaches for CAD of Quantum Circuits, the new research and development area that we try to establish by our research (Lukac et al. 2002; Lukac and Perkowski 2002; Al-Rabadi et al. 2002; Khan et al. 2003; Al-Rabadi 2002; Perkowski et al. 2001, pp. 245–252, 2001, pp. 119–138; Mishchenko and Perkowski 2002; Khlopotine et al. 2002; Negotevic et al. 2002; Dill and Perkowski 2001; Lukac et al. 2003; Yang et al. 2003; Lukac et al., in preparation; Perkowski et al. 2002).

So far, analytic and search approaches have been used in quantum logic synthesis. They are based on matrix decomposition, local circuit transformations, mapping from various types of decision diagrams, spectral approaches, and on adaptations of several EXOR logic, Reed-Muller, and other classical combinational circuit design methods. Our approach combines some of them but fundamentally belongs to the group of Evolutionary Algorithms. The approach has been also used to a similar problem of Reversible Circuit (RC) Logic synthesis; such circuits can be realized in CMOS, optical and nano-technologies (Vieri et al.).

Genetic Algorithms (GAs) are one of the well-known Evolutionary Algorithm problem solving approaches to Soft Computing (Goldberg 1989).

Their use is very popular in problems with no identified structure and high level of noise, because: (1) a big problem space can be searched, (2) the size of this space can be moderated by parameters, (3) a variety of new solutions can be produced, and (4) with long enough time a circuit can be obtained that is close to the optimal one. These advantages make GAs useful in the initial phases of research and investigations of the design problem space. GAs are very good candidates to be used in logic synthesis of new types of circuits, investigating the usefulness of new gate types and new circuit structures. The special cases of Evolutionary Algorithms include: Genetic Algorithms, Evolutionary Programming, Evolution Strategies and Genetic Programming. So far, to the best of our knowledge, only two of the four Evolutionary Algorithm types have been applied to the QC or RC synthesis. Genetic programming was used to synthesize EPR (Einstein-Podolsky-Rosen) pairs of qubits (Rubinstein 2001), while a growing number of works uses a classical GA for QC and RC synthesis. For instance (Lukac et al. 2002; Lukac and Perkowski 2002; Ge et al. 1998; Williams and Gray 1999; William and Clearwater 1998; Yabuki and Iba 2000; Spector et al. 1999) used GA to evolve quantum and reversible gates and circuits, like the teleportation circuits. A reversible circuit is one that has the same number of inputs and outputs and is a one-to-one mapping between vectors of input and output values. Such circuits are related to quantum circuits. An attempt at a general approach to encode both Quantum and Reversible Circuits was presented in Lukac et al. (2002; Lukac and Perkowski 2002). It is known that every quantum circuit is reversible (Williams and Clearwater 1998; Hirvensalo 2001; Nielsen and Chuang 2000), so the researches on classical binary reversible synthesis and quantum synthesis share many ideas. The Reversible Logic (RL) circuits (Fredkin and Toffoli 1982; Bennet 1973; Kerntopf 2000) are already technologically possible and have been implemented in CMOS technology (Vieri et al.). Thus, some of our results below are applicable also to such circuits. As described later, most of gates with more than one input used in QC are derived and originate from RC. Although this work is concerned with one approach to automated synthesis, it is important to notice that a parallel research on new gates is also explored by (Al-Rabadi et al. 2002; Kerntopf 2000; Smolin and DiVincenzo 1996; Miller 2002; Miller and Dueck 2003; Dueck and Maslov 2003; Iwama et al. 2002; Shende et al. 2002; Barenco et al. 1995; Khan et al. 2003; Al-Rabadi 2002; Kim et al. 2000; Lee et al. 1999; Price et al. 1999). The search of new Quantum gates (QG) and Reversible gates (RG) has two different aspects: invention and generalization. The invention of new gates is mainly aimed toward an optimization of a particular design in a specific technology or towards inventing new universal gates. The generalization approach is the use of already known gates and extending them to new gates of certain preferable properties. Usually there are also new particular synthesis methods that come together with proposed new gates (Perkowski et al. 2001, pp. 245–252, 2001, pp. 119–139; Mishchenko and Perkowski 2002; Khlopotine et al. 2002; Kerntopf 2001; Lukac et al., 2003; Lukac et al., in preparation; Perkowski et al. 2002).

This paper is organized as follows. Section 2 presents the minimal background in quantum circuit design necessary to understand the paper. Examples are used to illustrate the most important notions. Section 3 describes the general problem of synthesis of quantum circuits (called also quantum arrays) from primitive quantum gates, especially using evolutionary algorithms. Section 4 presents decomposition of gates to smaller primitives related to the cost of these gates. More realistic cost function for gates are one of main innovations of this paper. Section 5 presents our entire minimization methodology for quantum arrays synthesis. The sixth section describes local optimizing transformations used in the post-processing stage of GA. Section 7 presents our variant of GA and its settings for this work. Section 8 gives more details on the most important aspect: the fitness function design for GA and its role. Section 9 presents other aspects of the Genetic Algorithm that we applied. Section 10 describes experimental results and section 11 discusses issues and advantages of this approach as well as our current and future research and open questions. Section 12 concludes the paper.

2. Fundamentals of Quantum Logic

In quantum computation *quantum bits* (*qubits*) are used instead of classical binary bits to represent information. These information units are derived from the states of micro-particles such as photons, electrons or ions. These states are the basis states (basis vectors, eigenstates) of the computational quantum system. Assume an electron with two possible spin rotations: $+1/2$ and $-1/2$. Using Ket notation (Dirac 1930; von Neumann 1950) these distinguishable states will be represented as $|0>$ and $|1>$, respectively. Each particle in a quantum domain is represented by a wave function describing it as having both properties of a wave and of a particle as introduced by (Feynman 1996; Dirac 1930; von Neumann 1950; Einstein et al. 1935). Based on these properties, quantum computation inherited the powerful concept of *superposition of states*. Assume a particle p1 be represented by a wave function $\psi_1 = \alpha_1|0>$ $+ \beta_1|1>$. The coefficients α and β are complex numbers called the eigenvalues. They must be in general complex because only having complex values in wave functions allows to eliminate themselves in order to satisfy some experimentally observed properties of quantum world, such as for instance in the Two-Slit experiment (Klay 1988). It is not our goal here to explain such experiments or formulate fundaments of quantum mechanics (The reader can find information in literature, especially (Nielsen and Chuang 2000)). Our goal is only to introduce the formal calculus of quantum mechanics in order to explain the basic concepts of our CAD algorithms and especially the genetic algorithm for quantum circuit synthesis. Similarly as an engineer who has no understanding of electronic circuits, but knows only Boolean Algebra, is able to develop logic circuits synthesis software, one with no understanding of quantum phenomena and physics will be able to develop quantum CAD if he will only learn some fundamental quantum notation and associated algebraic properties and transformations. To teach these to the Reader is one of the goals of this paper.

The interpretation of wavefunction $\psi 1$ is that $|\alpha_1|^2$ is the probability of the particle being measured in state $|0>$ and $|\beta_1|^2$ is the probability of that particle being measured in state $|1>$. Thus, the measurement or observation process transforms the quantum world of complex wavefunctions to the macro-world of events that occur with standard probabilities. The superposition of states is represented by these facts: (1) each of these probabilities can be non-null, (2) $|\alpha_1|^2 + |\beta_1|^2 = 1$, and (3) if another particle p_2 with a wavefunction $\psi_2 = \alpha_2|0> + \beta_2|1>$ is added to the system, then the resulting wavefunction will be $|\psi_1\psi_2> = \alpha_1\alpha_2|00> + \alpha_1\alpha_2|01> + \alpha_1\beta_2|10> + \alpha_1\alpha_2|11>$. The system can be in any possible state of the wavefunction and will collapse to one of the eigenstates when measured (Wheeler 1983). The space of quantum computing is thus much larger than in classical computing, which causes that efficient algorithms for synthesis and analysis of quantum circuits are more difficult to develop. However, certain similarities exist which will be useful for us to explain the synthesis issues, especially to the readers with a digital design engineering background. The equations introduced above are the result of the Kronecker product (Graham 1981) on matrices of elementary quantum gates that operate "in parallel". The Kronecker Product of Matrices is defined as follows:

$$\begin{bmatrix} a & b \\ c & d \end{bmatrix} \otimes \begin{bmatrix} x & y \\ z & v \end{bmatrix} = \begin{bmatrix} a\begin{bmatrix} x & y \\ z & v \end{bmatrix} & b\begin{bmatrix} x & y \\ z & v \end{bmatrix} \\ c\begin{bmatrix} x & y \\ z & v \end{bmatrix} & d\begin{bmatrix} x & y \\ z & v \end{bmatrix} \end{bmatrix} = \begin{bmatrix} ax & ay & bx & by \\ az & av & bz & bv \\ cx & cy & dx & dy \\ cz & cv & dz & dv \end{bmatrix}$$

Mathematically, it is the Kronecker Product operation that allows the quantum logical system to grow dimensionally much faster than in classical logics. Observe that in a quantum system n qubits represent a *superposition* of 2^n states while in a classical system n bits represent only 2^n *distinct states*. Operations over a set of qubits are defined as matrix operations and map the physical states into the *Hilbert space* (von Neumann 1950; Hirvensalo 2001; Nielsen and Chuang 2000). The concept of Hilbert space will be used below only in the most elementary way necessary to understand the calculations. States of the wave function are eigenvalues of this space. Each matrix-operator represents a modification to the complex coefficients of the wave function. The new coefficients result in the modification of probabilities to find the system in a particular basic state. But we do not have to worry about classical probabilities in this paper since the operation of a quantum circuit is purely deterministic, so we will always deal with eigenvalues (complex probability) rather than standard probabilities in circuit design. Consequently a quantum gate will be a matrix having for input the vector of complex coefficients of the waveform and producing a vector of complex coefficients as the output. An illustrative example can be seen in equation 1.

$$\begin{bmatrix} a & b \\ c & d \end{bmatrix} \begin{bmatrix} \alpha|0> \\ \beta|1> \end{bmatrix} = \begin{bmatrix} a\alpha|0> + b\beta|1> \\ c\alpha|0> + d\beta|1> \end{bmatrix} \quad (1)$$

where a, b, c, and d are complex coefficients of the matrix indicating the (complex) probability to transit from one state to another, and $\alpha|0>$, $\beta|1>$ are the (complex) wavefunction coefficients to be propagated through the matrix-operator. Less formally, we can think about this matrix as a quantum equivalent of a classical gate such as AND which transforms its input states into output states. The designer of quantum algorithms has to deal with standard probabilities, but the designer of quantum circuits, which is our interest here, deals only with operations in quantum world because his input problem is described in such a way.

Assume j to be the square root of -1. Let us denote by U^+ a hermitian matrix of matrix U, which means the complex conjugate of the transposed matrix U (the matrix U is first transposed and next each of its complex numbers is replaced with its conjugate, thus $a - jb$ replaces $a + jb$). We say that gate U is unitary when $U * U^+ = I$, where I is an identity matrix. It can be shown that because the probabilities must be preserved at the output of the quantum gate, all matrices representing quantum gates are unitary. Thus every quantum gate, block and the entire circuit is described by a unitary matrix. Every quantum gate has therefore exactly one inverse matrix – quantum computing is reversible; quantum gate matrices represent logically reversible gates. Some of those gates are exactly the same as in classical reversible computing, which allows to use some results of binary reversible computing in quantum computing. While in general the coefficients in unitary matrices of quantum circuits are complex numbers, there are some special and important gates for which unitary matrices are just *permutation matrices* (Let us recall that a permutation matrix has exactly one "1" in every row and in every column and all other entries are zeros – it describes therefore an input-output permutation of value vectors). The gates whose unitary matrices are permutation matrices are called *permutation gates* and they correspond to gates of classical reversible logic. There exist, however, other important gates whose unitary matrices are not permutation matrices. Rotational operators (gates) and gates such as Hadamard gate (denoted by H) and Square Root of Not Gate (denoted by V) belong to this second category which we will call "*truly quantum primitives*". These gates are responsible for superposition, entanglement and all peculiarities of quantum computing, although they may constitute only a small fraction of gates in a quantum circuit. A Hadamard gate is an example of a gate that has a unitary matrix which is not a permutation matrix. Here are some useful matrices:

$$Hadamard = \frac{1}{\sqrt{2}}\begin{bmatrix} 1 & 1 \\ 1 & -1 \end{bmatrix}, Pauli\text{-}X = NOT = \begin{bmatrix} 0 & 1 \\ 1 & 0 \end{bmatrix},$$

$$Pauli\text{-}Y = \begin{bmatrix} 0 & -j \\ j & 0 \end{bmatrix}, Pauli\text{-}Z = \begin{bmatrix} 1 & 0 \\ 0 & -1 \end{bmatrix},$$

$$Phase = \begin{bmatrix} 1 & 0 \\ 0 & j \end{bmatrix}, V = \sqrt{NOT} = \frac{1+j}{\sqrt{2}}\begin{bmatrix} 1 & -j \\ -j & 1 \end{bmatrix}$$

We will denote these gates by H, X, Y, Z, S and V, respectively. Only X is permutative. It can be easily checked by the reader that multiplying the matrix of gate V by itself produces the matrix of Pauli-X which is an inverter or NOT gate. The reader can find interesting identities by multiplying these matrices by themselves and also by their inverse matrices. Such identities in quantum logic are used to simplify the quantum circuits that come directly from evolutionary algorithms (section 6). They play a role analogous to Boolean algebra identities such as De Morgan used in classical logic synthesis.

Below we will pay special attention to circuits composed of only permutation gates (these are the so-called *permutation circuits*, because their unitary matrices are permutation matrices). An example of a permutation gate is the Feynman gate (called also $CNOT$ gate or Quantum XOR). This gate can be described by the following logic expressions: (A, B) => (P, Q) = (A, A⊕B), see Figure 1a. This equation means that the output bit P is just the same as its input A, while on the output of the second wire Q the operation A⊕B is performed. This operation (EXOR of A and B) is a standard logic operation. Sometimes notation A' = P and B' = Q is used (Figure 1c, d, e) (In other papers notation A' = A, and B' = B is used to underlie the fact that the result occurs on the same quantum wire A or B). These expressions are also related to the quantum matrix. For instance, the permutation matrix and the equations from Figure 1c describe the Feynman operator from Figure 1a, b. The careful reader may verify that this is indeed a unitary matrix and that it satisfies the definition of a permutation matrix as well. Observe that each of its rows and columns corresponds to one of possible states of input and output values: 00,

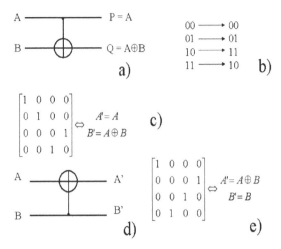

Figure 1. Feynman gates: (a) the circuit in quantum array notation, (b) the transformation between states executed by this gate, (c) the unitary matrix with binary enumeration of rows and columns and the corresponding Boolean equation of outputs, (d) Feynman EXOR up circuit schematics, (e) its unitary matrix and Boolean equations.

01, 10, and 11. The binary states are encoded to natural numbers as follows: $00 = 0, 01 = 1, 10 = 2, 11 = 3$. We will use natural numbers to address rows and columns in matrices. Let us observe that when $A = 0$ and $B = 0$, then $A' = B' = 0$. Thus, input combination 00 is transformed to output combination 00. This is represented by a value of "1" at the intersection of row 0 and column 0 of the matrix. Similarly, input combination 11 is transformed to output combination 10, which is represented by a value of "1' at the intersection of row 3 and column 2 (Figure 1c). Another variant of Feynman gate is in Figure 1d, e.

Figure 2 presents another permutation gate called a Swap gate. Observe that this gate just swaps the wires A and B, since input state 01 is transformed to output combination 10, input vector 10 is transformed to output vector 01 and the remaining combinations are unchanged (Figure 2a). Swap gate is not necessary in classical CMOS reversible computing, where it is just a crossing of connecting wires that can be for instance in two layers of metallization. However, in quantum computing technology like NMR, every unitary matrix other than identity is realized by NMR electromagnetic pulses so its realization has a cost. High cost of swap gates in quantum realization is one of main differences between quantum and reversible circuit synthesis (in reversible computing these gates are free). Thus quantum circuit design covers some aspects of not only logic synthesis but also physical design (placement and routing) of standard CAD.

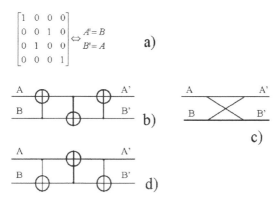

Figure 2. Swap gate: (a) unitary matrix and corresponding Boolean equations, (b) realization using Feynman gates, (c) a schematic, (d) another realization using two Feynman gates and one Feynman EXOR up gate.

Let us introduce now few more gates that we will use. There are several 2∗2 gates in binary reversible logic and they are all linear. A linear gate is one that all its outputs are linear functions of input variables. Let a, b, and c be the inputs and P, Q and R the outputs of a gate. Assuming *2∗2 Feynman gate*, $P = a$, $Q = a \oplus b$, when $a = 0$ then $Q = b$; when $a = 1$ then $Q = \neg b$. (Sometimes the negation of a is also denoted by a′). With $b = 0$ the 2∗2 Feynman gate is used as a *fan-out (or copying) gate*. It can be shown that a swap gate can be built as a cascade of three Feynman gates – Figure 2 (the reader can check it by multiplying matrices or by transforming Boolean equations of gates). Figure 2b shows realization of Swap gate with three Feynman gates, and Figure 2c its schematics. Observe that there is also another realization of the Swap gate (Figure 2d). Circuits from Figure 2b and Figure 2d are then called *equivalent* or *tautological*. Every linear reversible function can be built by composing only 2∗2 Feynman gates and inverters. There exist 8! = 40,320 3∗3 reversible logic gates, some of them with interesting properties, but here we are not interested in types of gates but in synthesis methods using *arbitrary* permutation gates, so we will restrict ourselves to only few gate types (many other gate types have been defined in our software). There exist two well-known universal 3∗3 reversible gates: Fredkin gate (Fredkin and Toffoli 1982) and Toffoli gate. Toffoli gate is also called 3∗3 Feynman gate or Controlled-Controlled-NOT (Figure 3). The 3∗3 Toffoli gate is described by these equations:

$$P = a, \ Q = b, \ R = ab \oplus c.$$

Toffoli gate is an example of *two-through* gates, because two of its inputs are given to the output. Similarly, the concept of *k-through gates* can be

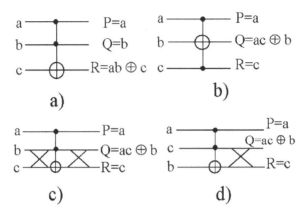

Figure 3. Toffoli gates: (a) a schematic of Toffoli gate (EXOR down), (b) Toffoli gate (EXOR middle), (c) realization of schematics from Figure 3b using the Toffoli EXOR down and two Swap gates, (d) realization of Toffoli EXOR middle with permuted input order.

introduced, as well as the concept of k∗k Toffoli Gates. In general, for a reversible gate with n inputs and n outputs, the matrix is of size $2^n * 2^n$.

The 3∗3 *Fredkin gate* (Figure 4) is described by these equations:

$$P = a, \ Q = if \ a \ then \ c \ else \ b, \ R = if \ a \ then \ b \ else \ c.$$

As we see, in terms of classical logic this gate is just two multiplexers controlled in a flipped (permuted) way from the same control input *a*. The symbol notation for a 3∗3 Fredkin Gate is shown in Figure 4a. Qubit *a* is the controlling qubit. A classical schematics of this gate that uses multiplexers is shown in Figure 4b. As we see, the 3∗3 Fredkin gates are permutation gates, they permute the data inputs *b*, *c* of the multiplexers under control of the control input *a* of these multiplexers that is also outputted from the gate. The fact that output *P* replicates input *a* is very useful because it allows to avoid fanout gates for the same level of the circuit (let us recall that fanout is not allowed in reversible logic). Copying of arbitrary signals is not allowed in quantum circuits (no cloning principle) (Nielsen and Chuang 2000), so the replication using Feynman gates can be applied only to basis states. Figure 4c presents a Fredkin gate controlled with qubit *c*. In Figure 4d qubit *b* is the controlling qubit. Realization of Fredkin gates using Toffoli and Feynman gates is shown in Figure 5.

Fredkin gates are examples of what we define here as *one-through gates*, which means gates in which one input variable is also an output.

Figure 6 presents the well-known realization of Toffoli gate from (Smolin and DiVincenzo 1996). According to the figure in right (Figure 6b), there are five 2-qubit primitives. Let us explain how the quantum primitives cooperate

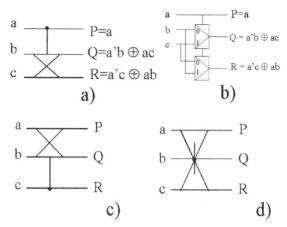

a) **b)**

c) **d)**

Figure 4. Fredkin gates: (a) Schematics with controlled qubit a, (b) the classical schematics of this gate using multiplexers, (c) Fredkin gate controlled by qubit c (Fredkin Down), (d) Fredkin gate controlled by qubit b (Fredkin middle).

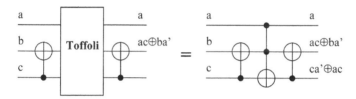

Figure 5. Fredkin gate realization using one Toffoli and two Feynman gates. (a) Toffoli shown schematically, (b) Toffoli with EXOR-down schematics.

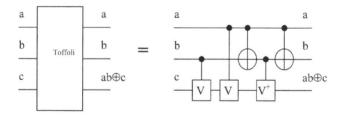

Figure 6. Toffoli gate synthesized using only 2-qubit primitives as proposed by Smolin.

to create correct gate behavior. The non-permutation matrices of quantum primitives are composed to create permutation behavior of the entire gate. Observe that Controlled-V and Controlled-V^+ quantum primitives are used. V is the unitary matrix called Square-Root-of-Not, thus a serial connection of gates V realizes an inverter: $V*V = NOT$. Matrix V^+ is the hermitian of V. As we remember from the definition of hermitian, we have that $V*V^+ = I$ which is the identity matrix (describing a quantum wire). The Controlled-U

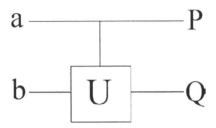

Figure 7. A general-purpose Controlled Gate. When a = 0 then Q = b, when a = 1, then Q = U(b).

gate (Figure 7) works as follows: when the controlling signal *a* (upper signal in gate Controlled-U) is 0, the output Q of the lower gate repeats its input *b*. When the controlling signal *a* is 1, the output of the lower gate is the operator of this gate (matrix U) applied to its input *b* (Figure 7). Therefore, in case of the Controlled-V gate, the output is operator V applied to the input. In case of Controlled-V^+ gate, the output is operator V^+ applied to the input of the gate. The reader can verify in Figure 6b that when *a* = 1, *b* = 0, the operation $V*V^+ = I$ is executed in line *c*, thus the output of this gate repeats the value of *c*. When *a* = 1 and *b* = 1, the operation $V*V = NOT$ is executed in line c, thus the lowest output is a negation of input *c*. Which is the same as *ab*⊕*c* executed on the corresponding qubit in Toffoli (Figure 6a). Other output values are also the same as in the Toffoli gate equations. The reader is asked to analyze the operation of this gate on the remaining input combinations to verify its correctness. Let us also remember that Feynman gate has two variants. The EXOR gate can be on an upper or on a lower wire. Similarly, there are three 3-qubit Toffoli gates, EXOR-up, EXOR-down and EXOR-middle (Figure 3). Observe that these gates have all different unitary matrices so they are formally different quantum gates.

3. Evolutionary Approaches to Synthesis of Quantum Arrays

There are basically two methods of designing and drawing quantum circuits. In the *first method* you draw a circuit from gates and you connect these gates by standard wires. This is like in classical circuit design, but *your gates are reversible or quantum gates*. The rules to design a reversible circuit using this approach are the following: (1) no loops allowed in the circuit and no loops internal to gates, (2) fan-out of every gate is one (which means, every output of a gate can be connected only to one input terminal). These rules preserve the reversible characteristic of gates thus the resulting circuit is also completely reversible. Next, when you draw the circuit, the gates are placed on a 2-dimensional space and the connections between them are routed. Every crossing of two wires in the schematics is replaced with the quantum Swap gate making the schematics planar, which means, no more two wires intersect in it. The schematics is thus rewritten to a quantum array notation illustrated below. It is relatively easy to transform a quantum array to its corresponding unitary matrix, as will be illustrated in the sequel. Kronecker and standard matrix multiplications are used in this transformation. A unitary matrix of a parallel connection of gates A and B is the Kronecker product A$\oplus$B of matrices A and B. A unitary matrix of a serial connection of gates A and B is the standard matrix product A$*$B of matrices A and B. Transformation of quantum array to its unitary matrix is done for analysis or verification and is a part of our evolutionary algorithm. The approaches that use this first design method of reversible circuits are closer to those of the classical digital CAD where the phases of logic synthesis and physical (geometrical) design are separated.

The *second design method for quantum circuits* is to synthesize directly the quantum array of a circuit that was initially specified by a unitary matrix. This is done without involving additional graph-based or equation-based representations. The synthesis is done by one of two approaches:

(a) composing matrices of elementary gates in series or in parallel until the matrix of the entire circuit becomes the same as the specification matrix,

(b) decomposing the specification matrix of the entire circuit to parallel and serial connections of unitary matrices until all matrices correspond to matrices of elementary gates directly realizable in the given technology. For simplification, from now on we will talk interchangeably about a gate and its unitary matrix.

In another synthesis variant, called the approximate synthesis, it is not required that the circuit specification matrix and the matrix of composed gates are exactly the same. They can differ by small allowed values or/and differ in some matrix coordinates only.

The synthesis problem of a quantum array has therefore a simple formulation as a *composition* or *decomposition* problem, but practical realization of synthesis algorithms for quantum circuits is extremely hard. Observe that even in the case of standard (irreversible) binary logic these composition and decomposition approaches are computationally very difficult and it is only very recently that they are becoming practical in CAD software, because of their high computational complexity. In the case of quantum logic the situation is much more difficult, because of the fast increase of data sizes and because so far the mathematics of description transformations is limited, heuristics are not known and there are no counterparts in quantum logic of such familiar notions as Karnaugh Maps, prime implicants or reductions to covering/coloring combinatorial approaches. Therefore most authors turned to evolutionary algorithms as the fast prototyping methods for quantum arrays (Lukac et al. 2002; Lukac and Perkowski 2002; Ge et al. 1998; Rubinstein 2001; Williams and Gray 1999; Yabuki and Iba 2000; Spector et al. 1999). These approaches seem to be good for introductory investigations of the solution space and its properties, with the hope that by analyzing solutions we will learn more about the search space and ultimately create more efficient algorithms. As we will see below, this phenomenon actually happened in case of our research.

In the evolutionary computation approach to quantum synthesis, two key elements strongly influence the convergence of the search. These are: (1) the evaluation function (called also the *fitness function*), and (2) the *encoding* of the individuals (Goldberg 1989). Genetic programming uses the encoding of data to trees so the operations on them are very time-consuming (Koza 1992). Here we propose a new way to calculate costs of gates and circuits as a part of the fitness function. Also proposed is a new method of QC and RL circuit encoding in a GA. Since we use the same encoding for QC and RL and for many variants of algorithms, design parameters and cost functions, we can easily compare respective synthesis results under strictly the same conditions. This has been not accomplished by previous authors.

The synthesis of quantum logic circuits using evolutionary approaches has two fundamental aspects.

First, it is necessary to find the circuit that either (A) exactly corresponds to the specification, or (B) differs only slightly from the specification. Case (A) is verified by a tautology of the specification function and solution function. In case of a truly quantum circuit this is done by a comparison of unitary matrices. In case of permutation functions this can be also done by comparing the truth tables. Observe that non-permutation matrices cannot be represented by truth tables which leaves the representation of unitary matrices as the only canonical function representation. This representation is responsible for less efficient tautology verification during fitness function calculations, which considerably slows down the software. Case (B) for permutation circuits is verified by an incomplete tautology (tautology with accuracy to all combinations of input values and with arbitrary logic values for don't care combinations). In some applications such as robot control or Machine Learning it is sufficient that the specification and the solution are close, like, for instance, differing only in few input value combinations.

Second fundamental aspect of quantum logic synthesis is that the size (cost) of the circuit has to be minimal, in order to allow the least expensive possible quantum hardware implementation (like the minimum number of electromagnetic pulses in NMR technology). The cost differs for various technologies, but some approximate costs functions will be formulated here that are still more accurate than those previously proposed in the literature.

Consequently, the fitness function of the GA should take into account both the above-mentioned aspects, which means the GA should be designed so as to minimize both the distance from the specification and the final circuit cost simultaneously. Sometimes we are not interested only in those solutions that meet the specification *exactly*. Usually we are not interested in the solution that would have the *exact minimal cost*. Multiple designs of a GA are tested here and their results are compared in order to select the minimum cost solution. This approach not only improves the cost of any solution found, but also helps us to develop gradually better generations of our evolutionary/algorithmic/heuristic programming tools for QC CAD.

In our software the fitness function is optimized for the minimal cost, speed of synthesis or precision of synthesis. In case of designing a reversible circuit, the fitness function can be also optimized for the speed of synthesis, the cost of the circuit or for the total number of wires (insertion of a minimum number of constants). These two methods of optimization for each technology and logic type allow to formulate precise constraints for the synthesis method in use.

In this paper we derive more precise cost functions for gates to be used in the optimization algorithm. We will follow the footsteps of some previous papers in quantum computing and we will realize all gates from 1∗1 gates (called also 1-qubit gates) and 2∗2 gates (i.e. 2-qubit gates). Moreover, according to (Smolin and DiVincenzo 1996) we will assume that the cost of every 2∗2 gate is the same. Although our methods are general, to be more specific we assume Nuclear Magnetic Resonance (NMR) quantum computers (Price et al. 1999; Kim et al. 2000, p. 022312, 2000, p. 032312; Lee et al. 1999; Price et al. 1999), for which we approximately assume costs of all 1-qubit gates to be 1, and 2-qubit gates to be 5. Thus, every quantum gate (particularly, every permutation quantum gate) will be build from only 1-qubit and 2-qubit primitives and its cost respectively calculated as the total cost of these primitives used in the circuit. Obviously, this approach can be used for both QC and RC, since in evolutionary approaches we can always arbitrarily restrict the gates used for synthesis, and in case of RC only a subset of permutation gates can be selected.

4. Costs of Quantum Gates

An important problem, not discussed so far by other authors, is the selection of the cost function to evaluate the QC (or RC) designs. Although the detailed costs depend on any particular realization technology of quantum logic, so that the cost ratios between for instance Fredkin and Toffoli gates can differ in NMR and ion trap realizations, the assumptions used in several previous papers (Lukac and Perkowski 2002; Ge et al. 1998; Price et al. 1999; Rubinstein 2001; Williams and Gray 1999; Yabuki and Iba 2000); that each gate costs the same, or that the cost of a gate is proportional to the number of inputs/outputs, are both far too approximate. These kinds of cost functions do not provide any basis for the synthesis optimization and as shown by us (Lukac et al. 2002; Lukac and Perkowski 2002) can lead to quantum sequences for NMR that are far from the minimum. In this section some of the well-known decompositions of known 3-qubit quantum permutation gates, as well as the new gates, will be discussed to illustrate this problem.

The cost of quantum gates in synthesis methods can be seen from as many points of view as there are possible technologies. Previous researches used specific cost to minimize the number of wires per gate, the number of gates in the circuit or other task-specific criteria. Assume a random function to be synthesized in QL and defined over a set of basic variables a, b, c and d. This function will use 4 qubits and the unitary matrix representation of the circuit will be of size 2^4*2^4. The matrix representation can be also called the *evolution matrix* of the system (Al-Rabadi et al. 2002) but this name has nothing to do with evolutionary synthesis algorithms. Defined as quantum evolutionary processes, there is theoretically an infinity of possible quantum gates. In the present work only well-known and widely used unitary quantum gates, as those listed above, will be used. The user of our software can however very easily extend the system by defining arbitrary gates as 256-character symbols together with their unitary matrices and natural numbers for costs. This applies also to multi-qubit gates and multi-valued quantum gates (Lukac et al. 2003).

Based on NMR literature (Kim et al. 2000, p. 022312; 2000, p. 032312; Lee et al. 1999; Price et al. 1999) we assume the 1-qubit gates to have a cost of 1 and the 2-qubit gates to have the cost of 5. This leads to 3-qubit gate costs of about 25. From this cost attribution it is evident that the optimization should be aimed toward the use of gates with as few qubits as possible. The cost of the circuit is the total cost of gates used. Of course, especially using GAs, the search might not always be successful but a solution will have always two evaluation criteria: the final circuit error and the final cost calculated after the optimizing "local equivalence" transformations being applied to the initial circuit produced by a GA. (Sometimes we use in GA more precise gate costs if they are known from the literature, but the costs given above will be used for illustration purposes here).

In the *ideal case* the GA will find an optimal circuit; i.e. the circuit with the smallest possible cost and with the correct matrix representation. Two *worse*

Table 1. Illustration of possible results using a combination of cost function per gate C_g and a global evaluation cost function C_t. The column in the middle indicates if the searched function was found or not found

Cost of the result in gates C_g	Function found	Total cost C_t
1. Min $\leq C_g$	Yes	Min $\leq C_t$
2. Min $< C_g <$ Max	Yes	Min $< C_t <$ Max
3. $C_g \leq$ Max	Yes	Min $< C_t <$ Max
4. Min $\leq C_g$	No	Min $< C_t <$ Max
5. Min $< C_g <$ Max	No	Min $< C_t <$ Max
6. $C_g \leq$ Max	No	$C_t \leq$ Max

alternatives are: (1) a circuit with a higher cost than the minimum but still with a correct matrix, (2) a circuit with not completely correct matrix but with a smaller cost. For future use let's refer to these three groups of alternatives as C_o (optimal cost), C_a (average cost) and C_w (worst cost), respectively. In the case of RL circuit synthesis the differences in results can occur on one more level: as mentioned before, input constants insertion is allowed in RL synthesis. However, we want to minimize the number of such constants in order to reduce the circuit's width. Once again, as will be shown below, this parameter can be also taken into account by our software as part of the cost function. Consequently, one can obtain a correct circuit with a small cost of gates but with a higher number of wires.

Table 1 systematizes possible results of a Quantum Circuit synthesis. In the table possible search solutions are characterized with respect to the combinations of the two cost functions.

The cost C_g is the cost of the circuit based on the sum of costs of all gates. C_t is the total cost where the correctness of the result is also included, based on one of the previously explained verification methods between the specification and the obtained circuit descriptions. As will be seen, both of these two cost functions can be handled in order to accelerate or moderate the speed of convergence of the entire synthesis process. A closer look at the table will reveal three categories of results as previously described. These are circuits with small, average and high total costs, respectively. These types can be again separated in two, depending whether the final solution was found or not. Groups 1, 2, and 3 are for exact design, groups 4, 5, and 6 for approximate. Groups 1 and 4 correspond to C_o, Groups 2 and 4 to C_a, and groups 3 and 6 to C_w. The category with the most circuits is the one with an average cost, and as will be seen later it is this category that needs particular attention while selecting and adapting a fitness function for a synthesis problem. Min and Max in the table are some parameters set by the user and based on experience.

Now let us continue the discussion of the circuit decomposition into primitives. Let us consider one practical example. The Toffoli or Fredkin gates introduced in section 2 are both universal quantum logic gates that are well-known. They have been built in several quantum and reversible technologies. The problem is to find an optimal decomposition of the universal gates into smaller parts, especially the directly realizable quantum primitives such as Feynman, NOT or Controlled-V (C_V) gates. As mentioned earlier, the gates with one and two qubits have costs 1 and 5, respectively. Consequently the assumption of using only 1-qubit and 2-qubits gates will be observed since only such gates are *directly realizable in quantum hardware*. Other gates are only abstract concepts useful in synthesis and not necessarily the best gates for any particular quantum technology (like NMR in our case). Figure 6 presented the well-known realization of Toffoli gate from (Smolin and DiVincenzo 1996). According to Figure 6b there are five 2-qubit primitives, so the cost is $5*5 = 25$.

Now that we learned in section 2 how Toffoli gate works internally based on the operation of its component quantum primitives, we will realize the Fredkin gate from the Toffoli gate. Using a GA (Lukac et al. 2002; Lukac and Perkowski 2002) or the synthesis method from this paper, we can synthesize the Fredkin gate using two Feynman gates and one Toffoli gate as in Figure 5. The cost of this gate is $2*5 + 25 = 35$.

Substituting the Toffoli design from Figure 6b to Figure 5 we obtain the circuit from Figure 8a. Now we can apply an obvious EXOR-based transformation to transform this circuit to the circuit from Figure 8b. This is done by shifting the last gate at right (Feynman with EXOR up) by one gate to left. The reader can verify that this transformation did not change logic functions realized by any of the outputs. Observe that a cascade of two 2*2 gates is another 2*2 gate, so by combining a Feynman with EXOR-up gate (cost of 5), followed by controlled-V gate (cost of 5) we obtain a new gate C_V with the cost of 5. Similarly gate C_V^+ with cost 5 is created. This way, a circuit from Figure 8c is obtained with the cost of 25 (This transformation is based on the method from (Smolin 1996) and the details of cost calculation of C_V and C_V^+ are not necessary here). Thus, the cost of Toffoli gate is exactly the same as the cost of Fredkin gate, and not half of it, as previously assumed and as may be suggested by classical binary equations of such gates.

Encouraged with the above observation that sequences of gates on the same quantum wires have the cost of only single gate on these wires, we used the same method to calculate costs of other well-known gates. Let us

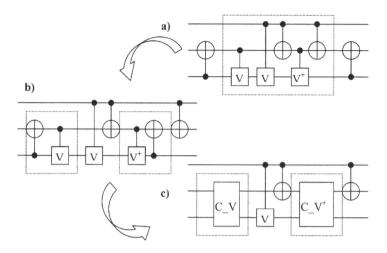

Figure 8. Example of reducing cost in the Fredkin gate realized with quantum primitives. Gates C_V and C_V$^+$ in Figure 8c are created by combining respective quantum primitives from Figure 8b which are shown in dotted lines.

first investigate a function of three majorities investigated by Miller (Miller 2002; Miller and Dueck 2003; Yang et al. 2003). This gate is described by equations: P = ab⊕ac⊕bc, Q = a′b⊕a′c⊕bc, P = ab′⊕ac⊕b′c. Where a′ is a negation of variable a. Function P is a standard majority and Q, R are majorities on negated input arguments a and b, respectively (Yang et al. 2003). We realized this function with quantum primitives, found it useful in other designs and thus worthy to be a stand-alone quantum gate. We call it the Miller gate (Yang et al. 2003). As seen in Figure 9a, the Miller gate requires 4 Feynman gates and a Toffoli gate, which would suggest a cost of 4∗5 + 25 = 45. However, performing transformations as in Figure 9b, we obtain a solution with cost 35. Another solution obtained by the same method has cost 35 and is shown in Figure 9c. It is also based on simple EXOR transformation (x⊕y)⊕ z = (x⊕z)⊕y applied to three rightmost Feynman gates from Figure 9a, with EXOR in the middle wire y. (More on optimizing equivalence-based transformations in section 6). Again, the Miller gate, based on its binary equations, looks initially much more complicated than the Toffoli gate, but a closer inspection using quantum logic primitives proves that it is just slightly more expensive.

5. Our Entire Minimization Methodology of Quantum Arrays Synthesis

Based on examples from section 4, let us observe that a new permutation quantum gate with equations:

$$P = a,$$
$$Q = a \oplus b,$$
$$R = ab \oplus c$$

can be realized with cost 20. It is just like a Toffoli gate from Figure 6b but without the last Feynman gate from the right. This is *the cheapest quantum realization* known to us of a complete (universal) gate for 3-qubit permutation gate (Figure 10). It is thus worthy further investigations. We found that the equation of this gate was known to Peres (Peres 1985), but it has not been used in practical designs of quantum or reversible circuits. We propose here to consider the Peres gate as a base of synthesis and transforms, similarly as 2-input NAND gates are used in technology mapping phase of classical Boolean CAD.

Observe that assuming the availability of Peres gate, the algorithms from literature (for instance Iwama et al. 2002; Shende et al. 2002) will not lead to a quantum array with minimum quantum costs formulated as above. When given the unitary matrix of the Peres gate, these algorithms would return the solution of cost 30 composed from one Toffoli gate and one Feynman gate, which would lead to clearly non-minimal electromagnetic pulse sequence. Thus, improved synthesis algorithms should be created to minimize the realistic quantum gate costs introduced in (Smolin and DiVincenzo 1996) and in this paper. Observe please also, that if a post-processing stage were added to the algorithms from (MIller 2002; Miller and Dueck 2003; Dueck and Maslov 2003; Iwama et al. 2002; Shende et al. 2002) (or to the result of the GA from section 7), then the optimum solution (of a single Peres gate and cost of 20) would be found for this example. Therefore our entire synthesis and optimization approach to quantum arrays is the following.

1. First create the circuit by some approximate synthesis method (evolutionary algorithm in our case),
2. Apply macro-generation of complex gates to quantum primitives,
3. Apply the optimizing equivalence-based transformations.

The application of this procedure to our example will have the following stages:

(a) the initial solution is a Toffoli gate followed by a Feynman gate (with EXOR down) on quantum wires 1, 2 (Figure 10b).

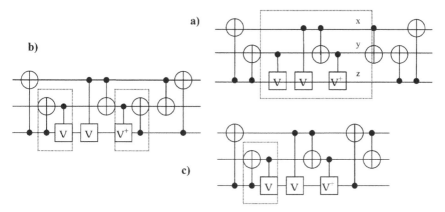

Figure 9. Reducing the cost of realization of Miller gate using quantum primitives: Feynman EXOR up, Feynman EXOR down, Controlled-V (V down), and Controlled V^+ (V^+ down).

(b) the macro-generation of the Toffoli gate leads to a Peres gate followed by the Feynman gate (with EXOR down) on quantum wires 1, 2.

(c) the equivalence-based transformation (presented below) will find a pattern of two Feynman gates of the same type (EXOR down) in sequence, which is replaced with two quantum wires, 1 and 2 (Figure 10c).

(d) thus the two right Feynman gates are cancelled, the same way as two inverters in series are cancelled in standard logic synthesis.

(e) the resultant circuit, a final solution, will have just one Peres gate of cost 20 (Figure 10d).

As can be seen from previous examples, the gates matrices have infinity of possible representations in QL. The above samples of synthesis are situated in the C_o category of Table 1, because if eventually a smaller circuit of a Fredkin gate were found, then the one presented here will remain very close to the minimal cost. Consequently the group that needs the closest inspection is the C_a because it includes circuits of great interest. As the number of circuits is infinite, the circuits in the last group can be considered as being too far from the searched solution. However the circuits in the group having average cost (not too far from the solution and not too big) can be simplified to better ones, and also they can already include parts of the optimal circuits.

6. Local Transformations

The transformations are grouped in 12 transformation sets. There are the following sets:

S1. 1-qubit transformations,
S2. 2-qubit transformations,
S3. 3-qubit transformations,
S4. 4-qubit transformations,
S5. n-qubit transformations,
S6. Ternary transformations,
S7. Mixed binary/ternary transformations,
S8. Macro-generations,
S9. Macro-cell creations,
S10. Peres Base transformations,
S11. Toffoli Base transformations,
S12. Controlled-V Base transformations,
S13. Input/Output permutting transformations.

Many transformations are shared between sets. In addition, in each of the above base sets, there are subsets to be chosen for any particular run of the optimizer program. Most of them are taken from (Miller 2002; Miller and Dueck 2003; Dueck and Maslov 2003; Iwama et al. 2002; Shende et al. 2002; Kim 2000, p. 0922312; Lomont 2003) but some other are based on our research or general quantum literature. Different groups of transformations are used in various stages of circuit optimization. The 1-qubit transformations are related to 1-qubit gates (Tables 2, 3 and 4). They can precede and also follow the 2-qubit, 3-qubit and other transformations. The 2-qubit transformations are for 2-qubit circuits or 2-qubit subcircuits of larger quantum circuits, similarly the 3-qubit transformations are for 3-qubit circuits or subcircuits of larger circuits (Table 5). The n-qubit transformations are general transformation patterns applicable to circuits with more than 3 qubits. They are less computationally efficient and they use internally transforms S1–S4. Ternary transformations (only 1-qubit, 2-qubit and 3-qubit) are used for ternary quantum logic synthesis and mixed transformations for mixed binary/ternary quantum logic synthesis (Lukac et al., in preparation). Macro-generations are transformations that convert higher order gates such as Fredkin, Margolus, DeVos, Kerntopf or Perkowski gates to standard bases. Macro-cell creations from set S9 are inverse to those from set S8. There are three standard bases of transformations: Toffoli Base, Peres Base, and Controlled-V Base. In Toffoli Base all permutation gates are converted to X (i.e. NOT), 3-qubit Toffoli and 2-qubit Feynman gates. This is the standard synthesis base used by all other authors in literature (Miller 2002; Miller and Dueck 2003; Dueck and Maslov 2003; Iwama et al. 2002; Shende et al. 2002; Nielsen and Chuang 2000). The Peres Base has been introduced originally by us based on the observation of superiority of this base in NMR realizations (and perhaps other realizations as well). It includes only X, 3-

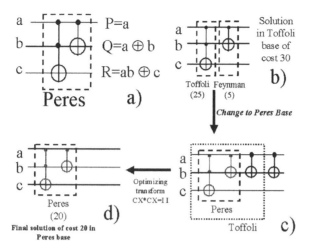

Figure 10. Peres gate: (a) Peres gate as a block using Toffoli Base notation, (b) Solution to Peres gate in Toffoli Base with cost 30, (c) Result of macro-generation of circuit from Figure 10b to Peres base, (d) After applying the optimizing transform that removes a successive pair of Feynman gates. The cost is now 20.

qubit Peres and 2-qubit Feynman gates. Controlled-V Base is another new base that is very useful to synthesize new low-cost permutation gates from truly quantum primitives of limited type (Lukac et al. 2003; in preparation). This base includes Controlled-V, Controlled-V$^+$, V, V$^+$, X and Feynman gates. In all bases the gates like Feynman, Toffoli, Controlled-V are stored in all possible permutations of quantum wires. Thus in 2-qubit base there are "Feynman EXOR up" and "Feynman EXOR down" gates and transformations respective to each of these gates. For simplification, in the tables below only some of the transformations are shown, for instance related only to "Feynman EXOR down" or "Toffoli EXOR down gates". Other transformations are completely analogous. The output permutting transformations lead in principle to a circuit that has an unitary matrix which is different from the original unitary matrix. Observe that each transformation can be applied forward or backward, so the software should have some mechanisms to avoid infinite loops of transformations. The matrix of the new circuit is the matrix of the original circuit with permuted output signals. In some applications the order of output functions is not important, so if the circuit is simplified by changing the output order, the output permutting transformation is applied.

In addition to operators defined earlier, we define now the following operators:

$$T = \begin{bmatrix} 1 & 0 \\ 0 & e^{\frac{j\pi}{4}} \end{bmatrix}, \ P(\phi) = e^{\frac{j\phi}{2}} I, \ X(\phi) = \cos\frac{\phi}{2} I - j\sin\frac{\phi}{2} X,$$

$$Y(\phi) = \cos\frac{\phi}{2}I - j\sin\frac{\phi}{2}Y, \; Z(\phi) = \cos\frac{\phi}{2}I - j\sin\frac{\phi}{2}Z$$

X, Y, Z defined earlier are Pauli spin matrices and $X(\phi)$, $Y(\phi)$, $Z(\phi)$ are the corresponding exponential matrices, giving rotations on the Bloch sphere (Nielsen and Chuang 2000). P is a phase rotation by $\phi/2$ to help match identities automatically (Lomont 2003).

The transformation software operates on sequences of symbols that represent gates and their parts. Symbol * is used to create sequences from subsequences. This symbol thus separates two serially connected gates or blocks. Numerically, it corresponds to standard matrix multiplication. Gate symbols within a parallel block may be separated by spaces, but it is necessary only if lack of space will lead to a confusion, otherwise space symbol can be omitted. So for instance symbols of macro-cells should be separated by spaces. There are four types of symbols: simple, rotational, parameterized and controlled. *Simple symbols* are just names (here – single characters, in software – character strings). The names of simple symbols (used also in other types of symbols) are the following: D – standard control point (a black dot in an array), E – control with negated input, F – control with negated output, G – control with negated input and output, X – Pauli-X, Y – Pauli-Y, Z – Pauli-Z, H – Hadamard, S – phase, T – Π/8 (although Π/4 appears in it). Symbols A, B and C are auxiliary symbols that can match several gate symbol definitions. They do not correspond to any particular gates but to groups of gates and are useful to decrease the set of rules and thus speed-up transformations. Other simple symbols will be explained below. As we see, characters are used here not only for gates but also for parts of gates, such as D – standard control used in gates. Thus we can combine these symbols to create gate descriptions: DX (Feynman with EXOR down), XD (Feynman with EXOR up), DDX (Toffoli or Toffoli with EXOR down), DXD (Toffoli with EXOR in middle), XDD (Toffoli with EXOR up), and so on. Names of macro-cells are for instance: FE (Feyman), TOD (Toffoli with EXOR down), SW (swap), FRU (Fredkin controlled with upper wire), MA (Margolus), KED (Kerntopf with Shannon expansion in lowest wire), etc. Symbols like ϕ, θ, π denote angles and other parameters. *Parameterized symbols* have the syntax: ***simple_name*** [***parameter***$_1$, ... ***parameter***$_n$], where ***parameteri*** are parameters. For instance, X[4Π/8], Y[3Π/2], Z[ϕ], etc. Rotational symbols are composed of the (simple) rotation operator symbol such as X, Y, Z, or P, and a number. X[ϕi], Y[ϕi], Z[ϕi], P[ϕi], where ϕi = 4Π/8*i, i = 1, 2, ... , 7. We assume here that all rotational operators have period 4Π and equal identity when their argument is 0, thus the choices for ϕi. In these operators the notation is like this, Xr = X[4Π/8*r], r = 1, 2, ... , 7, and so on for Yr, Zr and Pr. Controlled symbols have the syntax *simple_name* [*sequence of simple, rotational or parameterized symbols*]

Example 1. D[X*X] is a symbol of a controlled gate that is created from two subsequent Feynman gates. Observe that DX*DX = D[X*X] = D[I] = I I, which means that two controlled-NOT gates in sequence are replaced by two parallel quantum wires denoted by I I. *Example 2.* D[S*T*H] is a sequence S*T*H controlled by single qubit. *Example 3.* DD[X1*Y2*Y2] is a sequence of rotational gates controlled by a logic AND of two qubits (this is a generalization of a Toffoli gate). Observe that controlled symbols are created only as a transitional step during the optimization process – such gates do not physically exist. Using the concept of controlled symbols, n-qubit circuits can be optimized using 1-qubit transformations without duplicating all the 1-qubit identity rules. Similarly the parameterized and rotational symbols allow the reduction and hierarchization of the set of rules, which causes more efficient and effective run of the optimization software.

The simplified algorithm SA for performing rule-based optimization of 3-qubit quantum arrays is the following:

1. Apply all the 1-qubit transformations, until no more applications of such rules becomes possible.
2. Apply all the 2-qubit transformations and 1-qubit transformations induced by them (for instance using the controlled symbols).
3. Apply all the 3-qubit transformations until possible.
4. Iterate steps 1, 2 and 3 until no changes in the circuit.
5. Apply inverse transformations that locally optimize the array.
6. Repeat steps 1, 2, 3, 4 until possible.
7. Apply inverse transformations that do not worsen the cost of the array.
8. Repeat steps 1, 2, 3, 4 until possible.

Several similar variants of this heuristic algorithm can be created. In general, none of these versions gives a warranty of the optimal or even sub-optimal solution, as known from the theory of Post/Markov algorithms. As an example, we present 1-qubit transformation algorithm A1q:

A1. Combine modulo-8 the same types of rotational operators P, X, Y, Z. For instance X2*X3 becomes X5, X2*X3*X3 becomes I, and Y3*Y3*Y3 becomes I*Y1 = Y1.
A2. Apply directly applicable rules that do not include symbols A and B.
A3. Iterate 1 and 2 until no more changes possible.
A4. Starting from the left of the sequence, find a grouping pattern such as A2 = −BC
A5. Substitute symbols X, Y, Z for A, B and C from the pattern.

A6. Apply in forward directions the standard simplifying transformations such as $I*A = A$

A7. Repeat steps A1 to A6 until possible.

Example 4 (Lomont 2003). Given is an identity $Y = -X*Y*X$. Let us try to verify this identity using algorithm A1q. We have therefore to simplify the sequence $X*Y*X$. (A1) There are no patterns of the same rotational operators to combine, (A2) There are no directly applicable rules, (A4) We take pattern $-X*Y$ and match it with the rule $A2 = C*B$. This leads to $-Z2*X$. (A1) no, (A2) no, (A3) no, (A4) we find pattern $A2 = CB$ which leads to $-Y*X*X$. (A6) $X*X$ is replaced with I, $Y*I$ is replaced with Y. No further optimization steps are possible, so the sequence was simplified to $-Y$, proving that $Y = -Y*Y*X$.

Example 5. Simplify HXH. (A2) Use rule $H = X*Y1$ twice. This leads to $X*Y1*X*X*Y1$. (A2) Use rule $I = AA$ in reverse direction. This leads to $X*Y1*Y1$. (A1) $Y1*Y1$ is replaced by Y2. This leads to $X*Y2$. (A4) Use pattern $A = B*C2$. This leads to Z. No further optimization steps are possible. Thus we proved that $HXH = Z$. More optimization examples of algorithm SA will be given in the sequel.

In our runs of GA we look for solutions with the accuracy of: (1) permutation of quantum wires, (2) permutation of inputs, (3) permutation of outputs (transformation group S13). In addition we can also generate solution sets with accuracy of inverting input, output or input/output signals (the NPN classification equivalent circuits). Therefore, for each unitary matrix we generate therefore many logically equivalent solutions. We can generate solution sets also for the same set of Boolean functions, or for the same NPN classification class. One interesting aspect of such approach is that one can create new local equivalence transformations for circuits in each of these classes. Finding these transformations and applying them exhaustively to particularly interesting gates leads to levellized "onion-like" structures of gates, as the one shown in Figure 11. This Figure shows the layered structure of gates created by adding only Feynman gates to a seed composed of other gate types, in this case a Toffoli gate. Figure 12 shows the layered structure with Peres gate as a seed, in which Toffoli, Fredkin and Miller gates are created. These transformations are used to find efficient realizations of new gates from known gate realizations (Lukac et al. 2003, in preparation).

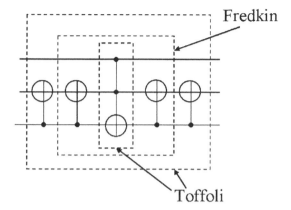

Figure 11. The layered structure of gates. Fredkin from Toffoli and Toffoli from Fredkin.

Table 2. 1-qubit transformations for I, A and X groups

		A1.	A = B*C2		
		A2.	A = B2*C		
		A3.	A = B3*A*B3		
		A4.	A = B3*C*B1	*X1.*	X = - H*Y3
I1.	I = A*A	*A5.*	A = - C*B2	*X2.*	X = S*X2*S
I2.	I = -A*C2*B	*A6.*	A = C1*B*C3	*X3.*	X = Y1*H
I3.	I = HH	*A7.*	A = - C2*B	*X4.*	X = Y3*H*Y2
I4.	I = - H*Y3*X	*A8.*	A = C3*A*C3	*X5.*	X = Z1*X*Z1
I5.	I = P2*A2*A	*A9.*	A = P2*A2	*X6.*	X = - Z3*Y*Z1
I6.	I = P2*H*X2*Y1				
I7.	I = - P2*H*Y3*X2	*A1.1.*	A1 = - B*A1*C	*X1.1.*	X1 = H*Z1*H
I8.	I = - P2*H*Z2*Y3	*A1.2.*	A1 = - B*A3*B	*X1.2.*	X1 = Z*X1*Y
I9.	I = P2*Y1*Z2*H	*A1.3.*	A1 = - C*A3*C		
I10.	I = P2*Y2*Y	*A1.4.*	A1 = P2*A3*A	*X2.1.*	X2 = H*Z2*H
I11.	I = - P2*Y3*H*Z2			*X2.2.*	X2 = P2*H*Y3
I12.	I = - P2*Y3*X2*H	*A2.1.*	A2 = - B*C	*X2.3.*	X2 = - S*X*S
I13.	I = P2*Z2*Z	*A2.2.*	A2 = CB	*X2.4.*	X2 = - Y3*H*Y
I14.	I = P3*Z3*S	*A2.3.*	A2 = - P2*A	*X2.5.*	X2 = Y3*Y*H
I15.	I = P4*A4				
I16.	I = - Y*X2*Z	*A3.1.*	A3 = B*A3*C	*X3.1.*	X3 = H*Z3*H
I17.	I = - Z*Y2*X	*A3.2.*	A3 = - C*A3*B	*X3.2.*	X3 = - S*H*S
I18.	I = - Z*Y3*H	*A3.3.*	A3 = - P2*A1*A	*X3.3.*	X3 = - Y*X1*Y
				X3.4.	X3 = - Z*X1*Z
		A4.1.	A4 = P4		

Table 3. 1-qubit transformations for Y and Z groups

Y1.	Y = - H*X2*Y3					
Y2.	Y = H*Y3*Z2	*Y2.1.*	Y2 = - H*Y2*H	*Z1.1.*	Z1 = H*X1*H	
Y3.	Y = - H*Z2*Y1	*Y2.2.*	Y2 = - H*Y3*Z	*Z1.2.*	Z1 = P2*Z3*Z	
Y4.	Y = P2*Y2	*Y2.3.*	Y2 = - P2*Y	*Z1.3.*	Z1 = - P3*S	
Y5.	Y = S*Y2*S	*Y2.4.*	Y2 = - S*Y*S	*Z1.4.*	Z1 = - X*Z3*X	
Y6.	Y = X1*Y*X1	*Y2.5.*	Y2 = - X*Y3*H	*Z1.5.*	Z1 = Y*Z1*X	
Y7.	Y = X2*Y3*H			*Z1.6.*	Z1 = - Y*Z3*Y	
Y8.	Y = X3*Y*X3	*Y3.1.*	Y3 = H*Z2*Y			
Y9.	Y = - X3*Z*X1	*Y3.2.*	Y3 = - H*X	*Z2.1.*	Z2 = H*X2*H	
Y10.	Y = -Y1*X2*H	*Y3.3.*	Y3 = - P2*H*X2	*Z2.2.*	Z2 = H*Y3*Y	
Y11.	Y = -Y3*H*X2	*Y3.4.*	Y3 = - P2*Y1*Y	*Z2.3.*	Z2 = P2*Y3*H	
Y12.	Y = - Y3*Z2*H	*Y3.5.*	Y3 = X2*H*Y	*Z2.4.*	Z2 = - P2*Z	
Y13.	Y = - Z1*X*Z3	*Y3.6.*	Y3 = Y*H*Z2	*Z2.5.*	Z2 = - P3*Z1*S	
Y13.	Y = Z1*Y*Z1	*Y3.7.*	Y3 = Y*X2*H	*Z2.6.*	Z2 = - Y*H*Y3	
Y14.	Y = - Z2*H*Y3	*Y3.8.*	Y3 = - ZH	*Z2.7.*	Z2 = Y1*Y*H	
Y15.	Y = Z3*Y*Z3					
		Z1.	Z = H*Y1	*Z3.1.*	Z3 = H*X3*H	
Y1.1.	Y1 = - H*Y*X2	*Z2.*	Z = P2*Z2	*Z3.2.*	Z3 = - P1*Z*S	
Y1.2.	Y1 = - H*Y2*X	*Z3.*	Z = S*S	*Z3.3.*	Z3 = - P2*Z1*Z	
Y1.3.	Y1 = - H*Y3*H	*Z4.*	Z = - X1*Y*X3	*Z3.4.*	Z3 = - P3*Z2*S	
Y1.4.	Y1 = H*Z	*Z5.*	Z = X1*Z*X1	*Z3.5.*	Z3 = - X*Z1*X	
Y1.5.	Y1 = P2*H*Z2	*Z6.*	Z = X3*Z*X3	*Z3.6.*	Z3 = X*Z3*Y	
Y1.6.	Y1 = P2*Y3*Y	*Z7.*	Z = Y2*H*Y3	*Z3.7.*	Z3 = - Y*Z1*Y	
Y1.7.	Y1 = X*H	*Z8.*	Z = - Y3*H			
Y1.8.	Y1 = Y*H*X2					
Y1.9.	Y1 = - Z*Y2*H					
Y1.10.	Y1 = Z2*H*Y					
Y1.11.	Y1 = - Z2*Y*H					

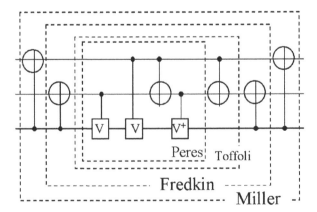

Figure 12. Internal part of a layered structure with Peres gate as a seed.

Table 4. 1-qubit transformations for S, H and parameterized rotation groups

		H1.	H = P2*Y1*Z2
S1.	S = P1*Z1	*H2.*	H = - P2*Y3*X2
S2.	S = P3*Z3*Z	*H3.*	H = S*X1*S
S3.	S = T*T	*H4.*	H = X*Y1
S4.	S = X1*S*Y1	*H5.*	H = - X1*H*Z3
S5.	S = X2*S*Y2	*H6.*	H = - X2*H*Z2
S6.	S = X3*S*Y3	*H7.*	H = - X2*Y3*Y
S7.	S = Y*S*X	*H8.*	H = - X3*H*Z1
S8.	S = - Y1*S*X3	*H9.*	H = Y*X2*Y3
S9.	S = - Y2*S*X2	*H10.*	H = - Y1*Y*X2
S10.	S = - Y3*S*X1	*H11.*	H = - Y1*Z

$X[\Pi] * Y[\phi] = Y[-\phi] * X[\Pi]$

$X[-\Pi] * Y[\phi] = Y[-\phi] * X[-\Pi]$

$X[\phi] * Y[\Pi] = Y[\Pi] * X[-\phi]$

$X[\phi] * Y[-\Pi] = Y[-\Pi] * X[-\phi]$

$X[\Pi/2] * Y[\phi] = Z[\phi] * X[\Pi/2]$

$X[-\Pi/2] * Y[\phi] = Z[-\phi] * X[-\Pi/2]$

$X[\phi] * Y[\Pi/2] = Y[\Pi/2] * Z[\phi]$

$X[\phi] * Y[-\Pi/2] = Y[-\Pi/2] * Z[-\phi]$

$X[3\Pi/2] * Y[\phi] = Z[-\phi] * X[3\Pi/2]$

$X[-3\Pi/2] * Y[\phi] = Z[\phi] * X[-3\Pi/2]$

$X[\phi] * Y[3\Pi/2] = Y[3\Pi/2] * Z[-\phi]$

$X[\phi] * Y[-3\Pi/2] = Y[-3\Pi/2] * Z[\phi]$

$Y[\Pi] * Z[\phi] = Z[-\phi] * Y[\Pi]$

H12. H = Y2*H*Y2
H13. H = Y2*X*Y3
H14. H = Y3*H*Y3
H15. H = -Y3*X
H16. H = Y3*Z*Y2
H17. H = Y3*Z2*Y
H18. H = - Z*Y3
H19. H = - Z1*H*X3
H20. H = - Z2*H*X2
H21. H = - Z2*Y1*Y
H22. H = - Z3*H*X1

7. Genetic Algorithm for Quantum Logic Synthesis

Now that we understand all the basic prerequisites for our approach to quantum array synthesis and we have a general plan for it, let us finally discuss the Genetic Algorithm itself. Similarities with genetic information and its transmission in Nature are applied and constitute the core of the evolutionary approach. The application of GA is well known in problems with high noise because their convergence can be moderated according to the fitness function. GAs are widely used optimization algorithms, however only few authors present their application programs for QC synthesis (Lukac et al. 2002; Lukac and Perkowski 2002; Ge et al. 1998; Willams and Gray 1999).

General steps of using our GA variant for QC (and RL) are the following:
1. Initiate a random population of individuals (here each individual is a Quantum or Reversible circuit). An individual is an encoding of a chromosome of a circuit for fitness function evaluation. The size of the population is generally between 50 and 500 individuals. A too small population will yield a too fast convergence while a too big population will uselessly search a too large problem space.

Table 5. Examples of 2-qubit and 3-qubit transformations: (a) 2-qubit transformations, (b) 3-qubit transformations; observe a space between X and DX in rule R3.57 that signifies that D control the lower X, (c) 4-qubit transformations

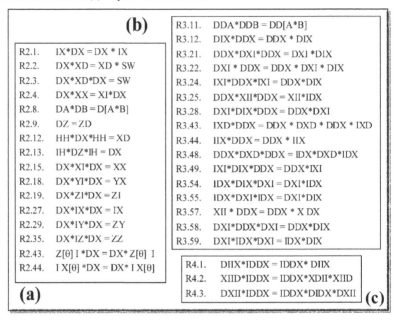

(b)

R2.1.	IX*DX = DX * IX
R2.2.	DX*XD = XD * SW
R2.3.	DX*XD*DX = SW
R2.4.	DX*XX = XI*DX
R2.8.	DA*DB = D[A*B]
R2.9.	DZ = ZD
R2.12.	HH*DX*HH = XD
R2.13.	IH*DZ*IH = DX
R2.15.	DX*XI*DX = XX
R2.18.	DX*YI*DX = YX
R2.19.	DX*ZI*DX = ZI
R2.27.	DX*IX*DX = IX
R2.29.	DX*IY*DX = ZY
R2.35.	DX*IZ*DX = ZZ
R2.43.	Z[θ] I *DX = DX* Z[θ] I
R2.44.	I X[θ] *DX = DX* I X[θ]

(a)

R3.11.	DDA*DDB = DD[A*B]
R3.12.	DIX*DDX = DDX * DIX
R3.21.	DDX*DXI*DDX = DXI *DIX
R3.22.	DXI * DDX = DDX * DXI * DIX
R3.24.	IXI*DDX*IXI = DDX*DIX
R3.25.	DDX*XII*DDX = XII*IDX
R3.28.	DXI*DIX*DDX = DDX*DXI
R3.43.	IXD*DDX = DDX * DXD * DDX * IXD
R3.44.	IIX*DDX = DDX * IIX
R3.48.	DDX*DXD*DDX = IDX*DXD*IDX
R3.49.	IXI*DIX*DDX = DDX*IXI
R3.54.	IDX*DIX*DXI = DXI*IDX
R3.55.	IDX*DXI*IDX = DXI*DIX
R3.57.	XII * DDX = DDX * X DX
R3.58.	DXI*DDX*DXI = DDX*DIX
R3.59.	DXI*IDX*DXI = IDX*DIX

R4.1.	DIIX*IDDX = IDDX* DIIX
R4.2.	XIID*IDDX = IDDX*XDII*XIID
R4.3.	DXII*IDDX = IDDX*DIDX*DXII

(c)

2. Steps describing the search using the GA are called generations. The search stops when a condition is attained, generally this is a result of having fitness greater than a limit, or that certain number of generations has been attained.
3. Pick n pseudo-randomly selected individuals. The rules of selecting individuals can speed up the convergence of the search, however if the selection operator is too "greedy" (selecting only the best individuals) the algorithm can get stuck in a local minimum. We experimented with several rules, some presented below. They are based on fitness function.
4. For all the individuals of the new population apply two genetic operators:
   a. Crossover – operator allows to recombine two individuals by exchanging parts of their respective chromosomes. This operation is the main power of the search with GAs.
   b. Mutation – operator introducing noise into the GA in order to avoid local minima. Generally applied with a very small probability [0.01, 0.1] (compared to the one of Crossover [0.3, 0.85]) modify randomly a chromosome of a randomly selected individual.

5. Replace the old generation of individuals by the new one.

6. If solution is found then exit, else go to step 2.

All individuals in the GA are quantum circuits partitioned into parallel blocks. Our representation allows describing any Quantum or Reversible circuit (Lukac et al. 2002; Lukac and Perkowski 2002). This partitioning of the circuit was in our case induced from the representation of any QC such as one in Figure 14. Each block has as many inputs and outputs as the quantum array width (for instance three in Figures 3–6). The chromosome of each individual is a string of characters with two types of tags. To indicate to the algorithm where a parallel block begins and ends, a special tag different than any gate was required. This special tag is named here R-tag or 'r'. Gates' tags are characters from the alphabet and in our examples tags are the first letter(s) of gates' names. An example of a chromosome can be seen on Figure 13. The circuit corresponding to the parent 1 from Figure 13 can be seen on Figure 14. In this particular encoding each space (empty wire or a gate) is represented by a tag. Our problem-specific encoding was applied to allow the definition of as simple genetic operators as possible. The advantage of these strings is that they allow encoding an arbitrary QL or RL circuit without any additional parameters known from previous research (Rubinstein 2001; Williams and Gray 1999). And only the possibility to move gates, remove and add them to the chromosome consequently make it possible to construct and arbitrarily modify a circuit.

To speed-up the search the crossover operator has only one restriction. Both crossover candidates are restricted to have the same number of wires. This considerably simplifies this operation because it allows to avoid the case when the two circuits are cut in incompatible locations. Consequently, for any circuit with a number of wires higher than one, the crossover is done only between parallel blocks. This assures that all children are well formed and eliminates also the necessity of a repair algorithm. In the case of individuals having only one wire in the circuit, the crossover is executed at a random location in the chromosome.

The mutation is more delicate. Its results should be: adding a gate, replacing a gate by another one, or a removal of a gate. All these operations can be easily implemented except for the replacement operator. Assume the Hadamard gate from circuit in Figure 14 is to be changed to a Feynman gate. It is a particular gate where we need to create a new parallel block because we want to preserve the remaining of the current circuit and also not to create a block with more wires than the circuit actually has. In the remaining cases: (1) the removing of a gate can easily be seen as replacing it by wires, (2) the adding of a gate implies a creation of a new parallel block. The results of genetic operators are shown in Figure 13. In the middle are two parents which

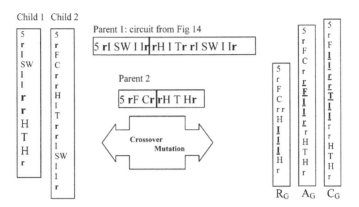

Figure 13. Chromosome representing the circuit on Figure 14 and genetic operations as example. On the left is a result of crossing-over and on the right is are results of mutation Removal of a gate (R_G), Adding of a gate (A_G) and Replacing of a gate (C_G).

yield two children on the left side as a result of a crossover. On the right side are the results of mutation (removal of a gate, adding a new gate and changing of a gate, respectively) on parent 2. As can be seen, the operation of adding a gate results in a creation of a new parallel block, but as will be shown later this operation is not equivalent to the operation of adding one block.

The evaluation of a QC is part of fitness function calculation. In the context of our encoding it is done in two steps. First for each parallel block a matrix representing the part of the circuit is calculated. Then each such a unitary matrix is multiplied with its neighbor so as at the end a matrix representing the whole circuit is obtained. An example of this calculation can be seen in Figure 15. Inside of each parallel block only Kronecker matrix products are used, while standard matrix multiplications are used to multiply these blocks.

In the case of our software variant for RC synthesis the evaluation is similar to the one used in QC but no matrix operations are required and truth tables of (in)complete multi-output functions are used instead. As mentioned earlier, constants insertion is possible in RC logic synthesis. Consequently a function defined over 3 wires can be searched on a circuit with 4, 5 or even more wires. The evaluation of a RC circuit consists in finding logical equation for each wire and then testing all input/output combinations on the searched wires. In this case the constants are alternatively tested for each wire.

As can be seen, the length of the chromosome is proportional to the size of the circuit. Also the time required to calculate each circuit is proportional to this size. A circuit with seven gates will have seven blocks if each gate cannot be connected in parallel with another one. In such case, the chromosome with seven blocks encoding a circuit for 3 qubits can encode maximally $7*3 = 21$ one-qubit gates.

8. The Fitness Function

The fitness function used was multiple times modified in order to observe improvements or changes in the search for the new optimized circuits. For dydactic reasons we will explain some of these variants. The fitness function evaluates each individual and assigns to it a fitness value, representing the quality of the encoded circuit, and the encoding is a direct mapping from genotype to phenotype. The original GA was defined as GA with Darwinian learning. It means, the fitness function evaluates the genotype. In Baldwinian GA the genotype is converted to the phenotype and the fitness function evaluates the phenotype. The objective of our experiments was consequently to modify the fitness function in order to observe the impact on the "evolution" of the desired solution. The fitness function does not modify the circuits but only biases the selection operator. This mechanism, according to the fitness function value of each circuit, will be able to select better or worse circuits. For example, a fitness function weighted too strongly towards the length of a circuit will prefer circuits with higher length, even if they are not correct. This kind of flexibility allows the GA to explore regions of the problem space that are inaccessible to classical computational methods without an extreme time and computational resources consumption.

The equation (8.1) represents the basic fitness function.

$$F = \frac{1}{1 + \text{Error}} \tag{8.1}$$

where Error is the evaluation of the correctness of the circuit. Although we found this fitness function very useful in space exploration, it is however too generic and does not take into account the cost of the circuit. An improved function is shown in equation (8.2). As can be seen in the equation, two parameters are sufficient to create a more sophisticated fitness function

$$F = \frac{1}{1 + \text{Error}} - \frac{\text{Min\_cost}}{\text{Cost}} \tag{8.2}$$

where the first right hand element is the evaluation of the correctness of the solution from equation (8.1) and the next one is an additional constraint forcing the selection operator to select circuits with a smaller number of gates or with gates having a smaller total cost. In the case of the QC search the error evaluation is based on the comparison of the resultant matrix with the unitary matrix of the evaluated circuit. The equation (8.3) describes the Error calculation

$$\text{Error} = \sum_i \sum_j |O_{ij} - S_{ij}| \tag{8.3}$$

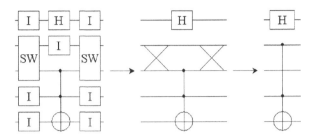

Figure 14. Transformation of a QC from the encoded chromosome (on the left) to a final quantum circuit notation representation of this QC (on the right). Here S is a Swap gate, H is a Hadamard gate and W is a wire. In the middle there is one CCNOT (Toffoli) gate.

where O_{xy} is an element from the expected synthesized unitary matrix and S_{xy} is an element from the matrix of the current circuit (for speed a truth table is used in case of RL). The error generated by this function is exact because it reflects any differences in the whole matrix of the circuit. In the case of an exact match, this error substituted in the equation (8.2) gives a fitness of 1. As can be seen in Table 6, the error has always value 128 in the denominator, since there are 64 ($2^3 \times 2^3$) units in the evaluation matrix of a 3-qubit gate and each of them is a complex number. The representation of complex numbers in computers is done via a definition of complex numbers as one number for the real part and one for the imaginary. Consequently one can scale the error either versus the complete set of real and imaginary parts of the matrix coefficients either versus the number of coefficients in the matrix of the resulting circuit.

Although useful in several runs of our program, the equation (8.2) has still a slight imperfection. In the case one wants to measure the fitness in the interval [0,1] the second element of the equation (8.2) will in some marginal cases output a fitness of 0 while the correct result was found. Consequently a modification of the fitness function is required. Equation (8.4) shows one possible form of a function producing fitness in the interval [0,1].

$$F = \frac{1}{\text{Error} + \frac{\text{Min\_Cost}}{\text{Cost}}} \tag{8.4}$$

However, this equation assumes that Min_Cost is the cheapest possible circuit. In the case a cheaper correct circuit is found the fitness will be of course smaller than 1.

This fitness function is centered in the point of the optimal circuit (Min_Cost and correct result) similarly to a Gaussian curve. It is clear from this example that the Min_Cost parameter should be also set to 1 which leads to the next version of the fitness function from equation (8.5):

$$F = \frac{1}{\text{Error} + \text{Cost}} \tag{8.5}$$

This equation then should be only maximized because the value of 1 is practically impossible to find. It is subsumed in the Cost parameter. The value of used gates varies as previously introduced in this paper. Consequently a circuit realized with only 1-qubit gates will always have higher fitness than one realized with gates of two and more qubits if for each more-than-one-qubit gate a cheaper variant with only 1-qubit gates is found. Important point to notice is the numerical non-compatibility of the Error and Length parameters. The Error is in the range $[0, 2^{1+(2*n)}]$ while the range of the Length parameter is in $[\text{Min}, \infty]$. This ill-scaled ratio would in consequence make the selection pressure mainly with respect to the length of the circuit and not to its correctness. As a solution to this problem, the final variant of the fitness function scales these two variables as shown in equation (8.6):

$$F = \alpha \left(1 - \frac{\text{Error}}{\text{Max\_Error}} \right) + \beta \frac{1}{\text{Cost}} \tag{8.6}$$

where α and β are scaling parameters that allow to modify the selection pressure either towards cheaper or more correct circuits.

Another possible approach induced from the previous fitness function was the introduction of the Pareto optimality (PO) fitness function (Goldberg 1989). PO is used for multiple objectives optimization problems such as here the Cost and the Error. A GA using PO uses a different evaluation for the fitness of each individual. This approach is the following:

- Each individual is compared with all others and is assigned with a rank.
- The rank of an individual depends on the number of wins while compared to all other individuals.
- The comparison is made on all parameters. Example: an individual with the highest rank will have all of its parameters higher (for maximization) or lower (for minimization) than most of all other individuals.

Such a comparison forces the selection process of the GA to reproduce mainly the globally optimal individuals. Similarly to equation (8.5) one needs to scale parameters of comparison when they are not equivalent.

9. Other Parameters of GA

In the following discussion of our results, Stochastic Universal Sampling (SUS) was used as the selection operator, although we experimented also with Roulette Wheel and other selection approaches. A decision for using SUS

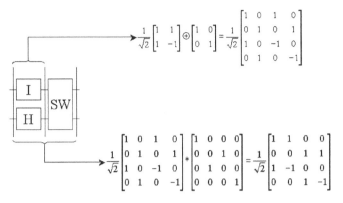

Figure 15. Examples of Kronecker product ⊕, and of Matrix product ∗ on a sample of a circuit.

Table 6. Results of sequential evaluation of Fredkin gate from Figure 5a

Circuit (blocks)	Error (3) ΣC_g	Cost $\alpha = 0.9$	I Fitness (8.6) $\beta = 0.1$
1	5/128	5	0.8815
2, 3, 4	18/128	12, 18, 24	0.78, 0.778, 0.777
5, 6	4/128	30, 36	0.875, 0.874
7 (circuit completed)	0/128	42	1

instead of Roulette Wheel (RW) was implied from the following demonstration of high non-linearity in circuit design. Consequently an operator allowing a selection of less locally optimal circuits such as SUS was needed in order to obtain good experimental results.

An implication of the previous part of this paper, one of the goals of the presented work was to "reinvent" gates such as those in Figures 2, 3 and 4, or to synthesize smaller ones with the same functionality. Consequently, the analysis of a circuit can give more precise explanation of possible non-monotonicity in the synthesis process evaluation. For this purpose let's have a closer look at the circuit from Figure 8a. This circuit has seven gates. The measurement of Error as from equation (8.3) and fitness from equation (8.6) are recorded in columns of Table 6.

A closer look at the above table will reveal non-monotonicity in the evaluation of this circuit. The first "trap" for GA is between the first gate and the adding of the second. Because both the error and the cost increase, the fitness is significantly reduced and creates a relatively big local minimum.

Consequently this step is difficult for any automated synthesis without heuristics. Two next "smaller" problems are that between steps 2, 3, 4 and 5, 6 the error remains constant while the cost of the circuit is linearly growing with each segment. A monotonic growth of the cost can be observed in the third column. It is important to notice that all gates have a cost including the wire gate (identity gate). Such a setting allows the algorithm to avoid some local minima and constructing a circuit with only empty wires. From the above example an important modification to the algorithm can be derived and is discussed below.

A solution to this non-monotonicity problem can be a modification of the above presented genetic operators. In this work an extension of the mutation operator was created. Its result is the addition of a complete parallel block to the circuit. Also the mutation is tested for each gate of the circuit because multiple mutations can occur on one chromosome. Such a "bitwise" mutation or noise introduction into the circuit can avoid such local minima as those in Table 6. The modification of the operator is required to override such traps as in Table 6. Moreover, as already mentioned, in each of the groups of steps {2, 3, 4} and {5, 6} the cost increases while the fitness decreases. It is possible to use a fitness function such as (8.1), however then the solution can have an arbitrary length and no parameters force the selection of a shorter circuit. The results and their comparison are described in the next section.

10. Experimental Results

As was shown earlier, several well-known gates can be synthesized in different manners depending on the cost function selection. There exist already a variety of implementations of universal gates, especially Fredkin or Toffoli. Our goal, however, is to obtain the gates of the smallest possible cost for realistic physical realizations. In section 2 of this paper few variations of Toffoli and Fredkin gates were given and we discussed possible minimization of their cost by realizing them with smaller, truly quantum, primitives. The two important properties allowing the minimization operation are the *commutability* and *concatenation* properties of gates in QL. The first property, whether order of two subsequent gates can be changed, is purely mathematical and can be studied in any literature on Quantum Computing. In general $AB \neq BA$, but some operators (gates) do commute. This can be used in synthesis, for instance as part of local optimization algorithms from section 6. The second property is more problematic because there are no any theoretical basis allowing or not this operation to be applied. It is only a heuristic useful to calculate approximate cost functions of composed

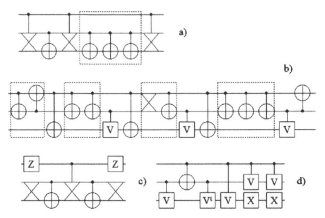

Figure 16. Various realizations of Toffoli gate obtained from GA before optimizing transformations.

gates (Smolin and DiVincenzo 1996) and the issue of concatenation should be further investigated.

All runs of the GA have these generic settings. The mutation probability was in the range of [0.01, 0.1], the crossover probability was in interval [0.75, 0.9], and the size of the population was [100, 200] individuals. All runs were done on a Pentium 500 Mhz with 64 megs of ram.

We reiterate again the non-monotonicity issue discussed previously. Because of two contrary flows of values in the evaluation function (growing a cost and reducing an error) the modifications to the fitness function and selection operator needed to be done. The first possible and already introduced modification is the scaling of the fitness function with the coefficients α and β. Such a use of scaling coefficients allows to eliminate the too small negative modifications due to the linearly increasing cost of the circuit. It is advised and experimentally proven in following results that the α coefficient should be in the interval of 0.9 to 0.99 while β should be a complement to it; $1 - \alpha$. Similarly these coefficients should be used in a PO selection operator to scale the ranking of individuals.

Before comparing available results one needs to define the input gates for the GA. In this case a distinction between a *complete starting set* and *a biased starting set* has to be made. A complete starting set of gates is one containing all available gates for synthesis, excluding the resulting gate. In a complete starting set of gates the following gates were available: Wire, Hadamard, Pauli X (NOT), Pauli Y, Pauli Z, Phase, V, Feynman, Swap, Controlled-V (C-V), Controlled-V-Hermitian (C-V$^+$), Fredkin, Toffoli, and Margolus. From these the resulting one was removed. A biased starting set of gates is one based on some constraints, for instance only 2-qubit gates are used, or only an

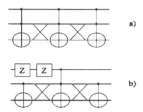

Figure 17. Results of synthesis of Fredkin gate. (a) is result of application of fitness (8.1) and a complete set and (b) the result of fitness (8.6) and a complete set.

arbitrary subset of gates is available. In figures representing results of the search following letters represent these gates: Z – Pauli-Z, V – controlled-V, V^+ – controlled V^+.

Let us compare results of different runs using various fitness functions. Figure 16 presents the results of search for the Toffoli gate. The first circuit (Figure 16a) is the result of a run where fitness from equation (8.1) was used and a complete starting set of gates. As can be seen, this result is very expensive because includes three Fredkin gates and four Feynman gates. This circuit can be simplified by removing two of the three subsequent identical Feynman gates based on the fact that Feynman gate is its own inverse. The next one, Figure 16b, is a result of the same setting as with the previous one but the starting set was a biased one. The available gates in this biased set were only the ones with number of inputs smaller or equal to 2 qubits. Similarly to the previous circuit, the group of gates in dash-squares can be concatenated in order to reduce the cost of the circuit. Moreover two pairs of consecutive identical Feynman gates (in dotted groups) can be removed. Thus, the second from left dotted group is removed entirely and the first from right dotted group is replaced by a single Feynman gate. The rightmost Feynman gate can be moved before the last Controlled-V gate and added to the dotted rectangle. Although this circuit is longer than the first one its cost is smaller. It is composed only from 2-qubit or 1-qubit gates. The first circuit has cost 55 (15∗3 + 5∗2) after minimization and the second circuit also after minimization has cost of 45 (9∗5). The two circuits below in Figure 16 were found using the improved fitness function from equation (8.6). The circuit from Figure 16c was found using a complete starting set and the one from Figure 16d using the biased set similar to the one used for the circuit from Figure 16b. The major improvement using the equation (8.6) as a fitness function is that the results are reproducible, because they are driven as already mentioned by two opposite flows of evaluation. Observe that by flipping over two Feynman gates from Figure 16c and removing the swap gates a solution with one Fredkin and two Feynman is used, which is close to a known realization of Toffoli gate from Fredkin gate. The only difference

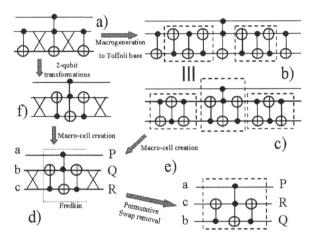

Figure 18. Simplification of Fredkin gate based on local transformations: (a) the gate found by the GA, (c) result of macro-generation to Toffoli Base, (d) circuit from Fig.18b rewritten to show left-side patterns, (d) result of macro-cell creation, (e) final circuit after permutative Swap removals, (f) another result of 2-qubit transformations in Toffoli Base applied to the initial circuit.

are two Pauli-Z rotation gates, which can be removed, as analysis shows. Interestingly, our software reinvented also the famous circuit of Smolin, since the sequence of two Pauli-X (NOT gates) can be removed, and the sequence of two Controlled-V is equivalent to Feynman gate. After these transformations, our gate is composed from the same basic quantum primitives as the Smolin's solution, but in a different order. The analysis of unitary matrices shows however that the circuits are equivalent. These examples show that not only can our software "reinvent" the realization of the known gates but it can also create similar minimum cost realizations of other gates, as will be presented below. We were not able to find a better solution to Toffoli and Fredkin gates from quantum primitives, since perhaps they do not exist.

Figure 17 presents two of many found realizations of the Fredkin gate. Again, observe that the consecutive Pauli-Z gates can be cancelled since this gate is its own inverse (a standard local transformation). Next, two Feynman gates can be flipped over and corresponding swap gates removed, leading to the known (minimal) realization Fredkin gate using one Toffoli and two Feynman gates. Figure 18a shows step-by-step simplification of the circuit from Figure 17b after removal of Pauli-Z gates. Applying macrogeneration of gates to Toffoli Base the circuit from Figure 18b is created which is the same as one from Figure 18c, in which gates were differently grouped to satisfy the left-hand-side patterns of rules. Then macro-cell creation leads to the schematics from Figure 18d. Finally, permutative transforms from set S13

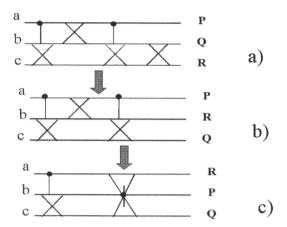

Figure 19. Margolus gate: (a) Original gate obtained directly from GA, (b) Removal of the right Swap gate permuted the order of output wires Q and R, (c) removal of the first Swap gate permuted output wires P and R.

that remove Swap gates lead to the classical realization of Fredkin gate from Figure 18e. Observe that the same solution is found when 2-qubit transformations are applied to Figure 18a (leading to the circuit from Figure 18f) and next the macro-cell generation is applied to the array from Figure 18f.

The circuit from Figure 17a is the result of using the equation (8.1) and a complete starting set. Figure 17b is the result of fitness equation (8.6) and the same starting set as in Figure 17a. The realization of the Margolus gate from Figure 19a is elegant and new. By applying the swap-related transformations of the Fredkin gates (Figure 19b, Figure 19c) two circuits with two Fredkin gates each are created that are not known from the literature. The 3-qubit Kerntopf gate family includes gates with one Shannon expansion (of the form v'u+vx, where v,u,x $\in$ {a,b,c,a',b',c'} and the expression cannot be simplified) and two Davio expansions (these expansions have the form v$\oplus$ux, where v,u,x $\in$ {a,b,c,a',b',c'} and the expression cannot be simplified). Figure 20a presents a gate from this family in Toffoli Base, and Figure 20b shows a gate from the Margolus gate family (NPN class), also realized in Toffoli Base. Based on these and other results, we can claim that our program invented several realizations of basic gates that have been not created yet by humans and are perhaps patentable.

Figure 21 shows the entanglement and teleportation circuits found also by the GP approach from (Rubinstein 2001). Finally, Figure 22 illustrates the using of transformations to prove that Fredkin gate is its own inverse. Transformations of this type lead often to significant reductions of long arrays found intially by the GA.

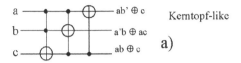

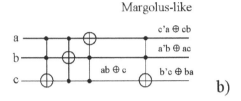

Figure 20. Kerntopf-like and Margolus-like gates in Toffoli Base after optimizing transformations. These are the best reported results for such gates.

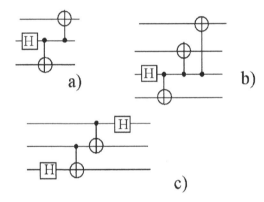

Figure 21. Entanglement and teleportation circuits from the literature found also by our program.

One of the striking observations of the obtained results is the respective presence in Fredkin or Toffoli gates in the synthesis of each of them. As can be seen in the results of synthesis using the complete starting set, the Fredkin gate is present in all Toffoli implementations and vice versa. It can be concluded that there is a local minimum that our algorithm was unable to override. Overriding this local minimum can be done using only basic quantum primitives such as controlled-V.

An interesting difference between our results of synthesis of Fredkin or Toffoli gate and those shown in Figures 5 and 6, is that our algorithm found circuits with generally slightly higher cost (in the case of improved fitness function). To understand this point one needs to look at the evolution of the search during a run of a GA. Figure 23 shows the recording of one run searching for a gate using the fitness function (8.1). For illustration the fitness

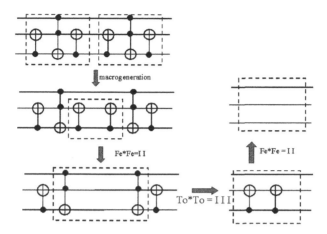

Figure 22. Using simplifying transformations to prove that Fredkin gate is its own inverse.

(8.6) is also drawn. The figure shows the best result of each hundred generations. First curve Error shows the evolution of the scaled error as used in the
fitness function. Second curve is the cost (1/Cost) and it is to be maximized
to increase the global fitness function. The larger a circuit the more the cost is
reduced. The last three curves show the values for fitness from equation (8.1)
and (8.6). Fitness function (8.6) is shown by two curves; first with parameters
$\alpha = 0.99$ and $\beta = 0.01$ then with $\alpha = 0.9$ and $\beta = 0.1$. Moreover for more
clarity all fitness functions are mapped on the secondary y-axis on the left.
Since there are many solutions to a gate at a higher cost, it is more likely
that GA finds those solutions, but hopefully running the local optimizing
transformations on each of these non-minimal solutions leads *to the same
optimal solution*, as was the case with the solutions discussed above.

Figure 23 illustrates the problem of fitness function used in quantum logic
synthesis. While both variants of the fitness function are globally stable or
stuck in a local minimum, the parameter having the greatest variation is the
cost of the circuit. This is because, as mentioned before, an infinity of circuits
exist theoretically in QL for a certain unitary matrix. This is illustrated by
the fact that while the error is constant all along the run, the cost oscillates.
Moreover, once the correct circuit is found, the fitness function (8.1) is at
its maximum but the fitness function (8.6) is relatively smaller because the
circuit is larger than the local minima shown in the graph. For this reason
while using the fitness function (8.6), once a correct circuit was found its
fitness was set up to 1 and the run was over. The presented curves capture
also different levels of detail while searching for a correct circuit. The Fitness
function (fitness (8.6) (0.9)) captures strongly the size of the circuit but the
final jump to the correct solution is very large in the value of fitness (≈ 0.1).

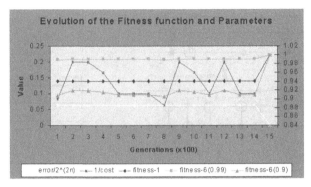

Figure 23. An example of evolution of the fitness function and its related parameters during a run. For more precision the two fitness function curves are shown on a secondary y-axis on the left. In parenthesis are values of α for fitness function 8.6.

Such a fitness function can be used to minimize the size constraints. However, the best fitness function to capture global properties during a search for a circuit is the fitness function (fitness (8.6) (0.99)), because on one hand it is mainly influenced by the correctness of the circuit and on the other hand it takes into account the size of it. Consequently, comparing it to the fitness function (fitness (8.1)) one can observe similar result but with small oscillations in the fitness function (fitness (8.6) (0.99)).

An important point to be discussed is the improvement of the results obtained with the hereby proposed cost functions and fitness functions variants. Intuitively one could think that the cost of gates reducing the global cost of the solution will only improve the results. However, as can be seen in Table 6 and Figure 23 the reasoning is not so simple. The duration of runs using normal fitness function as in equation (8.1) has generated solutions in less than 2000 generations. These results were not optimal; either too long and too expensive, or not correct at all. On the other hand the parameter that forces the selection operator to pick the individuals with smaller gate costs drives the evolutionary process deeper into the problem space. As the consequence, the search time using the improved fitness function increases. Also, when using the fitness function (8.6), the parameters α and β can be set to values in the interval [0,1]. Ultimately when using $\alpha = 1$ and $\beta = 1 - \alpha$, this leads to similar results as using the fitness (8.1). The problem remains the same: *"What are the correct parameters for an optimal relation between the error evaluation and the cost of the circuit in order to find the optimal representation of the searched function?"*

Table 7 presents a summary of some results in QC search. The first column is the function searched, the second column shows if the solution was found using either the non-optimized fitness function (8.1), or the optimized one

Table 7. Comparison of fitness function variants

Problem searched	Solution found 8.1/8.6	Time of search in number of generations
8.1/8.6		
Toffoli	Yes/Yes	<2000/<50000
Fredkin	Yes/Yes	<1000/<75000
Margolus	Yes/No	<1000/<100000

(8.6). The last column shows, for each circuit, the time in generations necessary to arrive at a solution. Each run was stopped after 100000 generations
in the case no solution was found previously in the run. The above presented
argument that the improved fitness function increases the required time to find
a solution can be argued here. As can be seen in two cases out of the three
problems, a significant time increase was observed. The presented results are
statistical averages over five runs for each problem. Out of them only the best
solutions are presented in Figures 16, 17, and 19. Two arguments support the
use of the improved fitness function. First, the solution was found in all cases
and it was found even when using a biased set. This implies that the cost
function allows the GA to explore parts of the problem space that were not
accessible to the GA which used the normal fitness function, i.e. without the
cost of gates. Also all results from Table 7 are from runs using the biased
starting set of gates.

The following observations can be made when using the GA proposed
here with variable length of chromosomes for quantum logic synthesis.

- No direct relation between the error and the cost values was observed.
  In general this means that there is almost an infinity of solutions to a
  given problem and the number of these solutions grows with increasing
  size of the circuit. A solution of non-minimal length can be found easier
  because of the variable length of chromosomes and next transformed to a
  shorter solution. It should be further investigated if this a better strategy
  than restricting a size to given value. When the length is short, such as
  5, the second strategy works, but for longer circuits the variable length
  of chromosomes seem to be more powerful.

- When drawing the fitness of the best individual each (let's say 100th)
  generation we observe that there is no tendency of optimization (convergence). This confirms our previous conclusion that the space of the
  fitness function is non-monotonic with lot of flat space (local optima)
  and some occasional peaks of good solutions.

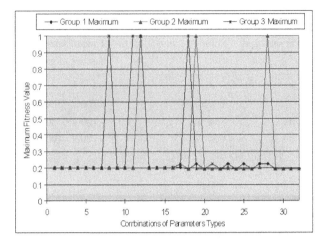

Figure 24. Results of the comparative tests with all 32 possible variations of the GA. Results are all normalized over 20 runs for each. The X axis represents each of the 32 possible combinations of the 5 concerned parameters. Y axis is the fitness function as from equation (8.6).

- The best type of Mutation is Bitwise, and the best type of Replication is SUS. The bitwise mutation operator seems to be more appropriate for such a high noise space as the one explored here. Also, this observation is based on an average measurement of fitness taken from the best individuals from each 100th generation but this approach does not necessary finds better solutions.
- Best type of GA observed from all previous results is Baldwinian, using SUS and bitwise mutation operator.

To illustrate these mentioned problems and conclusions we set up a small comparative experiment. In this experiment, three different parameterized GA were used. Two types of comparative experiments are set up and are used to explore the fitness landscape. In the first case all three configurations are using the same value of parameters (mutation, crossover, etc.) and five different parametric settings are modified. For this a binary encoding of each GA type was used. Each GA has a binary signature such as 00000 or 10110 corresponding to a particular configuration. Each digit in the binary signature represents if a certain feature is used or not. The five positions in the signature correspond respectively to the following options: fitness type, threshold, replication type, mutation type and Baldwinian. Fitness type means that if it is set to zero, the fitness will be calculated individually and if it is set to one the fitness will be calculated as a shared fitness in a group of 20 individuals. Threshold set to one means only individuals having fitness higher than threshold (0.6) will be used for replication. Replication type set to zero

Table 8. Options and parameters (group 1)

Population size	50–70–100	Calculation of fitness	Individual
Mutation probability	Variable [0.01–0.3]	Type of mutation	Normal
Crossover Probability	Variable [0.2—0.9]	Type of crossover	1 point
Factor alpha	Variable [0–1]	Type of replication	RW
Factor beta	Variable [0–1]	Type of fitness	Complex

means the GA will be using the Roulette Wheel type of selection and if it is set to one the stochastic universal sampling is used. The mutation type set to zero controls the GA to use normal mutation (once pre individual based on the mutation probability) while set to 1 selects the bitwise mutation (applied to each element of the chromosome). And finally Baldwinian bit indicates if the GA is using fitness calculated directly from the chromosome or calculated from a minimized chromosome. The results are presented on Figure 24. As can be seen the GA behaves differently for different parameters. In the case where no fitness = 1 is attained the GA was not always successful in finding the solution, while in the opposite case the GA found a good solution in at least half of runs. Each point on the figure is a result of a statistical mean over 20 runs. The main result from this experimentation is the fact that some configuration of the GA is more successful than others for the quantum search space search. This concerns mainly GAs of types 01000, 01011, 01100 and 11100. These configurations were experimental proofs of better adaptation and better search ability than others.

The second type of experiments comparing different GA settings is also based on three groups. In this approach the settings from Table 8 were used.

Three GA were created; each with 50, 70 and 100 individuals in the population. Each GA was configured as follows: no Baldwinian fitness calculation, normal (once per individual) mutation, single point crossover and individual fitness calculation were used. The parameters to be varied in this approach were: Alpha, Beta, mutation and crossover probability. Only a resume of the characteristics of Alpha and Beta factors and their best, worst and average values are listed in Table 9. We have chosen four sets of Alpha and Beta parameters, they are Alpha = 0.7 Beta = 0.3, Alpha = 0.8 Beta = 0.2, Alpha = 0.9 Beta = 0.1, and Alpha = 0.97 Beta = 0.03. Note that the relation between Alpha and Beta is that Alpha + Beta = 1.0.

We observe from the results in Table 9 that the fitness value increases when the Alpha value is increased. However, the best fitness of all generations is with the smallest from the proposed Alpha values, thus there is a two-fold (advantage and drawback) characteristic in increasing the factor Alpha. This confirms again the fact that the relation between the measure of error and the circuit cost is not direct. Based on solution analysis, it indicates that the size of the circuit should be controlled in a more direct way and not to be left to the evolution alone.

Next we analyzed also the influences of the Probabilities of Mutation and Crossover on the Fitness Value. Again, only a resume of the characteristics of Mutation and Crossover Probabilities and their best, worst and average values are to be listed in Tables 10 and 11, respectively.

Comparing the obtained results from Tables 9, 10 and 11, we arrive at the following conclusion: the quantum circuit synthesis is not extremely dependent on the values of such parameters as mutation or crossover probability. Rather, it depends on the types of selected operations on chromosomes that are used in various GA variants in experiments. This means the structure of the quantum space is not sensitive to the parameterization of genetic operators but is sensitive to the operations themselves. This is mainly because the size of the circuit is variable and the space of possible solutions available before and after one genetic operator is applied changes radically. In other words this means that for example having a circuit with 6 blocks that will have a group of solutions based on first four blocks equal to A, B, C, D with two additional blocks that are required. Now application of the mutation operator that will remove one block will completely modify the current outlook for success because it can be assumed that the same sequence of first four blocks A, B, C, D does not lead to a solution with 5 blocks. Consequently, the changes of length in the chromosome of individuals in the GA are one constraint more for the GA to overcome.

The attentive reader will remark the Pareto optimal GA results were not analysed. This results from a weak relation and non-equivalence of the two main parameters of our fitness function; the error and the cost. The Pareto optimal GA can be applied in the case the solution is on the intersection of best results of all parameters evaluation. This is not the case here because the cost of the circuit will never be minimal. A circuit representing a function will always have a cost equal to at least one parallel block with gates. Moreover, because in the ranking of individuals the error and cost are used to determine a win or a loss, the algorithm has the tendency to fall too soon into a local minimum with a relatively small error and a very small cost, from which there is no escape.

Table 9. Characteristic of fitness F with relation to factors alpha and beta

F	Group 1			Group 2			Group 3		
α	Best	Worst	Average	Best	Worst	Average	Best	Worst	Average
0.7	0.44	0.0777778	0.148206074	0.377778	0.0636364	0.144381594	0.188142	0.0333704	0.141694126
0.8	0.163663	0.0727273	0.159274043	0.163663	0.0652016	0.160634195	0.272727	0.0615385	0.161248625
0.9	0.181832	0.18	0.18007352	0.28	0.18	0.18110992	0.216788	0.18	0.18144116
0.97	0.224	0.194	0.19431102	0.194549	0.194	0.19403843	0.194549	0.194	0.19403294

Table 10. Characteristic of fitness F with relation to Mutation Probability (MP)

F	Group 1			Group 2			Group 3		
MP	Best	Worst	Average	Best	Worst	Average	Best	Worst	Average
0.01	0.194549	0.194	0.19401098	0.194549	0.194	0.19401098	0.194549	0.194	0.19401098
0.05	0.224	0.194	0.19431102	0.194549	0.194	0.19403843	0.194549	0.194	0.19403294
0.1	0.224	0.194	0.19503785	0.224	0.194	0.19491647	0.224	0.194	0.19432749
0.3	0.224	0.194	0.19491651	1	0.194	651	0.224	0.194	0.19465308

Table 11. Characteristic of fitness F with relation to Crossover Probability (CP)

F	Group 1			Group 2			Group 3		
CP	Best	Worst	Average	Best	Worst	Average	Best	Worst	Average
0.5	0.194549	0.194	0.19401098	0.194549	0.194	0.19401647	0.224	0.194	0.19491647
0.6	0.224	0.194	0.19403294	0.194549	0.194	0.19402749	0.194549	0.194	0.19403843
0.8	0.224	0.194	0.19432749	0.194549	0.194	0.19403843	0.194549	0.194	0.19403294
0.9	0.194549	0.194	0.19465308	0.194549	0.194	0.19402749	0.194549	0.194	0.19402196

11. Important Research Issues

The results of this work proved once again the "no free lunch theorem". The modifications of some parameters results in improvement in some runs but do not provide optimal solutions. Consequently the optimization of any algorithm needs to be specific to the proposed task. In our case, we have shown that the GA is well situated to explore the problem space of RC and QC logic synthesis. However few heuristics have been added to our algorithm and they provide us with a fresh look at this new area of logic synthesis.

An important issue to be discussed is the representation and transformations of circuits with more than 3 qubits. In the presented work we focused on 3-qubit universal gates. Consequently, the number of possible permutations of used gates is finite and a sufficient number of them can be defined in the starting set of gates. The definition of larger elementary units can be directly implied from the design of our individuals in the GA. As each circuit can be partitioned into parallel blocks, multiple optimizations can be used to improve the results. An example is to define blocks of predefined size bigger than two qubits, called complex blocks (CB). Consequently multi-gate blocks will be used as simple gates and will be manipulated by the classical genetic operators. One of such blocks can be seen in Figure 14. It was also illustrated in Figure 14 that synthesis of Toffoli on wires 2, 4 and 5 was too complicated to letting the GA to find such a gate. Particularly in QC there are many examples where more than 1-qubit gates are required to be connected to non-adjacent wires or even the wire order needs to be completely inverted. Consequently, our approach to solve this problem that evolved in the course of this research was to insert larger elementary units than simple gates. An example is the creation of a Feynman gate on wires 1 and 3. It can be argued that there are very many gates that could be defined such as this one, so should we define them all? The counter argument is that most gates can be synthesized from gates already used in our work. In our approach, the gate is connected to wires according to its position in the parallel block so the implementation of genetic operators is easy and can be executed quickly. The number of defined gates relates also closely to the number of respective equivalence transformations and the efficiency of rule-based optimizer decreases with too many rules. We believe that a middle way is the best one – there should be the major quantum primitives and basic gates defined on all wires and on all permutations of inputs. The penalty we pay however, is a big number of stored unitary matrices of basic gates. There is a solution though – declaring just subsets of available gates for every run. Thus only some of permutations of gates like Toffoli are declared in each run. With high speed of generating solutions and relatively high number of solutions for each matrix, our approach is competitive and allows various types of experimentations.

The approach presented here can be compared to a GP where each gate can be defined over a set of wires, with control bits and other parameters (Rubinstein 2001). In that case the mutation operator needs to be redefined to be able to modify any of the gate parameters. Consequently the number of possible variations of one gate grows quickly with the number of parameters of this gate and the number of wires it can be connected to. This makes genetic operators and evaluations of matrices in fitness function more complicated and slower. In our method all these parameters disappear and are replaced by the position of the gate itself in the parallel block. The negative point is that we need to define a particular gate for most of the variations of one gate in order to speed up the search. For 3-qubit and 4-qubit gates our approach is more efficient than having a highly complex parameterized gates such as in (Rubinstein 2001). A related important problem is how the GA representation affects the speed of fitness function calculations for larger circuits. The critical factor are definitely the calculations of resultant unitary matrices in fitness function, and especially the Kronecker matrix multiplications. In one approach, which is more flexible, the matrices are created in a run by decoding the GA or GP operators. In another approach, used by us, all matrices are calculated in advance and stored in the memory. They correspond to characters used in chromosomes and do not need to be created in a run. It results from our experiments that this approach is faster for small circuits since the gain of flexibility of the other approach is practically useless as leading to a too large time increase. On the other hand our approach does not scale well for a higher number of qubits than 5, because it requires to define and store many matrices. So, although more efficient than previous approaches, our encoding will not scale well for larger circuits and further research on synthesis of such circuits is necessary.

Another useful idea related to gate and circuit design that we acquired in the process of this work is the following. As can be seen from the examples, sometimes the solution can be found which is a permutation of output wires with respect to the initial specification. Because in many problems the order of output wires is not important, there should be some mechanism in the software that would automatically perform simplifications not to the circuit described by original order of output wires but to some of its permutations. A theory of respective local transformations should be developed to solve this problem.

12. Conclusions

In this paper we have presented an overview of a successful evolutionary approach to QC and RC synthesis. Even without heuristics we were able to find correct circuits representing universal gates, although perhaps non minimal ones. We find solutions to all quantum circuit synthesis problems presented by previous authors, such as entanglement and teleportation, and our programs reinvented several realizations of known gates such as Toffoli or Fredkin. In addition, we found very elegant solutions to gates such as Margolus, Miller and Peres. We believe that our solution to the Peres gate is minimal. It is now a challenge for future research to find a less expensive universal 3-qubit gate than the Peres gate using only 1-qubit and 2-qubit gates. The solution to this problem is not yet known, but in any case, based on our other studies, the Peres gate realization that we found seems to be very good, at least for NMR computing.

We hope that multiple circuit solutions will become eventually available from other authors for quantum circuit synthesis. We are compiling a library of *"Benchmark Quantum Functions and their Circuits"* (www.ece.pdx. edu/~lukacm/q-bench.html) so that the future quantum software developers will be able to compare their results with those of our programs, analogously as it is a standard in current binary CAD.

In the course of this research we found the importance and mutual relations of four problems: circuit encoding, fitness function, cost function and local optimizing transformations. It was thanks to running the program and analyzing its various solutions that we found the sets of equivalence transformations and the heuristics for their application (Lukac et al. 2003, in preparation). Although most examples were for 3-qubit circuits, our program allows in principle larger circuits. The RL variant is more efficient than the QL one.

Acknowledgments

The authors would like to thank Professor Soonchil Lee from Physics Department of KAIST in Korea for discussions about NMR computing and to Dr. Andrzej Buller from ATR Japan for his critical remarks on the first draft of this paper.

References

Al-Rabadi, A., Casperson, L., Perkowski, M. & Song, X. (2002). Canonical Representation for Two-Valued Quantum Computing. *Proc. Fifth Intern. Workshop on Boolean Problems*, 23–32. Freiberg, Sachsen, Germany, September 19–20.
Al-Rabadi, A. (2002). Novel Methods for Reversible Logic Synthesis and Their Application to Quantum Computing. Ph.D. Thesis, Portland State University, Portland, Oregon, USA, October 24.

Barenco, A. et al. (1995). Elementary Gates for Quantum Computation. *Physical Review A* **52**: 3457–3467.

Bennett, C. (1973). Logically Reversible Computation. *I.B.M. J. Res. Dev.* (17): 525–632.

Cirac, J. I. & Zoller, P. (1995). Quantum Computation with Cold Trapped Ions. *Physical Review Letters* (15 May) **74**: Issue 20, 4091–4094.

Dill, K. & Perkowski, M. (2001). Baldwinian Learning utilizing Genetic and Heuristic for Logic Synthesis and Minimization of Incompletely Specified Data with Generalized Reed-Muller (AND-EXOR) Forms. *Journal of System Architecture* **47**: Issue 6, 477–489.

Dirac, P. A. M. (1930). *The Principles of Quantum Mechanics*, 1st edn. Oxford University Press.

DiVincenzo, D. P. (1995). Quantum Computation. *Science* **270**: 255–256.

Dueck, G. W. & Maslov, D. (2003). Garbage in Reversible Designs of Multiple-Output Functions. *Proc. RM*: 162–170.

Einstein, A., Podolsky, B. & Rosen, N. (1935). *Physical Review* (47): 777.

Ekert, A. & Jozsa, R. (1996). Quantum Computation and Shor's Factoring Algorithm. *Review of Modern Physics* (July) **68**: Issue 3, 733–753.

Feynman, R. (1996). *Feynman Lectures on Computation. Addison Wesley*.

Fredkin, E. & Toffoli, T. (1982). Conservative Logic. *Int. J. of Theoretical Physics* (21): 219–253.

Ge, Y. Z., Watson, L. T. & Collins, E. G. (1998). Genetic Algorithms for Optimization on a Quantum Computer. In *Unconventional Models of Computation*, 218–227. London: Springer Verlag.

Goldberg, D. E. (1989). *Genetic Algorithms in Search, Optimization, and Machine Learning*. Addison Wesley.

Graham, A. (1981). *Kronecker Products and Matrix Calculus with Applications*. Chichester, UK: Ellis Horwood Limited.

Hirvensalo, M. (2001). *Quantum Computing*. Springer Verlag.

Iwama, K., Kambayashi, Y. & Yamashita, S. (2002). Transformation Rules for Designing CNOT-based Quantum Circuits. In *Proc. DAC*, 419–424. New Orleans, Louisiana.

Kerntopf, P. (2001). Maximally Efficient Binary and Multi-Valued Reversible Gates. In *Proceedings of ULSI Workshop*, 55–58. Warsaw, Poland, May.

Kerntopf, P. (2000). A Comparison of Logical Efficiency of Reversible and Conventional Gates. *Proc. of 3rd Logic Design and Learning Symposium (LDL)*. Portland, Oregon.

Khan, M. H. A., Perkowski, M. & Kerntopf, P. (2003). Multi-Output Galois Field Sum of Products Synthesis with New Quantum Cascades. In *Proceedings of 33rd International Symposium on Multiple-Valued Logic, ISMVL 2003*, 146–153. Meiji University, Tokyo, Japan, 16–19 May.

Khlopotine, A., Perkowski, M. & Kerntopf, P. (2002). Reversible Logic Synthesis by Gate Composition. *Proceedings of IWLS*: 261–266.

Kim, J., Lee, J-S. & Lee, S. (2000). Implementation of the Refined Deutsch-Jozsa algorithm on a Three-Bit NMR Quantum Computer. *Physical Review A* **62**: 022312.

Kim, J., Lee, J-S. & Lee, S. (2000). Implementing Unitary Operators in Quantum Computation. *Physical Review A*: 032312.

Klay, M. (1988). Einstein-Podolsky-Rosen Experiments: The Structure of the Sample Space I, II. *Foundations of Physics Letters 1*: 205–232.

Koza, J. (1992). *Genetic Programming. On the Programming of Computers by Means of Natural Selection*. The MIT Press.

Lee, J-S., Chung, Y., Kim, J. & Lee, S. (1999). A Practical Method of Constructing Quantum Combinational Logic Circuits. *arXiv:quant-ph/9911053v1* (12 November).

Lomont, Ch. (2003). Quantum Circuit Identities. *arXiv:quant-ph/0307111v1* (16 July).

Lukac, M., Pivtoraiko, M., Mishchenko, A. & Perkowski, M. (2002). Automated Synthesis of Generalized Reversible Cascades Using Genetic Algorithms. In *Proc. Fifth Intern. Workshop on Boolean Problems*, 33–45 Freiberg, Sachsen, Germany, September 19–20.

Lukac, M. & Perkowski, M. (2002). Evolving Quantum Circuits Using Genetic Algorithms. In *Proc. of 5th NASA/DOD Workshop on Evolvable Hardware*, 177–185.

Lukac, M., Lee, S. & Perkowski, M. (2003). Low Cost NMR Realizations of Ternary and Mixed Quantum Gates and Circuits. In preparation.

Lukac, M., Lee, S. & Perkowski, M. (2003). Inexpensive NMR Realizations of Quantum Gates. In preparation.

Miller, D. M. (2002). Spectral and Two-Place Decomposition Techniques in Reversible Logic. *Proc. Midwest Symposium on Circuits and Systems*, on CD-ROM, August.

Miller, D. M. & Dueck, G. W. (2003). Spectral Techniques for Reversible Logic Synthesis. *Proc. RM*: 56–62.

Mishchenko, A. & Perkowski, M. (2002). Logic Synthesis of Reversible Wave Cascades. In *Proc. IEEE/ACM International Workshop on Logic Synthesis*, 197–202. June.

Monroe, C., Leibfried, D., King, B. E., Meekhof, D. M., Itano, W. M. & Wineland, D. J. (1997). Simplified Quantum Logic with trapped Ions. *Physical Review A* (April) **55**: Issue 4, 2489–2491.

Monroe, C., Meekhof, D. M., King, B. E. & Wineland, D. J. (1996). A "Schroedinger Cat" Superposition State of an Atom. *Science* (May) **272**: 1131–1136.

Nielsen, M. A. & Chuang, I. L. (2000). *Quantum Computation and Quantum Information*. Cambridge University Press.

Negotevic, G., Perkowski, M., Lukac, M. & Buller, A. (2002). Evolving Quantum Circuits and an FPGA Based Quantum Computing Emulator. In *Proc. Fifth Intern. Workshop on Boolean Problems*, 15–22. Freiberg, Sachsen, Germany, September 19–20.

Peres, A. (1985). Reversible Logic and Quantum Computers. *Physical Review A* **32**: 3266–3276.

Perkowski, M., Kerntopf, P., Buller, A., Chrzanowska-Jeske, M., Mishchenko, A., Song, X., Al-Rabadi, A., Jozwiak, L., Coppola, A. & Massey, B. (2001). Regular Realization of symmetric Functions Using Reversible Logic. In *Proceedings of EUROMICRO Symposium on Digital Systems Design*, 245–252.

Perkowski, M., Jozwiak, L., Kerntopf, P., Mishchenko, A., Al-Rabadi, A., Coppola, A., Buller, A., Song, X., Khan, M. M. H. A., Yanushkevich, S., Shmerko, V. & Chrzanowska-Jeske, M. (2001). A General Decomposition for Reversible Logic. *Proceedings of RM*: 119–138.

Perkowski, M., Al-Rabadi, A. & Kerntopf, P. (2002). Multiple-Valued Quantum Logic Synthesis. *Proc. of 2002 International Symposium on New Paradigm VLSI Computing*, 41–47. Sendai, Japan, December 12–14.

Price, M. D., Somaroo, S. S., Tseng, C. H., Core, J. C., Fahmy, A. H., Havel, T. F. & Cory, D. (1999). Construction and Implementation of NMR Quantum Logic Gates for Two Spin Systems. *Journal of Magnetic Resonance* **140**: 371–378.

Price, M. D., Somaroo, S. S., Dunlop, A. E., Havel, T. F. & Cory, D. G. (1999). Generalized Methods for the Development of Quantum Logic Gates for an NMR Quantum Information Processor. *Physical Review A* (Octover) **60**(4): 2777–2780.

Rubinstein, B. I. P. (2001). Evolving Quantum Circuits Using Genetic Programming. In *Proceedings of the 2001 Congress on Evolutionary Computation (CEC2001)*, 144–151.

Shende, V. V., Prasad, A. K., Markov, I. K. & Hayes, J. P. (2002). Reversible Logic Circuit Synthesis. *Proc. 11th IEEE/ACM Intern. Workshop on Logic Synthesis (IWLS)*, 125–130.

Smolin, J. & DiVincenzo, D. P. (1996). Five Two-Qubit Gates are Sufficient to Implement the Quantum Fredkin Gate. *Physical Review A* (April) **53**(4), 2855–2856.

Spector, L., Barnum, H., Bernstein, H. J. & Swamy, N. (1999). Finding a Better-than-Classical Quantum AND/OR Algorithm Using Genetic Programming. In *Proc. 1999 Congress on Evolutionary Computation*, Vol. 3, 2239–2246. Washington DC, 6–9 July; IEEE, Piscataway, NJ.

Van Der Sypen, L. M. K., Steffen, M., Breyta, G., Yannoni, C. S., Sherwood, M. H. & Chuang, I. L. (2001). Experimental Realization of Shor's Quantum Factoring Algorithm Using Nuclear Magnetic Resonance. *Nature* (20/27 December) **414**: 883–887.

Vieri, C., Ammer, M. J., Frank, M., Margolus, N. & Knight, T. *A Fully Reversible Asymptotically Zero Energy Microprocessor*. MIT Artificial Intelligence Laboratory, Cambridge, MA 02139, USA

von Neumann, J. (1950). *Mathematical Foundations of Quantum Mechanics*. Princeton University Press.

Wheeler, J. A. & Zurek, W. H. (1983). *Quantum Theory and Measurement*. Princeton University Press.

Williams, C. W. & Gray, A. G. (1999). Automated Design of Quantum Circuits. ETC Quantum Computing and Quantum Communication, *QCQC '98*, 113–125. Palm Springs, California: Springer-Verlag, February 17–20.

Williams, C. P. & Clearwater, S. H. (1998)., *Explorations in Quantum Computing*. New York Inc.: Springer-Verlag.

www.ece.pdx.edu/~lukacm/q-bench.html

Yabuki, T. & Iba, H. (2000). Genetic Algorithms and Quantum Circuit Design, Evolving a Simpler Teleportation Circuit. In *Late Breaking Papers at the 2000 Genetic and Evolutionary Computation Conference*, 421–425.

Yang, G., Hung, W. N. N., Song, X. & Perkowski, M. (2003). Majority-Based Reversible Logic Gate. In *Proceedings of 6th International Symposium on Representations and Methodology of Future Computing Technology*, 191–200. Trier, Germany, March 10–11.

Conquering Uncertainty in Multiple-Valued Logic Design

DENIS V. POPEL

Computer Science Department, Baker University, P.O. Box 65, Baldwin City, 66006 Kansas, USA (E-mail: denis.popel@bakeru.edu); neoTROPY LCC, Lawrence, Kansas, USA

Abstract. In modern science, significant advances are typically made at cross-roads of disciplines. Thus, many optimization problems in *Multiple-valued Logic Design* have been successfully approached using ideas and techniques from *Artificial Intelligence*. In particular, improvements in multiple-valued logic design have been made by exploiting information/uncertainty measures. In this paper, we review well-known information measures in the multiple-valued domain and consider some methods of finding information measures for completely or incompletely specified functions with multiple-valued and continuous attributes. In this respect, the paper addresses the problem known as discretization and introduces a method of finding an optimal representation of continuous data in the multiple-valued domain. We also propose a technique for efficient calculation of different information measures using *Multiple-valued Decision Diagrams*. As one application of our technique, we outline an approach to synthesizing digital circuits derived from decision diagrams that can yield to reduction in power dissipation. The paper also shows the impact in several important areas of multiple-valued system design including (i) fuzzy logic, (ii) quantum computing systems, and (iii) data mining.

Keywords: decision diagrams, discretization, information theory, knowledge representation, machine learning, multiple-valued logic

> It will be seen that the processes which lose information are, as we should expect, closely analogous to the processes which gain entropy.
>
> *Norbert Wiener*

1. Introduction

The *entropy* principle of conquering uncertainty has a long history. It was pointed out in (Berger et al. 1996) that the concept of uncertainty can be traced back to Biblical times. In 1850, Rudolf Clausius, a German physicist, introduced the notion of entropy as a quantity that remains constant in the absence of heat dissipation. Since heat relates to the random movement of the individual particles that make up the system, entropy has been interpreted as the amount of disorder in the system. A hundred years later, in 1948, Shannon suggested a measure to represent the information by a numerical value, nowadays known as *Shannon entropy*. Since then, the term "uncertainty" is interchanged with the term "entropy." Here, we propose to broaden the concepts of information theory and direct them toward a variety of problems in multiple-valued logic design and beyond.

Over the years, many important problems in logic design and circuit synthesis have been approached using concepts from information theory. Thus, a number of promising results have been obtained in applying information theory methods to various optimization problems such as minimization of logic functions (Yanushkevich et al. 2000), testing of digital circuits (Agraval 1981), estimation of power dissipation (Marculescu et al. 1996), and quantum computing (Lo et al. 1998) to name just a few. The problem of employing information measures for uncertainty measuring in optimization, however, is far from being solved. For example, one of the most significant barriers to efficient utilization of information measures is that the majority of applications deal with Shannon entropy measures calculated for the tabular representations of discrete data although more efficient estimators can be found for different representations of discrete and continuous data.

This paper presents a convergence of two lines of research. In the first of these, the paper resolves the problem of finding a multiple-valued counterparts for continuous-valued signals. We introduce information density estimators on tabular representations of multiple-valued data and develop a unified discretization algorithm. In the second, this paper contributes to the pool of existing techniques for uncertainty measuring. We concentrate on employing graph structures and introduce a novel technique for calculating information measures on *Multiple-valued Decision Diagrams* (MDDs).

The purpose of this paper is twofold: (i) to cover uncertainty measuring for multiple-valued and continuous data representation, particularly concerning their use in circuit synthesis, logic design, and data mining; and (ii) to provide an introduction to some new techniques based on MDDs for information measuring and highlight their importance to a variety of problems in multiple-valued logic design. Thus, we present an entropy-driven algorithm for MDD ordering which is used to synthesize low-power circuits derived from decision diagrams. Among many emerging approaches, circuits derived from decision diagrams are of great interest to engineers and scientists as they compound the junction for the class of low-power and highly testable circuits. Concepts and techniques developed for such circuits can therefore be extended to other classes of digital circuits.

The breakdown of our efforts is given as follows. Section 2 reviews information measures for discrete tabular representations outlining their deficiencies for certain tasks of multiple-valued logic optimization. Section 3 introduces new information measures for continuous data and describes our new method of unsupervised and supervised discretization of continuous signals. It also discusses the algorithm and demonstrates the results of our study on mixed continuous and multiple-valued benchmarks. Section 4 intro-duces a new technique for calculating information measures using graph representations. Then, we point out two applications of this technique to MDD ordering and low-power synthesis in Section 5. Section 6 provides some ideas on possible applications of our discretization method and the information measuring technique to quantum computing and fuzzy logic. Each section contains a motivating example and a review of prior work in particular directions.

Tabular Representations

The prior research in information measuring focused on tabular represen-tations (truth tables, implicant tables, Karnaugh maps) of discrete data. In this Section, we consider a completely or incompletely specified m-valued function $f \colon B^n \to A$ over the variable set $\{X = x_1, \ldots, x_n\}$, where $B = \{0, \ldots, r - 1\}$ and $A = \{0, \ldots, m - 1\}$, and n is the number of r-valued variables. More formally, the function f can be represented by the sets C_i of cubes $\epsilon \colon C_i = \{\epsilon \in B^n | f(\epsilon) = i, = 0, \ldots, m - 1\}$; and $\mathcal{DC} = \{\epsilon \in B^n | f(\epsilon) = d\}$, where d is an undefined value. $k = \sum_{i=0}^{m-1} |C_i|$ is the number of cubes ϵ of the function f. It is understood that for the completely specified function, $\mathcal{DC}$ is an empty set, $\mathcal{DC} = \oslash$.

2. Information Measures: Discrete Domain

In order to quantify the content of information for a finite field of events $A = \{a_0, \ldots, a_{m-1}\}$ with the probability distribution $\{p(a_j)\}$, $j = 0, \ldots, m - 1$, Shannon introduced the concept of entropy (Shannon 1948). *Entropy* of the finite field A is given by

$$H(A) = - \sum_{j=0}^{m-1} p(a_j) \cdot \log p(a_j), \tag{1}$$

where the logarithm log is to base m, and the entropy is measured in bits ($m = 2$), dits ($m = 10$), etc. Suppose there are two finite fields of events A and B with probability distributions $\{p(a_j)\}$, $j = 0, \ldots, m - 1$, and $\{p(b_i)\}$, $i = 0, \ldots, r - 1$, respectively. Let $p(a_j, b_i)$ be the probability of the joint occurrence of a_j and b_i. The *joint entropy* of A and B is defined by

$$H(A, B) = - \sum_{j=0}^{m-1} \sum_{i=0}^{r-1} p(a_j, b_i) \cdot \log p(a_j, b_i), \tag{2}$$

note that

$$H(A) = -(\sum_{j=0}^{m-1}\sum_{i=0}^{r-1} p(a_j, b_i)) \cdot (\log \sum_{i=0}^{r-1} p(a_j, b_i)).$$

PROPERTY 1. *The uncertainty of a joint event is less than or equal to the sum of the distinct uncertainties:*

$$H(A, B) \le H(A) + H(B).$$

For any particular value a_j that A can take, there is a conditional probability $p(a_j|b_i)$ that B has a value b_i. It is expressed by

$$p(a_j|b_i) = \frac{p(a_j, b_i)}{\sum_{j=0}^{m-1}} p(a_j, b_i). \tag{3}$$

The *conditional entropy* of A given B is defined by

$$H(A|B) = -\sum_{j=0}^{m-1}\sum_{i=0}^{r-1} p(a_j, b_i) \cdot \log p(a_j|b_i). \tag{4}$$

This quantity determines how uncertain we are of A, on average, when specifying B. Using Equations (2) and (3) we obtain

$$H(A, B) = H(B) + H(A|B).$$

The *mutual information* between two finite fields A and B is given by

$$I(A; B) = H(A) + H(B) - H(A, B). \tag{5}$$

One can show that

$$I(A; B) = H(A) - H(A|B).$$

PROPERTY 2. *The average uncertainty of A is never increased by knowledge of B. It will be decreased unless A and B are independent events:*

$$H(A) \ge H(A|B).$$

In the study of multiple-valued functions, we deal with two finite fields: the set A denotes the values of the function f, and the set B denotes the values of the variable x. Equations (1)–(5) are used to calculate information measures of the function f and its variables X (Yanushkevich et al. 2000). In the majority of applications, the probability $p(f = a) = {}^{k(f=a)}/_k$, where $k(f = a)$ is the number of cubes ϵ for which $f = a$, and k is the total number of cubes ϵ of the function f. Other probabilities are calculated in the same way.

EXAMPLE 1. *Consider a 4-valued function given by its truth vector* $f =$ [0000023102130321]. *The probabilities of the function values are* $p(f = 0) = {}^7/_{16}$, $p(f = 1) = p(f = 2) = p(f = 3) = {}^3/_{16}$. *The entropy of the function* f *is* $H(f) = -{}^7/_{16} \cdot \log_4 {}^7/_{16} - 3 \cdot {}^3/_{16} \cdot \log_4 {}^3/_{16} = 0.94$. *The conditional entropy of the function* f *given* x_1 *is* $H(f|x_1) = -{}^4/_{16} \cdot \log_4 1 - 12 \cdot {}^1/_{16} \cdot \log_4 {}^1/_4 = 0.75$, *and given* x_2: $H(f|x_2) = -{}^4/_{16} \cdot \log_4 1 - 8 \cdot {}^1/_{16} \cdot \log_4 {}^1/_4 - 2 \cdot {}^2/_{16} \cdot \log_4 {}^1/_2 = 0.625$. *The mutual information of the function* f *and the variable* x_1 *is* $I(f; x_1) = 0.19$, *and of the function* f *and* x_2 *is* $I(f; x_2) = 0.315$.

The following applications take advantage of information measuring on tabular representations of discrete data:

Synthesis of low-power circuits: The power dissipation in integrated circuits is mainly caused by the *switching activity*, whose estimation is closely related to entropy measures described above. It implies that low-power consumption circuits should be designed taking into consideration the entropy criterion (Marculescu et al. 1996). Previous work relied heavily on tabular techniques for calculating entropy measures. A more detailed study on low-power synthesis for the circuits derived from decision diagrams is given in Section 5.

Testing of combinational circuits: It was shown in (Agraval 1981), that additional uncertainty, estimated by the information measures described above, leads to the coverage of primitive faults in digital circuits. It can be concluded that entropy measures are suitable criteria to construct circuits with high testability.

However, a number of other problems cannot be addressed by available information measuring techniques if initial data are specified in a non-tabular format, e.g. by a graph, or on a mixture of multiple-valued and continuous attributes. To cover possible tabular representations of continuous data, we introduce new information measures for the continuous domain in Section 3. A new technique for information measuring based on graph representations is given in Section 4. These trends are also depicted in Figure 1.

3. Information Measures: Continuous Domain

Continuous data analysis has gone through a dramatic change in the last decade and become an essential component in signal processing and knowledge discovery applications. This change and renewed interest in mixed analog/continuous/discrete systems have been facilitated by recent advances in continuous-discrete transformation techniques, where discrete in many cases implies multiple-valued representation. This problem, known as

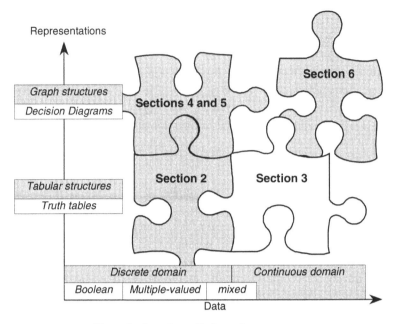

Figure 1. Overview of information measures.

quantization, covers a variety of applications ranging from circuit synthesis to data mining:

Circuit synthesis: The problem of quantization arises in many circuit applications including adaptive signal processing systems. The essential component of such systems, *long term analog memory*, is designed having quantizers (combinations of A/D and D/A converters) that store the discrete level nearest to analog input data. For example, the current-mode memory proposed in Baturone et al. (1997) is based on continuous algorithmic converters, which achieve an efficient trade-off between write/read speed and power consumption. The application in circuits also include *field programmable analog arrays* introduced in Pierzchala et al. (1994) for continuous, fuzzy, and multi-valued logic. Further extensions in this direction might include the development of multiple-valued, fuzzy and continuous reversible logic.

Logic design: The transition from continuous valued digits to discrete level multiple-valued digits is also important in the implementation of arithmetic operations using the theory of *analog digits* (Saed et al. 1999). Another area of applying discretization techniques and pattern finding for continuous data is the *decomposition* of logic functions.

Data mining: Substantial amount of work on quantization has been done by the data mining and database research communities. Some examples include the modifications of $ID3$ algorithm (Quinlan 1996) and the discretization of continuous attributes (Fayyad and Irani 1993; Dougherty et al. 1995). With the set of continuous variables as initial representation of database information, the problem is to find a technique that converts the database information into a multiple-valued function (Popel and Hakeem 2001).

3.1. *Multi-interval discretization*

Discretization is a process of transforming values of a continuous variable into a finite number of intervals, and associating with each interval a discrete numerical value. There are three different dimensions by which discretization approaches can be classified: global vs. local, static vs. dynamic, and unsupervised vs. supervised. We refer to (Dougherty et al. 1995) for the review of all discretization approaches. There are two basic techniques of discretization: the one which operates on the single continuous variable at a time is called *local*, in contrast to which, the *global* discretization processes the entire set of continuous variables. Some discretization approaches called *static* require an *a priori* set parameter η of the maximum number of intervals to produce. *Dynamic* approaches search through the space of all possible η values for all continuous variables simultaneously capturing differences in discretization. The first approach is important in multiple-valued representations, because of existing constrains on the number of intervals. In this paper, we work with local static discretization only. *Unsupervised discretization* does not utilize class labels in setting partition boundaries. *Supervised discretization* discretizes variables by taking into account the class labels assigned to examples. In our research, we consider a unified solution for both discretization approaches.

The major drawback of existing discretization techniques is that unsupervised discretization is considered without reference to supervised discretization. In addition to traditional information theory measures (see Equations (1)–(5)), some criteria such as *Gini Index* and *Information Gain* (Quinlan 1996) have been used. These criteria belong to the class of impurity measures designed to capture some aspects of partitioning relevant to "good" classification. We refer to the paper (Simovici and Jaroszewicz 2001) for detailed theoretical research on impurity measures. We suggest employing unified measures for both approaches based on an information density concept. Our goal is to develop a fast and efficient discretization algorithm which is applicable to any set of continuous data. The algorithm is supposed to maximize interdependence between the generated discrete intervals and generate the smallest number of intervals for a given continuous variable.

3.2. *Multiple-valued functions and intervals*

Let us use the following notation. Denote by s_i the value of a continuous variable x from the set of given values $S = \{s_1, \ldots, s_k\}$ assuming that x can take values in the range $T = [T^{min}, T^{max}]$, where $T^{min} < T^{max}$ and $\Delta T = T^{max} - T^{min}$. A partition $\mathcal{P}$ of the set S into r intervals $\mathcal{P}_u = [\mathcal{P}_u^{min}, \mathcal{P}_u^{max}]$, $u = 0, \ldots, r - 1$, can be characterized as

$$\begin{cases} \mathcal{P} = \bigcup_{u=0}^{r-1} P_u, \text{ and } \mathcal{P}_u \neq \oslash \text{ for } u = 0, \ldots, r - 1, \\ s_i \notin \mathcal{P}_v, \text{ if } s_i \in \mathcal{P}_u \text{ for } u \neq v. \end{cases}$$

Taking boundaries T^{min} and T^{max} into consideration, the partitions can be described by

$$\begin{cases} \mathcal{P}_0^{min} = T^{min}, \\ \mathcal{P}_u^{min} > T^{min} \text{ and } \mathcal{P}_u^{max} < T^{max}, \text{ for } u = 1, \ldots, r - 2, \\ \mathcal{P}_{r-1}^{max} = T^{max}. \end{cases}$$

We study below the transition of the continuous variable x into the r−valued variable χ. Let us define new information measures that are used in unsupervised (with no information about class labels) and supervised (with class labels attached) discretization.

3.3. *Unsupervised discretization*

We introduce the following class-independent measures.

DEFINITION 1. *The Entropy $H(T, S)$ of the partition $\mathcal{P}$ is defined by*

$$H(T, S) = -\sum_{i=0}^{k} \delta_i \cdot \log \delta_i, \tag{6}$$

where the probabilities δ_i are

$$\delta_i = \begin{cases} (s_1 - T^{min})/\Delta T, \text{ if } i = 0, \\ (s_{i+1} - s_i)/\Delta T, \text{ if } i = 1, \ldots, k - 1, \\ (T^{max} - s_k)/\Delta T, \text{ if } i = k. \end{cases}$$

The quantity $H(T, S)$ has a number of interesting properties which are utilized in our research.

PROPERTY 3. $H(T, S) = 0$ *if and only if the set S has no values: $S = \oslash$. Otherwise, $H(T, S)$ is positive.*

PROPERTY 4. *Given k, $H(T, S)$ is maximal and equal to $\log(k + 1)$ when all δ_i take equal values, that value being $1/(k + 1)$. This is, intuitively, the case of the most uncertainty for the uniform distribution of values.*

PROPERTY 5. *Any change toward equalization of probabilities $\delta_0, \ldots, \delta_k$ leads to increasing $H(T, S)$. Thus, if $\delta_0 < \delta_1$ and δ_0 is increasing, δ_1 is decreasing by the same amount, and if δ_0 and δ_1 are nearly equal, then $H(T, S)$ is close to its maximum value.*

DEFINITION 2. *The Information Density $D(T, S)$ of the partition $\mathcal{P}$ is given by*

$$D(T, S) = \begin{cases} H(T, S)/\log(k + 1), & if \ k > 0, \\ 0, & if \ k = 0. \end{cases} \tag{7}$$

Making the definition consistent, we set $D(T, S) = 0$ for the intervals where $S = \oslash$.

PROPERTY 6. *The terminal intervals $[T^{min}, s_1]$ and $[s_k, T^{max}]$ make no contribution to the information density $D(T, S)$. These segments of $D(T, S)$ have values of zero.*

3.4. Supervised discretization

Here we generalize the definitions presented above to cover supervised discretization. In supervised discretization, the dependencies between the set of class labels $C = \{c_0, \ldots, c_{m-1}\}$, where m is the number of classes, and the set of values $S = \{s_1, \ldots, s_k\}$ can be illustrated as a relation: $S \to C$. The notation for values in S can be adjusted as follows: $S^c = \{s^c | c \in C\}$. Thus, $S = \bigcup_{j=0}^{m-1} S^{c_j}$.

DEFINITION 3. *The Class Entropy $\mathcal{H}(T, S^{c_j})$ of the partition $\mathcal{P}$ is defined by*

$$\mathcal{H}(T, S^{c_j}) = -\sum_{i=0}^{k} \delta_i^{c_j} \cdot \log \delta_i^{c_j}, \tag{8}$$

where the probabilities $\delta_i^{c_j}$ are

$$\delta_i^{c_j} = \begin{cases} (s_1^{c_j} - T^{min})/\Delta T, & if \ i = 0, \\ (s_{i+1}^{c_j} - s_i^{c_j})/\Delta T, & if \ i = 1, \ldots, k - 1, \\ (T^{max} - s_k^{c_j})/\Delta T, & if \ i = k. \end{cases}$$

The class entropy $\mathcal{H}(T, S^{c_j})$ employs Properties 1-3. Hence, for the entropy of the partition $\mathcal{P}$ comprised of m classes, we have

$$\mathcal{H}(T, S) = \begin{cases} \frac{1}{m} \cdot \sum_{j=0}^{m-1} \mathcal{H}(T, S^{c_j}), & if \ \exists j \ |S^{c_j}| > 0, \\ H(T, S), & if \ \forall j \ |S^{c_j}| = 0. \end{cases} \tag{9}$$

DEFINITION 4. *The Class Information Density $\mathcal{D}(T, S)$ of the partition $\mathcal{P}$ is given by*

$$\mathcal{D}(T, S) = \begin{cases} \mathcal{H}(T, S)/\log(k+1), & if \ k > 0, \\ 0, & if \ k = 0. \end{cases} \tag{10}$$

PROPERTY 7. *For a single class of labels, $m = 1$, the measures of the entropy/information density and the class entropy/class information density are equal: $H(T, S) = \mathcal{H}(T, S)$ and $D(T, S) = \mathcal{D}(T, S)$.*

The measures of entropy/information density and class entropy/class information density can be classified as impurity measures: they are minimum if $\exists i$ such that $\delta_i = 1$; they are maximum if $\forall i: i = 0, \ldots, k, \delta = 1/(k+1)$; they are symmetric with respect to components of δ; and they are differentiable everywhere in their ranges.

3.5. *Conditional information density and information density gain*

For the given set of values S with the associated set of class labels C, an arbitrary cut point T^{cut} splits the partition $\mathcal{P}$ into two partitions $\mathcal{P}_1$ and $\mathcal{P}_2$. It has been proved in Fayyad and Irani (1993) that when searching for the best binary split by choosing a single cut point, we can restrict our attention to boundary points. Thus, the cut point T^{cut} preserves the following:

$$\begin{cases} \mathcal{P}_1^{min} = T^{min}, \\ \mathcal{P}_1^{max} = \mathcal{P}_2^{min} = T^{cut}, \\ \mathcal{P}_2^{max} = T^{max}. \end{cases}$$

DEFINITION 5. *The Conditional Information Density $\mathcal{D}(T, S|T^{cut})$ of the partition $\mathcal{P}$ given the cut point T^{cut} is defined by*

$$\mathcal{D}(T, S|T^{cut}) = p_1 \cdot \mathcal{D}(T_1, S_1) + p_2 \cdot \mathcal{D}(T_2, S_2), \tag{11}$$

where $\mathcal{D}(T, S)$ is the class information density (see Equation (10)), p_1 and p_2 are the following probabilities:

$$p_1 = (T^{cut} - T^{min})/\Delta T, \quad p_2 = (T^{max} - T^{cut})/\Delta T.$$

DEFINITION 6. *The Information Density Gain $\mathcal{I}(T, S; T^{cut})$ as the measure of "goodness" of the partitioning process is given by*

$$\mathcal{I}(T, S; T^{cut}) = \mathcal{D}(T, S|T^{cut}) - \mathcal{D}(T, S). \tag{12}$$

PROPERTY 8. *If the discretization process leads to increasing the information density of the formed partitions $\mathcal{P}_1$ and $\mathcal{P}_2$, then information density gain $\mathcal{I}(T, S; T^{cut}) > 0$. Otherwise, $\mathcal{I}(T, S; T^{cut})$ is not positive, and the cut point T^{cut} is worthless.*

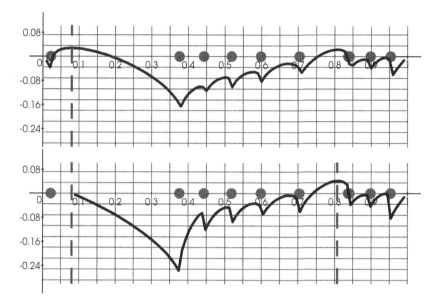

Figure 2. Unsupervised discretization for the set of values.

PROPERTY 9. *The information density gain $\mathcal{L}(T, S; T^{cut})$ is scale and shift invariant.*

3.6. *Optimization criterion*

Here we introduce a criterion based on the previously defined information density measures to control the number of intervals produced over the continuous space. The cut points are evaluated according to the *information density criterion* maximizing the distinction between classes assuming no gaps between intervals, intuitively,

$$T^{cut} = argmax\{\mathcal{D}(T, S|T^{cut}) - \mathcal{D}(T, S)\}. \tag{13}$$

Obviously, the lesser the interval, the lesser the density of values.

EXAMPLE 2. *Let us consider an example of the discretization process for a continuous variable x specified by the set of arbitrary values $S = \{0.022, 0.376, 0.443, 0.519, 0.598, 0.704, 0.837, 0.841, 0.899, 0.953, 0.954\}$ and the range $T = [0, 1]$. Figure 2 illustrates two steps of the unsupervised discretization. Shown on the left are the distributions of the information density gain $\mathcal{L}(T, S; T^{cut})$ for different cut points. Depicted on the right are the final cut points $T^{cut} = \{0.080, 0.807\}$ according to the optimization criterion (Equation (13)). Figure 3 illustrates several steps of the supervised discretization for the continuous variable x specified by*

the set of values and class labels $S = \{0.022^0, 0.376^0, 0.443^0, 0.519^1, 0.598^0,$
$0.704^1, 0.837^1, 0.841^2, 0.899^2, 0.953^2, 0.954^2\}$ *and the range* $T = [0, 1]$. *Eight cut*
points are detected $T^{cut} = \{0.097, 0.512, 0.559, 0.607, 0.833, 0.840, 0.842, 0.938\}$.
The result of the discretization process is the set of partitions and corresponding
values for their discrete representation (the radix of the multiple-valued variable
χ *is* $r = 2$ *for the unsupervised discretization and* $r = 8$ *for the supervised*
discretization).

3.7. *Discretization algorithm and experimental results*

Our discretization algorithm works in a greedy top-down manner: it starts
with the single partition $\mathcal{P}$ and splits it recursively. A sketch of the algorithm
for the continuous variable x is shown below:

Step1. Sort all given values from the set $S = \{s_1, \ldots, s_k\}$ in ascending order. Initialize the first
partition: interval boundaries T^{min}, T^{max}; and the set of class labels C.

Step 2. For the current partition, generate a set of possible cut points with a kernel $\Delta T / 100$.
Calculate the class information density: $\mathcal{D}(T, S) = \mathcal{H}(T, S)/\log(k + 1)$.

Step 3. Form two partitions $\mathcal{P}_1$ and $\mathcal{P}_2$. Calculate the resulting conditional information density
$\mathcal{D}(T, S|T^{cut}) = p_1 \cdot \mathcal{D}(T_1, S_1) + p_2 \cdot \mathcal{D}(T_2, S_2)$.

Step 4. Find a potential cut point according to the optimization criterion: $T^{cut} = argmax\{\mathcal{D}(T, S|T^{cut}) - \mathcal{D}(T, S)\}$.

Step 5. If the resulting conditional information density of two partitions $\mathcal{P}_1$ and $\mathcal{P}_2$ given the
potential cut point T^{cut} is greater than the information density of the initial partition $\mathcal{P}$
(see Property 8), then accept the cut point T^{cut}. Otherwise, terminate the recursion.

Step 6. Execute Steps 2–5 for the partitions $\mathcal{P}_1$ and $\mathcal{P}_2$ recursively.

Step 7. Sort all cut points, assign discrete values to formed partitions.

The algorithm is coded in Java and the experimental results are obtained on
a Pentium 1GHz workstation with 256Mb of memory (the same hardware
settings are used for other experiments reported below). The main purpose
of our experimental study is to compare different outcomes of the unsupervised and supervised discretization. The experiments are not intended to
estimate learning criteria or any restrictions for continuous data. The question
of applying certain restrictions of multiple-valued logic to assign values to
variables is beyond the scope of this paper (it can be resolved via algebraic
rules to refer to suitable systems of multiple-valued logic).

The experiments have been carried out on *Machine Learning* benchmarks[1]
each of which has at least one continuous variable. A fragment of our results
is shown in Tables 1 and 2, where r denotes the maximal radix of input
variables, m denotes the radix of a single-output function, [continuous] indicates the number of continuous variables, [discrete] indicates the number of
discrete variables. The columns labelled [Av. Radix] list the average radix

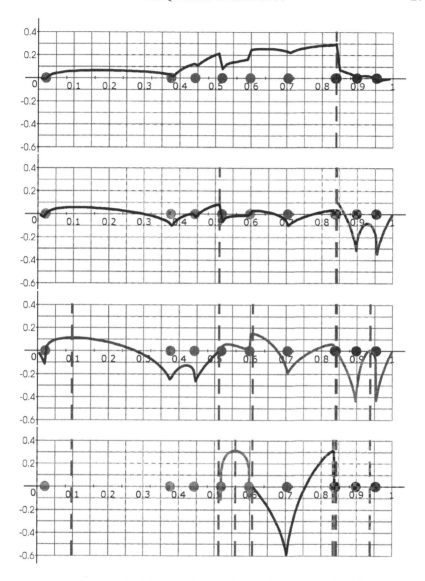

Figure 3. Supervised discretization for the set of values and class labels.

of continuous variables after discretization. The columns labelled [Error (%)] give the error rate of discretization algorithms. It is calculated as the ratio of the number of distinguished combinations to the total number of combinations.

In our experimental study, we compare unsupervised and supervised discretization approaches based on the information density criterion

Table 1. Results on the machine learning benchmark set (unsupervised discretization)

Dataset	r	m	Variables		Unsupervised	
			Continuous	Discrete	Av. radix	Error (%)
Adult	41	2	6	11	3.4	2.843
Allbp	6	2	7	22	2.5	10.250
Auto-mpg	14	5	5	2	4.0	4.117
Bupa	7	2	6	0	5.3	8.023
Clean1	476	2	166	2	5.3	0.935
Crx	14	2	6	9	3.8	4.822
Dis	6	2	7	22	2.9	12.540
Glass	7	7	9	0	3.1	15.567
Ionosphere	7	2	34	0	2.6	4.643
Iris	7	3	4	0	4.1	5.007
Letter	7	26	16	0	4.8	1.782
Pima-indians-diabetes	7	2	8	0	4.0	15.163
Post-operative	3	3	1	7	2.0	8.801
Sick	7	2	7	22	3.9	5.186
Spambase	7	2	57	0	2.1	2.257
Waveform	7	3	21	0	5.5	4.513
Wdbc	7	2	30	0	3.1	2.962
Yeast	7	10	8	0	3.7	5.320

discussed in Subsection 3.1. Not surprisingly, the experiments reveal that supervised discretization produces more accurate results than unsupervised discretization. The average radix of continuous variables is bigger for the supervised algorithm. For many benchmarks, our supervised discretization algorithm outperforms the algorithm based on information gain measures (column [Inf. Gain]) reported in Quinlan (1996). The information gain criterion from Quinlan (1996) has been integrated into our algorithm.

The extension of applicability of information measures towards the continuous domain, combined with the positive impact of traditional information measures, cannon however address crucial problems for different representations. Therefore, new techniques are required to compute information measures on graph structures.

Table 2. Results on the machine learning benchmark set (supervised discretization)

Dataset	Supervised		Inf. gain	
	Av. radix	Error (%)	Av. radix	Error (%)
Adult	4.8	2.572	4.5	2.238
Allbp	5.1	6.560	4.7	8.121
Auto-mpg	7.0	1.005	5.5	1.899
Bupa	6.6	6.087	4.7	11.950
Clean1	6.9	0.210	7.3	4.741
Crx	5.0	2.174	4.4	3.338
Dis	5.1	6.560	4.0	10.502
Glass	5.9	10.748	3.6	22.747
Ionosphere	4.7	1.139	4.9	1.250
Iris	6.0	4.667	6.5	3.667
Letter	6.4	0.025	7.2	4.093
Pima-indians-diabetes	5.3	6.771	6.0	1.903
Post-operative	3.0	7.78	3.0	7.780
Sick	5.3	4.546	4.3	9.153
Spambase	3.6	1.304	4.8	2.020
Waveform	7.0	1.660	6.1	3.677
Wdbc	4.53	1.757	4.8	1.535
Yeast	5.3	5.054	5.0	10.011

Graph Representations

Over the past decade, many important problems in computer science and engineering have been approached using graph structures: decision trees, decision diagrams, fractals, lattices, etc. The superior example is a *Binary Decision Diagram* (BDD) which has become an advanced structure in computer-aided design because of its simplicity, canonical nature, and effective algorithms. Thus, a large variety of decision diagrams has been introduced since the late 80s. There are almost 40 different abbreviations for them.

As well-known, Boolean functions $\{0, 1\}^n \rightarrow \{0, 1\}$ can be represented by BDDs (Bryant 1986). A BDD is a directed acyclic graph with the node set where: (i) each *non-terminal* node *node* is labeled by a variable x and assigned as a decision variable with two successors $node_0$ and $node_1$; (ii) each *terminal* node is labeled with the leaf value *leaf* and has no successors. Obviously, BDDs can be extended to represent multiple-valued functions $\{0, \ldots, r - 1\}^n \rightarrow \{0, \ldots, m - 1\}$, and the resulting graphs are denoted as *Multiple-valued Decision Diagrams* (MDDs). For this, each non-terminal node has r successors $node_0, \ldots, node_{r-1}$ and v predecessors $_0node, \ldots, _{v-1} node$.

A MDD is called *ordered* if the variables X appear in the same order $x_{l_1} \prec \ldots \prec x_{l_n}$ in each path from the root to a terminal node. Otherwise, a MDD is called *free*. In the following, only ordered MDDs are considered. It is known that the order of variables can be changed to reduce the MDD size as the number of nodes. We employ this fact and integrate information measures into our MDD ordering algorithm in Section 5.

4. Calculation of Information Measures on MDDs

In this Section, we extend the scope of the techniques presented above for tabular representations of continuous and multiple-valued data to encompass decision diagram structures. The first and natural step of calculating information measures on decision diagrams is to determine the set of probabilities (see Equations (1)–(4)) of the function f and its sub-functions $f_{|x=0}, \ldots, f_{|x=r-1}$.

4.1. *Computation of probabilities on BDDs*

The most distinguishing characteristic of decision diagrams is their ability to support recursive algorithms. Such algorithms are linear in the size of the corresponding graph (the size of the graph, however, may be exponential in the number of variables). Thus, an exact algorithm to compute output probabilities based on the traversal of decision diagrams and called the *probability assignment algorithm* has been reported in Miller (1998). In the case of a completely specified Boolean function f: $p(x = 0) = p(x = 1) = {}^1/_2$, and the output probability of the constant '1' is $p(leaf = 1) = 1$ and the output probability of '0' is $p(leaf = 0) = 0$. Since each node of the designed diagram is an instance of Shannon expansion, the probability of the function f can be calculated by the following recursive technique:

$$p(f) = \frac{1}{2} \cdot p(f_{|x=0}) + \frac{1}{2} \cdot p(f_{|x=1}). \tag{14}$$

An alternative is to use the top-down strategy. The probability assignment algorithm introduced in Thornton and Nair (1995) can be described as:

$$\begin{cases} p(node) = 1, \ if \ node = root \\ p(node) = p(_0node) + \ldots + p(_{v-1}node), \ if \ node \neq root. \end{cases}$$

top-down   **down-top**

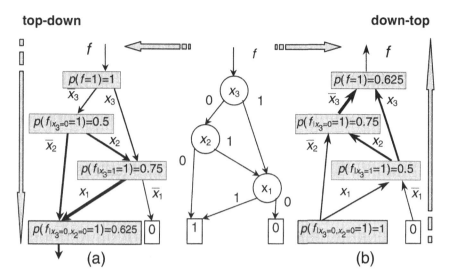

Figure 4. Computation of output probabilities.

The output probabilities are taken from the terminal nodes: $p(f = 1) = p(leaf = 1)$ and $p(f = 0) = p(leaf = 0)$.

EXAMPLE 3.   *Let us consider an example of computation of output probabilities for the function $f = x_1 \vee \overline{x}_2 \cdot \overline{x}_3$. The top-down strategy based on assigning $p(root) = 1$ gives $p(f = 1) = p(leaf = 1) = 0.625$, as illustrated in Figure 4(a). The down-top strategy based on assigning $p(leaf = 1) = 1$ computes $p(f = 1) = p(root) = 0.625$, as shown in Figure 4(b).*

4.2.   *Computation of probabilities on MDDs*

To extend the down-top strategy discussed above to the multiple-valued domain, we suggest the following modification of the probability assignment algorithm (Equation (14)):

$$p(node) = \sum_{i=0}^{r-1} p(x = i) \cdot p(node_i). \qquad (15)$$

To compute the output probabilities of the multiple-valued function f, proper probability values need to be assigned to all terminal nodes. Thus, assigning $p(leaf = i) = 1$ and $p(leaf \neq i) = 0$ leads to $p(f = i) = p(root)$. In fact, this modification allows us to compute conditional and joint probabilities which are required for the calculation of the conditional entropy $H(f|x)$ according to Equation (4). For example, it is necessary to set $p(x = 1) = 1$,

$p(x \neq 1) = 0$, $p(leaf = 1) = 1$ and $p(leaf \neq 1) = 0$ before the MDD traversal to get the joint probability $p(f = 1, x = 1)$. This modification is especially noteworthy in light of the fact that only one MDD traversal is required to calculate the whole spectrum of probabilities.

4.3. *Efficient calculation of entropies on MDDs*

Once the output probabilities have been computed and stored, implementation of the technique for exact computation of information measures is straightforward. The entropy $H(f)$ of the function f is calculated by

$$H(f) = -\sum_{j=0}^{m-1} p(f = j) \cdot \log p(f = j). \tag{16}$$

To calculate the conditional entropy $H(f|x)$ of the function f given the variable x, we exploit the fact that Equation (4) can be simplified:

$$H(f|x) = \sum_{i=0}^{r-1} p(x = i) \cdot H(f_{|x=i}). \tag{17}$$

In technical terms, this means that for the calculation of the conditional entropy $H(f|x)$ we need to know the entropy $H(f_{|x=i})$ of each sub-function $f_{|x=i}$.

EXAMPLE 4. *(Continuation of Example 3) The entropy of the function f is $H(f)$ = 0.96. The conditional entropy of the function f given x_2: $H(f|x_2) = {}^1/_2 \cdot H(f_{|x_2=0}) + {}^1/_2 \cdot H(f_{|x_2=1}) = 0.41 + 0.5 = 0.91$ (Figure 5). The same manipulation yields: $H(f|x_1) = 0.41$ and $H(f|x_3) = 0.91$. The conditional entropy of the function f given a set of variables $\{x_1, x_2\}$ is $H(f|x_1x_2) = 0.25$.*

We have implemented the technique within the CUDD package[2] and carried out experiments on several sets of ISCAS85[3] and ISCAS89[4] benchmarks (combinational and sequential circuits). A fragment of our results is shown in Table 3, where some information measures are calculated on BDDs and MDDs (the entropy $H(f_0)$ of the first output f_0; conditional entropies $H(f_0|x_1)$ and $H(f_0|x_2)$ given the first x_1 and second x_2 inputs). Since in case of multiple-valued functions there are no accepted benchmark sets, we interpret ISCAS85 and ISCAS89 benchmarks as multiple-valued circuits by transforming AND gates into 4-valued MIN gates and OR gates into 4-valued MAX gates.

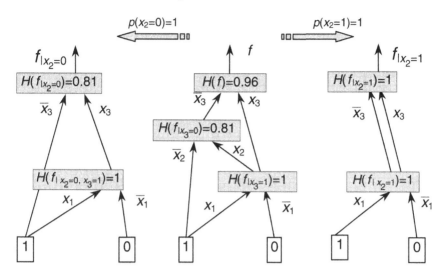

Figure 5. Calculation of information measures.

Table 3. Some values of information measures.

| | Information measures | | | | | |
| | BDDs | | | MDDs | | |
Dataset	$H(f_0)$	$H(f_0|x_1)$	$H(f_0|x_2)$	$H(f_0)$	$H(f_0|x_1)$	$H(f_0|x_2)$
C1908	1.000	0.017	1.000	1.000	0.129	1.000
C3540	0.337	0.272	0.272	0.533	0.160	0.160
C432	0.385	0.378	0.378	0.582	0.214	0.214
C6288	0.811	0.500	0.500	0.169	0.500	0.500
C880	0.544	0.406	0.406	0.389	0.027	0.027
S1196	0.896	0.842	0.883	0.306	0.711	0.309
S1512	0.954	0.905	0.749	0.344	0.195	0.033
S344	0.811	0.500	0.749	0.169	0.500	0.500
S4863	0.982	0.922	0.922	0.631	0.272	0.272
S635	0.811	0.811	0.500	0.169	0.169	0.500

5. Applications of Information Measuring on MDDs

As mentioned earlier in Section 4, the order of variables can be changed to reduce the size of a decision diagram.

5.1. *MDD ordering algorithm*

Here we introduce a MDD ordering algorithm. The *information criterion* is used to choose a decomposition variable x for the arbitrary level of a decision diagram, so that the conditional entropy $H(f|x)$ of the function f given the variable x has to be minimal:

$$x = argmin\{H(f|x_l), l = 1, \ldots, n\}. \tag{18}$$

A sketch of our MDD ordering algorithm *Info* for the function f is shown below:

Step 1. Initialize the variable ordering lexicographically: $x_1 \prec \ldots \prec x_n$ (x_1 is the topmost variable).

Step 2. Run Steps 3-6 for all levels ($level = 1, \ldots, n$) of the decision diagram.

Step 3. Run Step 4 for all variables $x_l, l = 1, \ldots, n - level$.

Step 4. Calculate $p(f = 0|x_l = 0), p(f = 0|x_l = 1), \ldots, p(f = m - 1|x_l = r - 1)$ (see Section 4). Calculate $H(f|x_l)$ according to Equation (17).

Step 5. Reorder variables according to the criterion (Equation (18)): $H(f|x_{l_1}) \leq \ldots \leq H(f|x_{l_{n-level}})$. Select the variable x_{l_1} as the decomposition variable for the current level.

Step 6. Rebuild MDD for the new variable ordering $x_{l_1} \prec \ldots \prec x_{l_{n-level}}$.

The experimental results have been obtained on several sets of ISCAS85 and ISCAS89 benchmarks (see Table 4). BDDs and MDDs are built for both combinational and sequential circuits using different ordering approaches (window, genetic, sifting and our *Info*). We observe that the algorithm $Info^{BDD}$ demonstrates better results in terms of diagram size for some testing functions compared to the existing ordering approaches. Finally, we show that our approach to MDD ordering can yield to reduction in the power dissipation for the circuits derived from decision diagrams.

5.2. *Low-power synthesis*

With the progress made in such areas as portable computing and wireless communication, power dissipation is becoming a crucial factor in circuit design. Such a consideration has resulted in the growing need for minimizing power consumption in today's digital systems. Considering the impact of design strategies on power dissipation, it is important to integrate the technology-independent and technology-dependent stages. In other words, logic specification should relate to the final topological structures and physical implementation of circuit. To move in this direction, we address the problem of automatic synthesizing transistor-level implementations starting from technology-independent structures like decision diagrams. Diagram-based circuits, implemented as networks of transistors, are of great interest

Table 4. Different BDD/MDD ordering algorithms

Dataset	Size [number of nodes]				
	Window^{BDD}	Genetic^{BDD}	Sifting^{BDD}	Info^{BDD}	Info^{MDD}
C17	11	11	11	9	15
C1908	36007	36007	36007	65032	210781
C432	1733	1733	1733	2425	4029
C880	346660	346660	21965	319071	455173
S1423	98454	98454	98454	73680	90244
S298	125	125	125	175	257
S349	206	206	206	141	296
S382	168	168	168	171	352
S820	2651	2651	2651	2283	3978

to researchers as they address the class of low-power dissipation circuits. Concepts and techniques developed for low-power circuits derived from decision diagrams can be used in the study of other classes of digital circuits.

The dominant part of power dissipation in CMOS circuits is the dynamic power dissipation, which is directly proportional to the signal switching activity (charging and discharging of load capacitances of logic gates). It is quite natural and useful to consider switching activity in terms of information measures (Marculescu et al. 1996) and use entropy as a complexity characteristic of the optimization process (Popel 2001). The entropy characterizes the uncertainty of a sequence of applied vectors and thus is intuitively related to switching activity. It is shown in Marculescu et al. (1996) that the average switching activity is bounded above by half of its entropy. The distinctive idea of our approach is to minimize switching activity for each level while designing BDDs. This will ultimately decrease switching activity of synthesizing circuits derived from BDDs.

EXAMPLE 5. *(Continuation of Example 4) According to the information criterion (Equation (18)), the optimal order of variables in the decision diagram for the function f is $x_1 \prec x_2 \prec x_3$ (Figure 4(b)). The power dissipated by circuits derived from BDDs with different variable ordering is*

Order			Power [μW]
$x_1 \prec x_2 \prec x_3$	*or*	$x_1 \prec x_3 \prec x_2$	31.87
$x_2 \prec x_1 \prec x_3$	*or*	$x_3 \prec x_1 \prec x_2$	45.00
$x_3 \prec x_2 \prec x_1$	*or*	$x_2 \prec x_3 \prec x_1$	37.50

Table 5. Power dissipation

| Dataset | Power [μW] | | | |
	Window^{BDD}	Genetic^{BDD}	Sifting^{BDD}	Info^{BDD}
C17	1 25	125	125	105
C1908	–	–	–	80583
C432	25860	25860	25860	28470
S1238	30268	30268	30268	19643
S1423	–	–	–	87082
S27	182	182	182	162
S298	1688	1688	1688	1363
S349	2865	2865	2345	2899
S526	3216	3216	3216	2395
S713	18215	18215	18215	15341
S953	18464	18464	18464	15661

The power consumption for the best case ($x_1 \prec x_2 \prec x_3$) is 1.5 times smaller than for the worst case taking into account the same size of BDDs for both cases.

We have used SIS[5] as an environment for modeling and simulating power dissipation. Since the environment does not permit the simulation of multiple-valued circuits derived from MDDs, we have carried out experiments on several sets of ISCAS85 and ISCAS89 benchmarks (combinational and sequential circuits). Table 5 shows the results of experiments, where BDDs have been built for circuits using different ordering approaches (window, genetic, sifting and our $Info^{BDD}$). Results show that power dissipation for the circuits derived from BDDs ordered by $Info^{BDD}$ algorithm are on an average 14% better than the corresponding circuits derived from BDDs where ordering is done by other approaches.

6. Potential Applications

Quantum computing: Quantum computers are devices that process information in a way that preserves quantum coherence. The most common model of quantum computation deals with coherent logic operations on two-state quantum variables called qubits. Quantum computation can also be performed on variables with three or more states, and where possible degrees of freedom are continuous (Lloyd and

Braunstein 1999). Hybrid quantum computers operate on both discrete and continuous quantum variables. The reason for investigating hybrid computers is that nature contains both discrete quantum variables such as nuclear spin, photon polarization, and atomic energy levels, and continuous variables such as position, momentum, and the amplitudes of the electromagnetic field. The design of such a processor is based on a hybrid register (continuous variables for different qubit rotations and discrete ones for two-gates). This paper presents new information measures for mixed continuous/multiple-valued data as well as the discretization method (see Section 3) which can be used to simplify the design of hybrid quantum computers.

Fuzzy sets: Existing data mining systems discretize numerical domains into intervals and treat intervals as nominal values during induction. Problems occur if test examples are not covered in training data (no match or multiple match). The solution is a hybrid approach using fuzzy intervals for the no-match problem. The application of fuzzy logic can also be extended to one-dimensional interval matching. Fuzzy matching is an interval match where the interval width determines the fuzziness of the match. In the induction of fuzzy decision trees, the continuous variables need to be partitioned into several fuzzy sets prior to the tree induction (Peng and Flach 2001). This paper gives an idea how to perform discretization, which can be treated as "soft" partitioning for fuzzy systems. In addition to potential utilization of the discretization method (see Section 3), we envisage some applications of the information measuring technique (see Section 4) on fuzzy decision diagrams (Strehl et al. 1999).

7. Summary

In different areas ranging from computer-aided design to machine learning, it is natural and beneficial to use information measures as complexity characteristics of optimization problems. In this paper, we reviewed well-known uncertainty measuring approaches in the multiple-valued domain based on tabular representations and discussed their applicability to some optimization problems. To contribute to the pool of existing information measures and to cover an important class of continuous data, we introduced new information density measures and formulated their properties. An efficient discretization algorithm is developed which utilizes the information density criterion. It is worth noting that the suggested discretization algorithm is robust, unaffected by data shift and scaling.

Extending information measuring even further, we developed a technique for computing information measures based on decision diagram representations. This technique features a novel algorithm for ordering of decision diagrams which contributes to an improvement in the diagram size in contrast to existing ordering algorithms. The ordering algorithm is shown to be of interest for application to different problems, especially in the area of low-power design of digital circuits derived from decision diagrams. The power reduction is achieved by decreasing the switching activity in a circuit while paying close attention to information measures as an optimization criterion.

The paper also addressed the importance of information measuring in the design of multiple-valued systems covering such areas as (i) fuzzy sets, (ii) quantum computing and (iii) data mining. It is hoped that this paper will help to broaden the understanding of practical applications of information measuring in the multiple-valued domain.

Acknowledgements

We express our sincere appreciation to Nawar Hakeem, Vlad Shmerko, Radomir Stankovich and Svetlana Yanushkevich for valuable discussions, as well as to Anita Dani and Christine Wacker for critical comments and data preparation. We thank unknown reviewers, Robert Fraga and Alexei V Nikitin for their input and useful remarks. We also thank Rolf Drechsler for useful discussions on incompletely specified functions and Claudio Moraga for providing references on fuzzy decision diagrams.

Notes

[1] http://www.ics.uci.edu/mlearn/MLRepository.html.
[2] F. Somenzi, University of Colorado Decision Diagrams Package, 1996.
[3] http://www.cbl.ncsu.edu/CBL_Docs/iscas85.html
[4] http://www.cbl.ncsu.edu/CBL_Docs/iscas89.html
[5] SIS Release 1.2., UC
Berkeley, 1994. The power is estimated by the power estimator available in SIS at 20MHz executing power_estimate (zero delay model).

References

Agraval, V. (1981). An Information Theoretic Approach to Digital Fault Testing. *IEEE Transactions on Computers* **C-30**(8): 582–587.
Baturone, I., Sanchez-Solano, S. & Huertas, J. (1997). A Self-Checking Current-Mode Analogue Memory'. *Electronics Letters* **33**(16): 1349–1350.

Berger, A., Pietra, S. D. & Pietra, V. D. (1996). A Maximum Entropy Approach to Natural Language Processing. *Computational Linguistics* **22**(1): 39–71.

Bryant, R. (1986). Graph – Based Algorithm for Boolean Function Manipulation. *IEEE Transactions on Computers* **C-35**(8): 667–691.

Dougherty, J., Kohavi, R. & Sahami, M. (1995). Supervised and Unsupervised Discretization of Continuous Features. In *Proc. Int. Conference on Machine Learning*, 194–202.

Fayyad, U. & Irani, K. (1993). Multi-interval Discretization of Continuous-valued Attributes for Classification Learning. In *Proc. Int. Conference on Artificial Intelligence*, 1022–1027.

Lloyd, S. & Braunstein, S. (1999). Quantum Computation Over Continuous Variables. *Physical Review* **82**: 1784–1787.

Lo, H., Spiller, T. & Popescu, S. (1998). *Introduction to Quantum Computation and Information*. World Scientific.

Marculescu, D., Marculesku, R. & Pedram, M. (1996). Information Theoretic Measures for Power Analysis. *IEEE Transactions on CAD of Integrated Circuits and Systems* **15**(6): 599–610.

Miller, D. (1998). An Improved Method for Computing a Generalized Spectral Coefficient. *IEEE Transactions on CAD of Integrated Circuits and Systems* **17**(3): 233–238.

Peng, Y. & Flach, P. (2001). Soft Discretization to Enhance the Continuous Decision Tree Induction. In *Integrating Aspects of Data Mining, Decision Support and Meta-Learning (PKDD'01 Workshop)*, 109–118.

Pierzchala, E., Perkowski, M. & Grygiel, S. (1994). A Filed Programmable Analog Array for Continuous, Fuzzy, and Multi-Valued Logic Applications. In *Proc. IEEE Int. Symposium on Multiple-Valued Logic*, 148–155.

Popel, D. (2001). Towards Efficient Calculation of Information Measures for Reordering of Binary Decision Diagrams. In *Proc. IEEE International Symposium on Signals, Circuits and Systems*, 509–512.

Popel, D. & Hakeem, N. (2001). Improving Web Database Access Using Decision Diagrams. In *Proc. IEEE Int. Conference on Computer Systems and Applications*, 519–525.

Quinlan, J. (1996). Improved Use of Continuous Attributes in C4.5. *Journal of Artificial Intelligence Research* **4**: 77–90.

Saed, A., Ahmadi, M. & Jullien, G. (1999). Arithmetic Circuits for Analog Digits. In *Proc. IEEE Int. Symposium on Multiple-Valued Logic*, 186–193.

Shannon, C. E. (1948). A Mathematical Theory of Communication. *Bell Systems Technical Journal* **27**: 379–423, 623–656.

Simovici, D. Jaroszewicz, S. (2001). An Axiomatization of Generalized Entropy of Partitions. In *Proc. IEEE Int. Symposium on Multiple-Valued Logic*, 259–264.

Strehl, K., Moraga, C., Temme. K.-H. & Stanković, R. (1999). Fuzzy Decision Diagrams for the Representation, Analysis, and Optimization of Rule Bases. Technical Report TIK-77, Gloriastrasse 35, CH-8092 Zurich.

Thornton, M. and Nair, V. (1995). Efficient Calculation of Spectral Coefficients and Their Applications. *IEEE Transactions on Computer-Aided Design of Integrated Circuits and Systems* **14**(11): 1328–1341.

Yanushkevich, S., Popel, D., Shmerko, V., Cheushev, V. & Stanković, R. (2000). Information Theoretic Approach to Minimization of Polynomial Expressions over GF(4). In *Proc. IEEE Int. Symposium on Multiple-Valued Logic*, 265–270.

Evolutionary Synthesis of Logic Circuits Using Information Theory

ARTURO HERNÁNDEZ AGUIRRE[1] and CARLOS A. COELLO COELLO[2]

[1]*Center for Research in Mathematics (CIMAT), Department of Computer Science, Guanajuato, Gto. 36240, Mexico (E-mail: artha@cimat.mx);* [2]*CINVESTAV-IPN, Evolutionary Computation Group, Depto. de Ingeniería Eléctrica, Sección de Computación, Av. Instituto Politécnico Nacional No. 2508, Col. San Pedro Zacatenco, México, D. F. 07300, Mexico (E-mail: ccoello@cs.cinvestav.mx)*

Abstract. In this paper, we propose the use of Information Theory as the basis for designing a fitness function for Boolean circuit design using Genetic Programming. Boolean functions are implemented by replicating binary multiplexers. Entropy-based measures, such as Mutual Information and Normalized Mutual Information are investigated as tools for similarity measures between the target and evolving circuit. Three fitness functions are built over a primitive one. We show that the landscape of Normalized Mutual Information is more amenable for being used as a fitness function than simple Mutual Information. The evolutionary synthesized circuits are compared to the known optimum size. A discussion of the potential of the Information-Theoretical approach is given.

Keywords: circuit synthesis, computer-aided design, evolutionary algorithms, evolvable hardware, genetic programming, information theory

1. Introduction

The implementation of Boolean functions using the minimum number of components is important for ASIC circuit designers and programmable devices since silicon surface on a chip is a limited resource. Classic graphical methods such as Karnaugh Maps become harder to use when the number of variables increases, and it becomes impossible to use for a relatively small number of variables. Tabular methods such as Quine-McCluskey, although amenable for digital computers, have been proved to need memory in the order of 3^n. Modern minimization CAD tools, like Espresso, and hybrid methods based on binary decision diagrams have proved very powerful for circuit design, therefore finding near optimal solutions to complex circuits. Such automated design tools make use of *a priori* knowledge that human experts have extracted from the problem domain. Knowledge is stored and represented in the form of axioms and laws for immediate use. For example, knowledge of Boolean algebra is incorporated to Espresso in the form of three heuristics: reduce, expand, and redundant, aiming to construct the minimum set cover. In decision diagrams-based methods, rules for graph reduction preserving Boolean equivalence are used for minimization. For many design domains, human knowledge encoded in this form suffices to automate the design process. For some other areas such as logic circuit minimization, humans have not yet derived the whole set of rules that would allow to find the smallest circuit that implements an arbitrary Boolean function. Therefore, computation of the smallest circuit is achieved by searching for a solution in a combinatorial space spawned by operators of Boolean algebra.

As noted, the search space humans know (in some domains) is constructed by a deduction process, or repetitive application of the rules of the problem domain over an initial seed. But the search space could contain other elements not known to humans if it is created in some other way. Evolutionary computation methods build the search space in a bottom-up fashion by combining only some sampling elements (Thompson et al. 1999) (called individuals of the population). Hence, solutions found in this space challenge human designers since the deduction rules that lead to them are not common and, in many cases, such rules are in fact unknown.

Hybrid methods combining evolutionary computation algorithms and heuristics have proved very powerful. Nonetheless, purer forms of evolutionary algorithms deserve attention since the avoidance of human assistance is tantalizing.

In this article we use a pure form of Genetic Programming to synthesize logic circuits using binary multiplexer(s) ("mux" or "muxes"). That is, no knowledge is incorporated to the guiding search mechanism other than a fitness function based on entropy. We show that the Shannon expansion of a Boolean formula implementing a circuit is a sound basis for our evolutionary method. We also review several possibilities from information theory domain useful to fitness function design. The conclusions drawn here are applicable to any entropy-based fitness function for Boolean circuit design, regardless of the evolutionary technique in use.

The organization of this paper is the following: In Section 2, we describe some previous related work from both hybrid and pure areas. Section 3 provides the detailed description of the problem that we wish to solve. Section 4 provides the mathematical foundations for justifying the use of multiplexers as universal logic elements. Some basic concepts of information theory and genetic programming are provided in Sections 5 and 6, respectively. The use of entropy for circuit design is discussed in Section 7. With these ideas in mind, we introduce fitness functions for evolutionary circuit design in Section 8. A set of experiments illustrating the design of logic circuits is described in

Section 9. In Section 10, we provide some final remarks and our conclusions and, finally, in Section 11, we briefly discuss some possible paths for future research.

2. Previous Work

Most of the previous work in this domain has been done using Genetic Algorithms (GAs). Genetic Programming (Koza 1992) is also included in this review, since this technique is really an extension of GAs in which a tree-based representation is used instead of the traditional linear binary chromosome adopted with GAs.

Louis (1993) introduced the use of GAs for the design of combinational circuits, and the use of a matrix array inside of which a circuit is evolved. A layer or stage of his circuits was constrained to get its inputs only from the previous stage. Thus, he defined a new operator called *masked crossover* that exploits information unused by standard genetic operators over the matrix representation. After Louis, several authors followed his representation, for example, Miller et al. (1998) also evolved logic circuits in a matrix but the position of the circuit output is also considered a variable. In their approach, Miller et al. (1998) encode a set of complex Boolean expressions instead of simple gate functions aiming to design more complex circuits given the set of more powerful primitives. In principle, the approach is sound but, unfortunately, its main drawback is its lack of flexibility to handle a large number of inputs. Miller et al. (2000, 2000a), Louis and Johnson (1997), and Islas et al. (2003) studied the combination of GAs with Case-Based Reasoning tools that incorporate and preserve knowledge about the problem domain.

Our own previous work using GAs for circuit design (Coello Coello et al. 1996, 1997, 2000, 2001), shows successful results when small circuits are evolved inside the previously mentioned matrix representation. From our research using this sort of encoding, we concluded that the matrix causes a strong representation bias since some inputs and gates are favored by the genetic operators in a probabilistic sense. In an attempt to deal with such a biased representation, in more recent work, we proposed another approach based on genetic programming and multiplexers that seems to have a more neutral representation (Hernández Aguirre et al. 1999, 2000, 2000b).

Genetic programming (GP) is a natural alternative to genetic algorithms for circuit synthesis since its main goal is *program construction*, rather than vector optimization of the latter. Therefore, for GP, constructing a program is similar to constructing a circuit, with either domain being defined by its own set of terminals and functional symbols. Koza (1992) has used Genetic Programming to design combinational circuits, but his emphasis has been in the generation of functional circuits (in symbolic form) rather than their optimization. Dill et al. (1997) used GP to produce logic equations from partial information, covering up to 98.4% of the target function. Iba et al. (1997), have studied circuit synthesis at gate-level on FPGAs, concluding that intrinsic (on chip) evolution is harder than extrinsic (as our approach), because of the fine grained architecture of the FPGA. Note however, that we have showed that it is in fact possible for evolutionary algorithms to learn Boolean functions if we correctly estimate the so-called Vapnik-Chervonenkis dimension of our design tool (Hernández Aguirre et al. 2000a, 2001).

To establish the niche of this paper we should consider the two main approaches to Boolean function synthesis described in Section 1: pure methods, and hybrid methods. Pure methods are extended with procedures to cope with the poor scalability of evolutionary algorithms. Kalganova (2000) proposed a two-way strategy called "bidirectional incremental evolution", in which circuits are evolved in top-down and bottom-up fashion. Vassilev et al. (2000) proposed the use of predefined circuit blocks which can improve both the convergence speed and the quality of the solutions. Although these results are promising, a big problem remaining is defining what a good block is for the problem in turn. Recently, Torresen (2002) proposed a scalable alternative with limited success called "increased complexity evolution". Here, training vectors are partitioned, or the training set is partitioned, solving a problem in a divide and conquer fashion.

Hybrid methods have had more success, particularly binary decision diagram based methods. Droste (1997) used GP and two heuristics: deletion and merging rules, to reduce directed acyclic graphs. His approach solved the 20-multiplexer problem for the first time ever. Another important hybrid method has been developed by Drechsler et al. (1995, 1996, 1998). This GP system uses directed acyclic graphs for representing decision diagrams. Two heuristics that are representative of Dreschler's work are: sifting and inversion.

Claude Shannon suggested the use of information entropy as a measure of the amount of information contained within a message (Shannon 1948). Thus, entropy tells us there is a limit in the amount of information that can be removed from a random process without information loss. For example, music can be (losslessly) compressed and reduced up to its entropy limit. Further reduction is only possible at the expense of information loss (Weaver and Shannon 1949). In a few words, entropy is a measure of disorder and it constitutes the basis of Information Theory.

The ID3 algorithm for the construction of classifiers (based on decision trees) probably is the best known classification algorithm among computer scientists that relies on entropy measures (Quinlan 1983). For ID3, an attribute is more important for concept classification if it provides greater "information gain" than the others.

Information Theory (IT) was early used by Hartmann et al. (1982) to convert decision tables into decision trees. Boolean function minimization through IT techniques has been approached by several authors (Kabakcioglu et al. 1990; Lloris et al. 1993). Some work related to the proposal presented in this article can be found in Luba et al. (2000, 2000a), whom address the synthesis of logic functions using a genetic algorithm and a fitness function based on conditional entropy. Their system (called *EvoDesign*) works in two phases: first, the search space is partitioned into subspaces via Shannon expansions of the initial function. Then the GA is started in the second phase. The authors claim that the partition of the space using entropy measures is the basis for their success. In their domain, a fitness function based on Mutual Information apparently worked well. Note however, that in our case such an approach did not produce good results. Conditional entropy has also been used by Cheushev et al. (2001) in top-down circuit minimization methods. In fact, the formal result of Cheushev et al. (2001) indicates that a Boolean function can be synthesized by using entropy measures, thus, providing a sound ground for our approach.

We aim to explore the use of entropy-based fitness functions in a pure GP framework. Since no other knowledge than entropy will be used in the fitness function, only limited size circuits can be tested. However, as noted before, the conclusions we draw are applicable to any evolutionary system for Boolean function synthesis based on entropy measures.

3. Problem Statement

For the purposes of this article, let us consider a target Boolean function T specified by its truth table. The problem of interest to us is the design of the smallest possible logic circuit, with the minimum number of binary multiplexers (described in Section 4), that implements the target function. The multiplexer is the only functional unit replicated, each one being controlled by a variable of the target function. Note that only 1s and 0s are fed into the multiplexers (the analog of the Shannon's expansion shown in Figure 1). This strategy allows the implementation of the synthesized circuits by means of pass transistor logic (Scholl et al. 2000). The design metric driving the implementation is the number of components. Therefore, the best among a set of circuits with the same functionality is the one with the lowest number of components. Our goal is to find 100% functional circuits, specifying compo-

$f=a'b'c + a'bc' + ab'c'$

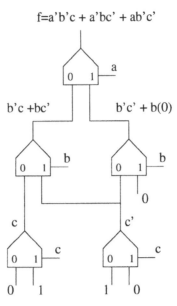

Figure 1. Shannon expansion implemented with binary multiplexers.

nents and connections, instead of a symbolic representation of it. Thus, the approach of this paper could be classified as "gate-level synthesis".

Since the number of circuit components is unknown for most circuits, the use of an stochastic method such as Genetic Programming seems appropriate. Also, the tree-like structure of the circuits makes Genetic Programming the most appropriate evolutionary technique.

4. Multiplexers as Universal Logic Elements

The use of binary muxes is a sound approach to circuit synthesis since they can form a basis for Boolean functions. That is, muxes are "universal generators".

A binary multiplexer is a logic circuit with two inputs a and b, one output f, and one control signal s, logically related as follows:

$$f = as + bs'$$
(1)

In other words, the output is the value a when the selector is "high", and b when the selector is "low".

The Shannon expansion is the representation of a Boolean function through the residues of a Boolean function.

DEFINITION 1. **Residue of a Boolean function**: *The residue of a Boolean function* $f(x_1, x_2, \ldots, x_n)$ *with respect to a variable* x_j *is the value of the function for a specific value of* x_j. *It is denoted by* f_{x_j}, *for* $x_j = 1$ *and by* $f_{\bar{x}_j}$ *for* $x_j = 0$. *The Shannon expansion in terms of residues of the function is,*

$$f = \bar{x}_j f|_{\bar{x}_j} + x_j f|_{x_j} \tag{2}$$

For mapping Boolean expansions into circuits using binary multiplexers, the variable x_j in Equation 2 takes the place of the control variable s in Equation 1. For the sake of an example consider the function $f(a, b, c) = a'b'c + a'bc' + ab'c'$.

The residue of the expansion over the variable a is:

$$
\begin{aligned}
f(a, b, c) &= a' f|_{a=0} + a f|_{a=1} \\
&= a' \cdot (b'c + bc') + a \cdot (b'c')
\end{aligned}
$$

The factor $b'c + bc'$ must be taken by the mux when the selector a is "low", and $b'c'$ when a is "high". These factors could also be expanded in the same way. The expansion of the first factor over the variable b is:

$$
\begin{aligned}
b'c + bc' &= b'(b'c + bc')|_{b=0} + b(b'c + bc')|_{b=1} \\
&= b' \cdot c + b \cdot c'
\end{aligned}
$$

And the expansion of the second factor over b is:

$$
\begin{aligned}
b'c' &= b'(b'c')|_{b=0} + b(b'c')|_{b=1} \\
&= b' \cdot c' + b \cdot 0
\end{aligned}
$$

Since in our approach the only valid inputs to the muxes are "0" and "1", the variable "c" has to be fed to the circuit through a mux. This is done by the two muxes at the bottom of the "tree". The circuit implementing the function of our example is shown in Figure 1.

5. Basic Concepts of Information Theory

Uncertainty and its measure provide the basis for developing ideas about Information Theory (Cover and Thomas 1991). The most commonly used measure of information is Shannon's entropy.

DEFINITION 2. **Entropy**: *The average information supplied by a set of* k *symbols whose probabilities are given by* $\{p_1, p_2, \ldots, p_k\}$, *can be expressed as,*

$$H(p_1, p_2, \ldots, p_k) = -\sum_{s=1}^{k} p_k log_2 p_k \tag{3}$$

The information shared between a transmitter and a receiver at either end of a communication channel is estimated by its Mutual Information,

$$MI(T; R) = H(T) + H(R) - H(T, R) = H(T) - H(T|R) \qquad (4)$$

The conditional entropy $H(T|R)$ can be calculated through the joint probability, as follows:

$$H(T|R) = -\sum_{i=1}^{n}\sum_{j=1}^{n} p(t_i r_j) log_2 \frac{p(t_i r_j)}{p(r_j)} \qquad (5)$$

An alternative expression of mutual information is

$$MI(T; R) = \sum_{t \in T}\sum_{r \in R} p(t, r) log_2 \frac{p(t, r)}{p(t)p(r)} \qquad (6)$$

Mutual information, Equation 4, is the difference between the marginal entropies $H(T) + H(R)$, and the joint entropy $H(T, R)$.

We can explain it as a measure of the amount of information one random variable contains about another random variable, thus it is the reduction in the uncertainty of one random variable due to the knowledge of the other (Cover and Thomas 1991).

Mutual information is not an invariant measure between random variables because it contains the marginal entropies. Normalized Mutual Information is a better measure of the "prediction" that one variable can do about the other (Studholme et al. 1999):

$$NMI(T; R) = \frac{H(T) + H(R)}{H(T, R)} \qquad (7)$$

The joint entropy $H(T, R)$ is calculated as follows:

$$H(T, R) = -\sum_{t \in T}\sum_{r \in R} p(t, r) log_2 p(t, r) \qquad (8)$$

Normalized MI has been used in image registration with great success (Maes et al. 1997).

EXAMPLE: *We illustrate these concepts by computing the Mutual Information between two Boolean vectors f and c, shown in Table 1. Variable c is an argument of the Boolean function $f(a, b, c) = a'b'c + a'bc' + ab'c'$.*

We wish to estimate the description the variable c can do about variable f, that is, $MI(f; c)$.

Table 1. Function $f = a'b'c + a'bc' + ab'c'$ used to compute MI(f;c)

a	b	c	$f = a'b'c + a'bc'$ $+ab'c'$
0	0	0	0
0	0	1	1
0	1	0	1
0	1	1	0
1	0	0	1
1	0	1	0
1	1	0	0
1	1	1	0

We use Equations 4 and 5 to calculate $MI(f; c)$. Thus, we need the entropy $H(f)$ and the conditional entropy $H(f|c)$.

Entropy requires the discrete probabilities $p(F = 0)$ and $p(F = 1)$ which we find by counting their occurrences

$$H(f) = -(\frac{5}{8}log_2\frac{5}{8} + \frac{3}{8}log_2\frac{3}{8}) = 0.9544$$

The conditional entropy, Equation 5, uses the joint probability $p(f_i, c_j)$, which can be estimated through conditional probability, as follows: $p(f, c) = p(f)p(c|f)$. Since either vector f and c is made of two symbols, the discrete joint distribution has four entries. This is:

$$p(f = 0, c = 0) = p(f = 0)p(c = 0|f = 0) = \frac{5}{8} \times \frac{2}{5} = 0.25$$

$$p(f = 0, c = 1) = p(f = 0)p(c = 1|f = 0) = \frac{5}{8} \times \frac{3}{5} = 0.375$$

$$p(f = 1, c = 0) = p(f = 1)p(c = 0|f = 1) = \frac{3}{8} \times \frac{2}{3} = 0.25$$

$$p(f = 1, c = 1) = p(f = 1)p(c = 1|f = 1) = \frac{3}{8} \times \frac{1}{3} = 0.25$$

Now, we can compute the conditional entropy by using Equation 5. The double summation produces four terms:

$$H(f|c) = -(\frac{1}{4}log_2\frac{1}{2} + \frac{3}{8}log_2\frac{3}{4} + \frac{1}{4}log_2\frac{1}{2} + \frac{1}{8}log_2\frac{1}{4})$$
$$H(f|c) = 0.9056$$

Therefore, $MI(f; c) = H(f) - H(f|c) = 0.9544 - 0.9056 = 0.0488$. In fact, for the three arguments of the example function we get $MI(f; a) = MI(f; b) = MI(f; c)$. The normalized mutual information between either argument and the Boolean function is $NMI(f; a) = NMI(f; b) = NMI(f; c) = 1.0256$. Although no function argument seems to carry more information about the function f, we show later that the landscape of NMI contains a region implying information sharing. This region is hard to find on the MI landscape.

6. Genetic Programming Concepts

Genetic Programming is a variation of the genetic algorithm in which a tree-based representation is adopted. Its main aim is to solve automatic programming problems and symbolic regression problems (Koza 1992). Problems such as the exponential growth of the search space even for some very specific problem domains, and the representation or encoding of computational structures of the objective language, remained unsolved for several years. John R. Koza (1992), used genetic algorithms to tackle the search space problem, and S-expressions which are naturally encoded as trees, to represent programs. The evolutionary operations applied over trees always produce valid trees and, therefore, syntactically correct programs. One of the earliest genetic programming systems ran in LISP, as to take advantage of the native parser provided by this programming language. Thus, the language interpreter "runs" the evolutionary algorithm, and at the same time is the evaluator of the S-expressions produced by the evolutionary algorithm.

Genetic Programming evolves functions encoded as trees. We should see a tree as the abstract semantic view of a program, that is, a parse tree. Therefore, nodes and leaves of the tree represent non-terminal and terminal grammar elements of the objective language. Furthermore, genetic programming has to be initialized with a set of operators and functions that work as a basis for the evolutionary synthesis of programs. For example,

- Arithmetic operations (e.g., $+$, $-$, $\times$, $\div$)
- Mathematical functions (e.g.,sine, cosine, log, exp)
- Boolean operations (e.g., AND, OR, NOT)
- Conditionals (IF-THEN-ELSE)
- Iterators (DO-UNTIL)

These are operators and functions commonly adopted in genetic programming. For the evolution of logic circuits we define a pertinent set of grammar elements (note the relationship with Figure 1):

- Terminals= {0, 1}
- Non-terminals= {a multiplexer}

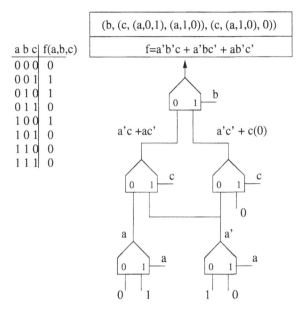

a b c	f(a,b,c)
0 0 0	0
0 0 1	1
0 1 0	1
0 1 1	0
1 0 0	1
1 0 1	0
1 1 0	0
1 1 1	0

(b, (c, (a,0,1), (a,1,0)), (c, (a,1,0), 0))

f=a'b'c + a'bc' + ab'c'

Figure 2. Logic function specification, encoding of the circuit using lists, and the circuit.

Therefore, a circuit is represented as a tree using binary multiplexers as node functions, and 0s and 1s for the leaves. An illustration of this kind of tree is shown in Figure 2 (the circuit was derived using the technique described in this article). The circuit is 100% functionally equivalent to the one derived by Shannon expansions in Figure 1. Note that the circuits are also structurally similar but the muxes are controlled by different variables.

The main body of the standard Genetic Programming algorithm is the following:

Genetic Programming Algorithm
$t = 0$;
$P_t \leftarrow$ initial population
$P_t \leftarrow fitness(P_t)$
while (stopcondition $=$ false) {
    $S \leftarrow selection(P_t)$;
    $C \leftarrow crossover(S)$;
    $M \leftarrow mutation(C)$;
    $t = t + 1$;
    $P_t \leftarrow fitness(M)$;
}

Several issues regarding the application of Genetic Programming as a problem solving tool for Boolean function synthesis are discussed next.

- **Implementation language**: Although the implementation language is not relevant for the results, we chose Prolog because lists are natural structures of this language and allow the representation of trees. The evaluation of either circuit just requires one predicate that translates a list into a tree.

- **Initial population**: In genetic programming, the size of the trees plays an important role in the search. In our work, we adopted the following setting for the size of the trees which was experimentally derived: maximum tree height should not exceed half the number of variables of the Boolean function. These trees could be required to be complete binary trees but our strategy was to randomly create them as to have a rich phenotypic blend in the population.

- **Representation**: A circuit is represented using binary trees, and trees are encoded as Prolog lists. This representation is less flexible than directed acyclic graphs (used in Drechsler et al. 1998) but still suitable to generate circuits for pass transistor logic. A circuit tree is a recursive list of triplets of the form: $(mux, left-child, right-child)$. Mux is assigned a control variable, and either child could be a subtree or a leaf. The muxes are treated as "active high" elements, therefore the left subtree is followed when the control is 0, and the right subtree otherwise. The tree captures the essence of the circuit topology allowing only the children to feed their parent node, as shown in Figure 2. The representation also captures with no bias the requirement for the leaves of being only 0 or 1.

- **Crossover operator**: The exchange of genetic information between two parent trees is accomplished by exchanging subtrees as shown in Figure 3. Crossover points are randomly selected, therefore, node-node, node-leaf, and leaf-leaf exchanges are allowed since they produce correct circuits. The particular case when the root node is selected to be exchanged with a leaf is disallowed, so that invalid circuits are never generated.

   In Figure 3, two parents exchange genetic information (subtrees circled) and produce two children in the way used in this paper.

- **Mutation operator**: Mutation is a random change with very low probability of any gene of a chromosome. The mutation of a tree can alter a mux or a leaf. If a mux is chosen then a random variable is generated anew and placed as a new gene value. The mutation of a leaf is as simple as the changing of a 0 to 1, and 1 to 0. The mutation of both a node and a leaf of the tree is shown in Figure 4.

- **Fitness function**: The design of the fitness function using entropy principles is explained in Section 7. Nevertheless, every fitness func-

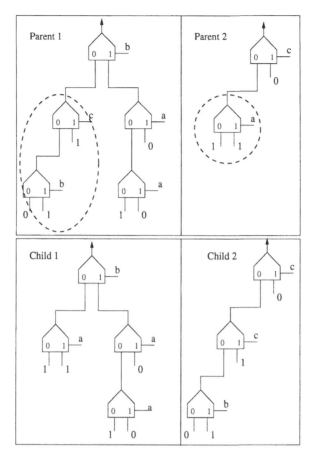

Figure 3. The crossover operator over trees encoding circuits.

tion used in our experiments works in two stages since the goal is twofold: the synthesis of 100% functional circuits, and their minimization. At stage one, genetic programming explores the search space and builds improved solutions over partial solutions until it finds the first 100% functional circuit. The fitness function for this stage uses entropy concepts in order to reproduce the truth table. Once the first functional circuit appears in the population, a second fitness function is activated for measuring the fitness of the new circuit. Thus, if a circuit is not 100% functional its fitness value is estimated through entropy; if the circuit is 100% functional its fitness value denotes its size and smaller circuits are preferred over larger ones. The fitness value of a 100% functional circuit is always larger than the value of a non functional one, so that circuits are protected from fitness conflicts.

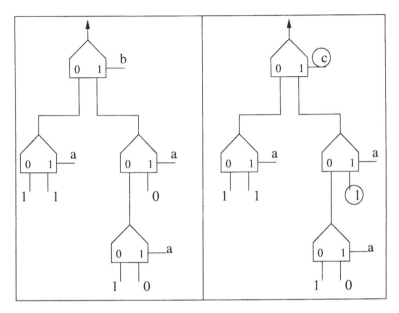

Figure 4. Mutation operation over a tree (two mutations are shown).

7. Entropy and Circuits

Entropy has to be carefully applied to the synthesis of Boolean functions. Let us assume any two Boolean functions, $F1$ and $F2$, and a third $F3$ which is the one's complement of F2, then

$$F3 \neq F2$$

For these complementary functions,

$$H(F2) = H(F3)$$

Also Mutual Information shows a similar behavior.

$$MI(F1, F2) == MI(F1, F3)$$

The implications for Evolutionary Computation are important since careless use of entropy-based measures can nullify the system's convergence. Assume the target Boolean function is T. Then, $MI(T, F2) = MI(T, F3)$, but only one of the circuits implementing $F2$ and $F3$ is close to the solution since their Boolean functions are complementary. A fitness function based on mutual information will reward both circuits with the same value, but one is better than the other. Things could get worse as evolution progresses because the mutual information increases when the circuits are closer to the solution, but in fact, two complementary circuits are then given larger rewards.

The scenario is one in which the population is driven by two equally strong attractors, hence convergence is never reached.

The fitness function of that scenario is as follows. Let us assume T is the target Boolean function (must be seen as a Boolean vector), and C is the output vector of any circuit in the population. Our fitness function is either the maximization of mutual information or the minimization of the conditional entropy term. This is,

$$bad\,fitness\,function\#1 = MI(T, C) = H(T) - H(T|C) \qquad (9)$$

The entropy term $H(T)$ is constant since T is the target vector. Therefore, instead of maximizing mutual information, the fitness function can only minimize the conditional entropy,

$$bad\,fitness\,function\#2 = H(T|C) \qquad (10)$$

We called *bad* to these entropy-based fitness functions because we were not able to find a solution with them. Although mutual information has been described as the "common" information shared by two random processes, the search space is not amenable for evolutionary computation. In Figure 5 we show this search space over mutual information for all possible combinations with two binary strings of 8 bits. The area shown corresponds to about $\frac{1}{4}$ ($[1, 150] \times [1, 150]$) of the whole search space of ($[1, 254] \times [1, 254]$) (the values 0 and 255 were not used). Horizontal axes are decimal values of 8 bit binary strings, and height represents mutual information.

The mutual information space, clearly full of spikes, does not favor the area of common information. For any two equal vectors, their Mutual Information lies on the line at $45°$ (over points $\{(1, 1), (2, 2), (3, 3) \ldots (n, n)\}$). In the next Section we continue this discussion and design fitness functions whose *landscape* seems more promising for exploration.

8. Fitness Function based on Normalized Mutual Information

So far we have described the poor scenario where the search is driven by a fitness function based on the sole mutual information. We claim that fitness functions based on Normalized Mutual Information (NMI) should improve the performance of the genetic programming algorithm because of the form of the NMI landscape. This is shown in Figure 6 for two 8-bit vectors (as we did for MI in Section 7, Figure 5). Note on the figure how the search space becomes more regular and, more important yet, note the appearance of the *wall* at $45°$ where both strings are equal.

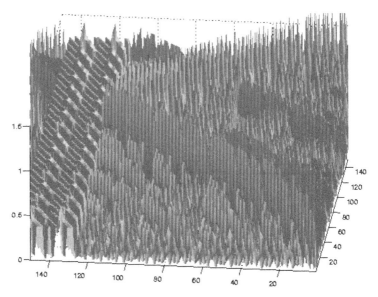

Figure 5. The search space of mutual information (Equation 9).

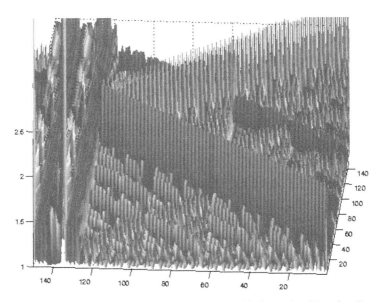

Figure 6. The search space of normalized mutual information (Equation 7).

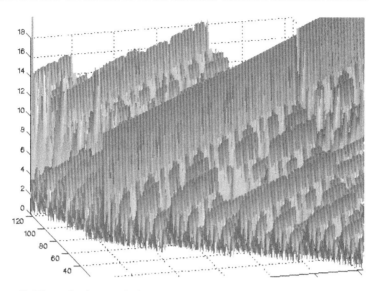

Figure 7. Fitness landscape of: $f = (Length(T) - Hamming(T, C)) \times NMI(T, C)$.

We propose three new fitness functions based on Normalized Mutual Information (Equation 7) and report experiments using the following three fitness functions (higher fitness means better).

Let us assume a target Boolean function of m attributes $T(A_1, A_2, \ldots, A_m)$, and the circuit Boolean function C of the same size. In the following, we propose variations of the basic fitness function of Equation 11, and discuss the intuitive idea of their (expected) behavior.

$$fitness = (Length(T) - Hamming(T, C)) \times NMI(T, C) \qquad (11)$$

We tested Equation 11 in the synthesis of several problems and the results were quite encouraging. Thus, based on this primary equation we designed the following fitness functions. In Figure 7 we show the *fitness landscape* of Equation 11.

$$fitness1 = \sum_{i=1}^{m} \frac{fitness}{NMI(A_i, C)} \qquad (12)$$

$$fitness2 = \sum_{i=1}^{m} fitness \times NMI(A_i, C) \qquad (13)$$

$$fitness3 = (Length(T) - Hamming(T, C)) \times (10 - H(T|C)) \,(14)$$

The function fitness, Equation 11, is driven by NMI(T,C) and adjusted by the factor $Length(T) - Hamming(T, C)$. This factor tends to zero when T

and C are far in Hamming distance, and tends to $Length(T)$ when T and C are close in Hamming distance. The effect of the term is to give the correct rewarding of NMI to a circuit C close to T. Equation 11 is designed to remove the convergence problems described in the previous section.

*Fitness*1 and *Fitness*2, Equations 12 and 13, combine NMI of T and C with NMI of C and the attributes A_k of the target function. Thus, *fitness*1 and *fitness*2 look for more information available in the truth table in order to guide the search. *Fitness*3 is based on conditional entropy and it uses the mentioned factor to suppress the reproduction of undesirable trees. Since conditional entropy has to be minimized we use the factor $10 - H(T|C)$ in order to maximize fitness. Equations 10 and 12 use the conditional entropy term. Nevertheless, only Equation 12 works fine. As a preliminary discussion regarding the design of the fitness function, the noticeable difference is the use of Hamming distance to guide the search towards the aforementioned *optimum wall* of the search space. The Hamming distance favors the destruction of individuals on one side of the wall, and favors the other side. Thus, in principle, there is only one attractor in the search space.

9. Experiments

In the following experiments we find and contrast the convergence of the genetic programming system for the three fitness functions of Equations 12, 13, 14.

9.1. *Experiment 1*

Here we design the following (simple) Boolean function:

$$F(a, b, c, d) = \sum(0, 1, 2, 3, 4, 6, 8, 9, 12) = 1$$

We use a population size of 300 individuals, a crossover rate (p_c) of 0.35, a mutation rate (p_m) of 0.65, and a maximum of 100 generations. The optimal solution has 6 nodes, thus we find the generation in which the first 100% functional solution appears, and the generation number where the optimal is found. The problem was solved 20 times for each fitness function. Table 2 shows the results of these experiments.

9.2. *Experiment 2*

Our second test function is:

$$F(a, b, c, d, e, f) = ab + cd + ef$$

Table 2. Generation number where the first 100% functional circuit is found, and the generation where the optimum is found, for our three proposed fitness functions

Event	Gen. at fitness1	Gen. at fitness2	Gen. at fitness3
100% functional	13 ± 5	14 ± 7	18 ± 6
Optimum solution	30 ± 7	30 ± 10	40 ± 20

Table 3. Generation number where the first 100% functional circuit is found, and the generation where the optimum is found, for our three proposed fitness functions

Event	Gen. at fitness1	Gen. at fitness2	Gen. at fitness3
100% functional	39 ± 12	40 ± 11	50 ± 12
Optimum Solution	160 ± 15	167 ± 15	170 ± 20

In this case, we adopted a population size of 600 individuals, $p_c = 0.35$, $p_m = 0.65$, and a maximum of 200 generations. The optimal solution has 14 nodes. Each problem was solved 20 times for each fitness function. Table 3 shows the results of these experiments.

A solution found to this problem is shown in Figure 8. The evolutionary solution is equivalent to the optimum reported by Reduced Order Binary Decision Diagram techniques.

9.3. *Experiment 3*

Our third problem is related to partially specified Boolean functions (Hernández Aguirre et al. 2000). With this experiment we address the ability of the system to design Boolean functions with "large" numbers of arguments and specific topology. For this, we have designed a synthetic problem where the topology is preserved when the number of variables increases.

Boolean functions with $2k$ variables are implemented with $(2 * 2k) - 1$ binary muxes *if* the truth table is specified as shown in Table 4.

We ran experiments for $k = 2, 3, 4$, thus 4,8, and 16 variables and we contrasted these results with the best known so far for this problem (reported in Hernández Aguirre et al. 2000). For completeness, all previous results are reported together with the results of the new experiments in Table 5, where we use the three fitness functions proposed before (Equations 12, 13, 14).

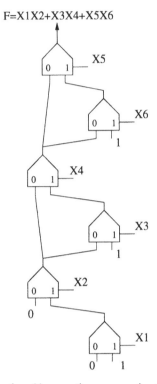

F=X1X2+X3X4+X5X6

Figure 8. Solution found by genetic programming to Experiment 2.

Table 4. Partially specified function of Example 3 needs $(2*2k) - 1$ muxes

A	B	C	D	F(ABCD)
0	0	0	0	0
0	0	0	1	1
0	0	1	0	1
0	1	0	0	1
1	0	0	0	1
0	1	1	1	1
1	0	1	1	1
1	1	0	1	1
1	1	1	0	1
1	1	1	1	0

Table 5. Generation number where the first 100% functional circuit is found, and the generation where the optimum is found, for the three proposed fitness functions

k	Variables	Size	Average (previous)	Aveage (fitness1)	Average (fitness2)	Average (fitness3)
2	4	7	60	60	60	60
3	8	15	200	190	195	194
4	16	31	700	740	731	748
5	32	63	2000	2150	2138	2150

All parameters are kept with no changes for similar experiments, average is computed for 20 independent runs. In previous work we used a fitness function based on the sole Hamming distance between the current solution of an individual and the target solution of the truth table (Hernández Aguirre et al. 2000). One important difference is the percentage of correct solutions found. Previously, we reported that in 90% of the runs we found the solution (for the case of fitness based on Hamming distance). For the three fitness functions based on entropy we found the solution in 99% of the runs.

9.4. *Experiment 4*

Our fourth (and last) problem, is the synthesis of even 4-parity circuits (the output of the circuit is 1 if the number of ones assigned to the input variables is odd). This experiment is harder to solve because only XOR gates are used in the optimum solution. Since our approach will need to implement XOR gates by using muxes, or make some abstraction of the overall circuit, interesting behaviors on the fitness functions will be observed. In this case, we adopted a population size of 810 individuals, $p_c = 0.35$, $p_m = 0.65$, and a maximum of 300 generations. Each problem was solved 30 times for each fitness function. The optimal solution has 15 nodes, which after removing similar branches gets its final form shown in Figure 9.

For this experiment we report the number of optimum solutions found as a percentage of trials (30). The results are shown in Table 6.

The first three columns are similar to previous experiments, the column labeled "Hamming" indicates the use of a fitness function optimizing Hamming distance, and the column labeled "H-NMI" indicates the use of the fitness function described in Equation 11. Except for function Fitness3, all fitness functions found functional circuits in all cases. It is important to remember that Fitness3 is based on conditional entropy and Hamming

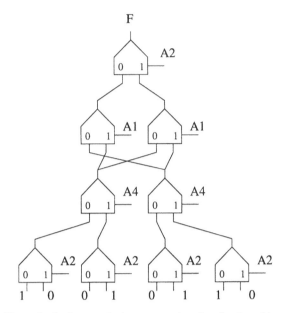

Figure 9. Optimum solution to even 4-parity circuit problem.

Table 6. Percentage of optimum solutions found in 30 runs, for our three proposed fitness functions

Event	Fitness1	Fitness2	Fitness3	Hamming	H-NMI
% opt. solutions	54.5	44.1	0.0	36.6	36.8

distance; the detailed results are: about 50% of the runs strayed from convergence showing an ever increasing number of nodes. In the other 50%, functional solutions were found but showing an erratic behavior. No circuit with optimum fitness solution was found.

Experiments show that Fitness H-NMI is quite similar to fitness Hamming, but Fitness2 and Fitness1 improve H-NMI and Hamming, most likely due to the normalized mutual information measured between the variables of the target and evolving functions.

10. Final Remarks and Conclusions

We have tested in this paper a fitness function using only conditional entropy for circuit synthesis, with no success at all. We believe this is a clear indication of a fitness function that does not take into account entropy properties.

Therefore, the evolutionary algorithm cannot converge because there is more than one attractor in the search space. Figure 5 reveals an amorphous search MI landscape with a quite weak wall at 45°. The left handside of the wall seems more regular than the right handside. Although it is hard to derive any conclusions from this figure, it is clear that no attractor dominates the area and it could explain the failure of the fitness function based on MI only.

The landscape of Normalized Mutual Information seems less chaotic and more regular. The great advantage of a fitness function designed over NMI is the appearance of the wall at 45°. It is clear that the wall must appear when the random vectors are equal; as the intersection of the vectors increases so it does the MU. What we have shown in this paper is that, in spite of all the credit given to MI as the "real information" shared between two random processes, the NMI landscape is more suitable for searching than the MI landscape. In the landscape of the fitness function of Figure 7, we can see the wall due to equal vectors is preserved, so we believe it is part of the landscape of three fitness functions using Equation 11.

In the first three experiments, the three fitness functions proposed in this paper worked quite well. All of them found the optimum in most cases, thus they are comparable to other fitness functions based on Hamming distance (Hernández Aguirre et al. 2000). Nonetheless, Experiment 4, which is harder since the optimum solution is implemented by only using XOR gates, tell us a different story. Remember that Fitness1 and Fitness2 are based on NMI, and that their design hypothesis is that some relevant information shared between the Boolean variables of the target function and the target function itself, could be extracted and used to guide the search. This seems to be the case of Experiment 4, since the best results are obtained by fitness functions based on NMI.

A final remark goes to the convergence time and quality of results for Experiment 4 previously reported in the specialized literature. Miller et al. (2000a), solved this problem using a genetic algorithm whose evolution is contained by the matrix representation used (called cartesian genetic programming). They found the optimum in 15% of the runs, each run made 4 million fitness function evaluations. In our case, we only need 240,000 fitness function evaluations, and we get the perfect fitness in 54.5% of the trials. It is not possible to derive firm conclusions from the comparison because the representation and the evolutionary technique of each approach is different, but it is worth to note how our GP based approach needs less computational resources to find perfect fitness circuits. From Tables 2 and 3 we can give some advantage to normalized mutual information over simple mutual information because it is less biased. Results from Table 5 and 6 could imply that mutual information is able to capture some relationship between the data that the sole Hamming distance cannot convey to the population.

11. Future Work

There are two immediate lines for further research in which we are already working. These are the following:

1. **Use of entropy-based measures to estimate the complexity of a tree**: In the first three experiments of this article, we reported the generation at which the first 100% functional solution was found. We also counted the number of muxes in that first functional solution (not reported here), and also the size of the best tree produced along all the evolutionary process. Comparing the size of trees for fitness functions based and not based on entropy, we found shorter trees when using entropy (NMI to be precise). A detailed analysis is in process in order to explain this behavior. It is quite important because shorter trees usually denote shorter convergence time to the solution. Thus, our hypothesis is that entropy could improve the measure of the complexity of a tree and that it could help to relate tree size with target goal. So far, the experiments show that trees tend to grow but evolution keeps their entropy low and with about the same value. Certainly, many different trees can share the same entropy value, but if this is low then it could mean some trees are replicating a branch with low probability of success. Entropy could also be used to keep diversity high. This is of major importance during the exploration phase of the evolutionary algorithm and, therefore, deserves to be explored in more depth.

2. **Circuits with more than one output**: The generation of common or shared branches is the main issue in this type of circuits. Our initial approach has consisted on measuring the mutual information between the output vectors, and to relate the circuit outputs to the argument vectors of the Boolean function. Our preliminary results are encouraging but more work is still necessary in this direction

Acknowledgements

The authors gratefully acknowledge the valuable comments provided by the anonymous reviewers which greatly helped them to improve the contents of this paper.

The first author acknowledges partial support from CONCyTEG project No. 03-02-K118-037. The second author acknowledges support from CONACyT through project no. 32999-A.

References

Cheushev, V., Moraga, C., Yanushkevich, S., Shmerko, V. & Kolodziejczyk, J. (2001). Information Theory Method for Flexible Network Synthesis. In *Proceedings of the IEEE 31st. International Symposium on Multiple-Valued Logic*, 201–206. IEEE Press.

Coello Coello, C. A., Christiansen, A. D. & Hernández Aguirre, A. (2000). Use of Evolutionary Techniques to Automate the Design of Combinational Logic Circuits. *International Journal of Smart Engineering System Design* **2**: 299–314.

Coello Coello, C. A., Christiansen, A. D. & Hernández Aguirre, A. (2001). Towards Automated Evolutionary Design of Combinational Circuits. *Computers and Electrical Engineering* **27**: 1–28.

Coello Coello, C. A., Christiansen, A. D. & Hernández Aguirre, A. (1996). Using Genetic Algorithms to Design Combinational Logic Circuits. In Dagli, C. H., Akay, M., Chen, C. L. P., Benito, R., Farnández & Ghosh, J. (eds.) *Intelligent Engineering Systems Through Artificial Neural Networks. Volume 6. Fuzzy Logic and Evolutionary Programming*, 391–396. St. Louis, Missouri, USA: ASME Press, Nov.

Coello Coello, C. A., Christiansen, A. D. & and Hernández Aguirre, A. (1997). Automated Design of Combinational Logic Circuits Using Genetic Algorithms. In Smith, D. G., Steele, N. C. & Albrecht, R. F. (eds.) *Proceedings of the International Conference on Artificial Neural Nets and Genetic Algorithms*, 335–338. University of East Anglia, England: Springer-Verlag, April.

Cover, T. M. & Thomas, J. A. (1991). *Elements of Information Theory*. New York: John Wiley & Sons.

Dill, K. M., Herzog, J. H. & Perkowski, M. A. (1997). Genetic Programming and Its Application to the Synthesis of Digital Logic. In *Proceedings of the 1997 IEEE Pacific Rim Conference on Communications, Computers, and Signal Processing*, Vol 2, 823–826. IEEE Computer Society.

Drechsler, R., Becker, B. & Göckel, N. (1995). A Genetic Algorithm for Minimization of Fixed Polarity Reed-Muller Expressions. In *Proceedings of the International Conference on Artificial Neural Networks and Genetic Algorithms*, 392–395. Ales, April.

Drechsler, R., Göckel, N. & Becker, B. (1996). Learning Heuristics for OBDD Minimization by Evolutionary Algorithms. In Voigt, H.-M., Ebeling, W., Rechenberg, I. & Schwefel, H. P. (eds.) *Proceedings of the Conference Parallel Problem Solving from Nature PPSN-IV*, 730–739. Berlin: Springer.

Drechsler, R. & Becker, B. (1998). *Binary Decision Diagrams: Theory and Implementation*. Boston, USA: Kluwer Academic Publishers.

Droste, S. (1997). Efficient Genetic Programming for Finding Good Generalizing Boolean Functions. In Koza, J. R., Deb, K., Dorigo, M., Fogel, D. B., Garzon, M., Iba, H. & Riolo, R. L. (eds.) *Genetic Programming 1997: Proceedings of the Second Annual Conference*, 82–87. San Francisco, CA, USA: Morgan Kaufmann

Hartmann, C. R. P., Varshney, P. K., Mehrotra, K. G. & Gerberich, C. L. (1982). Application of Information Theory to the Construction of Efficient Decision Trees. *IEEE Transactions on Information Theory* **28**(5): 565–577.

Hernández Aguirre, A., Buckles, B. P. & Coello Coello, C. A. (2001). Ga-Based Learning of $kdnf_n^s$ Boolean Formulas. In *Evolvable Systems: From Biology to Hardware*, 279–290. Tokyo, Japan: Springer Verlag, 3–5 October.

Hernández Aguirre, A., Buckles, B. P. & Coello Coello, C. A. (2000). Evolutionary Synthesis of Logic Functions Using Multiplexers. In Dagli, C., Buczak, A. L. et al. (eds.) *Proceed-

ings of the 10th Conference Smart Engineering System Design, 311–315. New York: ASME Press.

Hernández Aguirre, A., Buckles, B. P. & Alcántara, A. M. (2000a). The PAC Population Size of a Genetic Algorithm. In *Twelfth International Conference on Tools with Artificial Intelligence*, 199–202. Vancouver British Columbia, Canada: IEEE Computer Society, 13–15 November.

Hernández Aguirre, A., Buckles, B. P. & Coello Coello, C. A. (2000b). Gate-Level Synthesis of Boolean Functions Using Binary Multiplexers and Genetic Programming. In *Conference on Evolutionary Computation 2000*, 675–682. IEEE Computer Society.

Hernández Aguirre, A., Coello Coello, C. A. & Buckles, B. P. (1999). A Genetic Programming Approach to Logic Function Synthesis by Means of Multiplexers. In Stoica, A., Keymeulen, D. & Lohn, J. (eds.) *Proceedings of the First NASA/DoD Workshop on Evolvable Hardware*, 46–53. Los Alamitos, California: IEEE Computer Society.

Iba, H., Iwata, M. & Higuchi, T. (1997). Gate-Level Evolvable Hardware: Empirical Study and Application. In Dasgupta, D. & Michalewicz, Z. (eds.) *Evolutionary Algorithms in Engineering Applications*, 259–276. Berlin: Springer-Verlag.

Islas Pérez, E., Coello Coello, C. A. & Hernández Aguirre, A. (2003). Extracting and Re-Using Design Patterns from Genetic Algorithms Using Case-Based Reasoning. *Engineering Optimization* (April) **35**(2): 121–141.

Kabakcioglu, A. M., Varshney, P. K. & Hartmann, C. R. P. (1990). Application of Information Theory to Switching Function Minimization. *IEE Proceedings, Part E* **137**: 387–393.

Kalganova, T. (2000). Bidirectional Incremental Evolution in Extrinsic Evolvable Hardware In Lohn, J., Stoica, A., Keymeulen, D. & Colombano, S. (eds.) *Proceedings of the Second NASA/DoD Workshop on Evolvable Hardware*, 65–74. Los Alamitos, California: IEEE Computer Society.

Koza, J. R. (1992). *Genetic Programming. On the Programming of Computers by Means of Natural Selection*. Massachusetts, USA: MIT Press.

Lloris, A., Gomez-Lopera, J. F. & Roman-Roldan, R. (1993). Using Decision Trees for the Minimization of Multiple-Valued Functions. *International Journal of Electronics* **75**(6): 1035–1041.

Louis, S. J. (1993). *Genetic Algorithms as a Computational Tool for Design*. PhD thesis, Indiana University, Indiana, USA.

Louis, S. J. & Johnson, J. (1997). Solving Similar Problems using Genetic Algorithms Case-Based Memory. In Bäck, T. (ed.) *Proceedings of the Seventh International Conference on Genetic Algorithms*, 283–290. San Francisco, California: Morgan Kaufmann Publishers.

Luba, T., Moraga, C., Yanushkevich, S., Shmerko, V. & Kolodziejczyk, J. (2000). Application of Design Style in Evolutionary Multi-Level Network Synthesis. In *Proceedings of the 26th EUROMICRO Conference Informatics:Inventing the Future*, 156–163. IEEE Press.

Luba, T., Moraga, C., Yanushkevich, S., Opoka, M. & Shmerko, V. Evolutionary Multi-Level Network Synthesis in Given Design Style. In *Proceedings of the 30th IEEE International Symposium on Multiple valued Logic*, 253–258. IEEE Press.

Maes, F., Collignon, A., Vandermeulen, D., Marchal, G. & Suetens, P. (1997). Multimodality Image Registration by Maximization of Mutual Information. *IEEE Transactions on Medical Imaging* (April) **16**(2): 187–198.

Miller, J. F., Thomson, P. & Fogarty, T. (1998). Designing Electronic Circuits Using Evolutionary Algorithms. Arithmetic Circuits: A Case Study. In Quagliarella, D., Périaux, J., Poloni, C. & Winter, G. (eds.) *Genetic Algorithms and Evolution Strategy in Engineering and Computer Science*, 105–131. Chichester, England: Morgan Kaufmann.

Miller, J. F., Job, D. & Vassilev, V. K. (2000). Principles in the Evolutionary Design of Digital Circuits – Part I. *Genetic Programming and Evolvable Machines* (April) **1**(1/2): 7–35.

Miller, J. F., Job, D. & Vassilev, V. K. (2000a). Principles in the Evolutionary Design of Digital Circuits – Part II. *Genetic Programming and Evolvable Machines* (July) **1**(3): 259–288

Quinlan, J. R. Learning Efficient Classification Procedures and Their Application to Chess Games. (1983). In Michalski, R. S., Carbonell, J. G. & Mitchell, T. M. (eds.) *Machine Learning: An Artificial Intelligence Approach*, 463–482. Berlin, Heidelberg: Springer.

Shannon, C. E. (1948). A Mathematical Theory of Information. *Bell System Technical Journal* (July) **27**: 379–423.

Scholl, C. & Becker, B. (2000). On the Generation of Multiplexer Circuits for Pass Transistor Logic. In *Proceedings of Design, Automation, and Test in Europe*.

Studholme, C., Hill, D. L. G. & Hawkes, D. J. (1999). An Overlap Invariant Entropy Measure of 3D Medical Image Alignment. *Pattern Recognition* **32**: 71–86.

Torresen, J. (2002). A Scalable Approach to Evolvable Hardware. *Genetic Programming and Evolvable Machines* **3**(3): 259–282.

Thompson, A., Layzell, P. & Zebulum R. S. (1999). Explorations in Design Space: Unconventional Design Through Artificial Evolution. *IEEE Transactions on Evolutionary Computation* (September) **3**(3): 167–196.

Vassilev, V. K. & Miller, J. F. (2000). Scalability Problems of Digital Circuit Evolution In Lohn, J., Stoica, A., Keymeulen, D. & Colombano, S. (eds.) *Proceedings of the Second NASA/DoD Workshop on Evolvable Hardware*, 55–64. Los Alamitos, California: IEEE Computer Society.

Weaver, W. & Shannon, C. E. (1949). *The Mathematical Theory of Communication*. Urbana, Illinois: University of Illinois Press.

Three-Dimensional Feedforward Neural Networks and Their Realization by Nano-Devices

VLAD P. SHMERKO and SVETLANA N. YANUSHKEVICH

Department of Electrical and Computer Engineering, University of Calgary (E-mail: {shmerko;yanush}@enel.ucalgary.ca)

Abstract. The three-dimensional (3D) model of a feedforward neural network (NN) based on so called $\mathcal{N}$-hypercube topology is proposed. The $\mathcal{N}$-hypercube is different from the classical hypercube used in communication theory, and in Boolean algebra. This new structure has been created based on a novel algorithm for embedding a binary decision tree and binary decision diagram into a $\mathcal{N}$-hypercube. It is shown that $\mathcal{N}$-hypercube topology is a reasonable solution to implement NN of threshold gates, in particular, on the single-electron devices. The 3D design methodology of feedforward NN is oriented to technology mapping to nanodevices. Results of extensive experimental study of feedforward networks consisting of over 3500 $\mathcal{N}$-hypercubes are presented.

Keywords: decision tree, hypercube, single-electron logic, three-dimensional neural network

1. Introduction

It has been justified in a number of papers that 3D topologies of computer elements are optimal, physically realistic models for scalable computers (Frank et al. 1998). Nanoelectronic devices are expected to be 3D. This has motivated search for new design methodologies to meet expected computing demands in the nanometric era. This paper contributes in design methodologies of such tools, namely, we focus on feedforward NNs.

One of the primary problems in the development of large nanoICs circuits is the lack of appropriate design methodologies that effectively utilize CAD tools.

Recent results indicate that the most powerful NNs have 10^2–10^3 inputs and about 10^3–10^4 neuron-like gates of 10^2–10^3 inputs each (Fiesler and Beale (eds.) 1997). It corresponds to about 10^8 connections and processing speed about 10^{11} "connections per second". These results are based on the state-of-the-art technology and 2D design methodology. It should be noted that biological NN realize 10^{14} connections with a speed of 10^{16} connections per second in 3D space. It is expected that only breakthrough can change such drastic limitations. This is the motivation to develop a method of mapping NN into 3D space.

313

We study a computational framework for models of a linear threshold neuron in nanometric space. The crucial idea is to utilize a hypercube topological data structure that has been well studied in communication, supercomputer design, and logic circuit synthesis. However, direct application of the traditional hypercube model to3D representation of a set of simple computational nanoelements leads to serious difficulties. We propose to utilize the unique properties of a hypercube topology by modifying the hypercube structure. It is based on the similarities between the topological structure of binary decision trees and binary decision diagrams, and of Single Electron Transistor (SET).

The problem of mapping 2D models of SET into 3D space corresponds to the problem of embedding a binary decision tree or binary decision diagram into appropriate 3D model. This problem is only solved, for simple forms of decision trees. Binary decision tree are relevant to functional formulation, and thus, are very different from conventional hypercubes used in communication and supercomputer design. Hence, it is critical to modify the hypercube in order to embed binary decision trees and binary decision diagrams into 3D hypercube space.

The main feature of the methodology described in this paper is the application to SET technology. The second important feature is the development of a library of 3D threshold elements and 3D macro components, allowing the design of nanocircuits of arbitrary complexity. Finally, the proposed methodology is generic because it does not limit the types of decision trees, so it is possible to design technology independent 3D structures.

In this paper we develop a new data structure, $\mathcal{N}$-hypercube, based on the modified classical hypercube. The differences between the classical hypercube and the proposed $\mathcal{N}$-hypercube is in functions of the nodes. However, there are many common features: classical hypercube and $\mathcal{N}$-hypercube have the same rules for encoding the terminal nodes (Gray code) and the same measurements, except special evaluations of the effectiveness of a resulting 3D architecture.

It is worthy to note that the hypercube data structure together with the Reduced Ordered Binary Decision Diagram (ROBDD) are the state-of-the art tools of advanced logic circuit design. The spatial topology of the hypercube is suitable for mapping network models into 3D space, and, thus, is an appropriate model for nanotechnologies.

We refer to the most important recently reported results as follows.

(a) The concept of nanocomputing based on Josephson junction technology is discussed in (Likharev 1999; Takahashi et al. 2000; Wasshuber 2001). In Takahashi et al. (2000) applications of SET to neuron modeling are reported.

(b) Mapping a ROBDD into a hexagonal nanowire network controlled by Schottky wrap gates have been introduced in Asahi et al. (1997) and Yamada et al. (2001). In our paper we utilize and generalize the unique correspondence of SET topology and ROBDD topology.

(c) Survey on NN hardware implementation and its trends is given in Fiesler and Beale (eds.) (1997) and Webster (1999).

(d) Boolean functions computing by feedforward NN is introduced in Siu et al. (1991). In our approach, we study the 3D feedforward network (without feedback) that consists of linear threshold gates.

(e) Topological properties of hypercubes have been introduced in Saad et al. (1988). In Roth (1999), Boolean function manipulation based on hypercubes is discussed. We extend the classical hypercube topological data structure toward the $\mathcal{N}$-hypercube.

The results of our study are introduced as follows. In the next section we give the basic definitions starting with graph theory and classical hypercube. Embedding binary tree into hypercube is discussed in section 3. In section 4 a new hypercube data structure, $\mathcal{N}$-hypercube, is introduced. The method of 3D NN design via embedding a decision tree into a $\mathcal{N}$-hypercube is discussed in section 5. We show in section 6 that $\mathcal{N}$-hypercube is a reasonable model for a SET technology. We focus on library of nanoneurons, and introduce our approach to 3D NN design in section 7. The results of experimental studies are presented in section 8.

2. Basic Definitions

The theoretical basis of our study combines

(i)   NN design methodology, that focus on feedforward threshold network,

(ii) Graph data structures, including decision trees, decision diagrams, and hypercubes, and

(iii) SET nano-technology.

Definition 1. *The threshold gate is a computational element that calculates the sum of n inputs $x_i \in (0, 1)$ weighted with weights w_i. The sum is converted into digital output y by comparing the sum with a given threshold θ*

$$y = f\left(\sum_{i=1}^{n} w_i x_i - \theta\right) = \begin{cases} 1, & \text{if } \sum_{i=1}^{n} w_i x_i \geq \theta, \\ 0, & \text{if } \sum_{i=1}^{n} w_i x_i < \theta. \end{cases}$$

Adapting the weights and threshold, a linearly threshold gate computes any linear separable Boolean function of n inputs. The feedforward network is defined as a network of interconnected linear threshold gates, and NN to be a feedforward network of these gates. The tree is an appropriate model to describe a set of threshold gates.

Definition 2. *A* tree *is a rooted acyclic graph in which every node but the root has indegree 1.*

The form of representation of a tree is critical to the design methodology that is based on principle of embedding of the tree into a hypercube. In a tree, for any vertex v there exists a unique path from the root to v. The length of this path is called the *depth* or *level* of v. In logic design, the path corresponds to sum-of-product (SOP) terms of the Boolean function. The height of a tree is equal to the largest *depth* in the tree. A node with no children is a *terminal* (*external*) node or *leaf*. Terminal nodes carries information such as, for example, the function value (in ROBDD). A nonleaf node is called an *internal* node. The function of a node can be given in abstract form (communication problems) or specified exactly (logic design problems).

This edge model is suitable, to example, to describe transmission of data, so that data are sent in only one direction at a time, up or down. The *size* of the tree is the number of nodes and it often measures of software or hardware requirements.

If all nodes are grouped with the same depth together, the network is to be arranged in *layers*. Hence, the depth of the network can be interpreted as time of parallel computations.

Example 1. *A feedforward network is defined as an acyclic labeled directed graph with distinguished input nodes of indegree zero, internal nodes of arbitrary indegree, and distinguished output nodes.*

Definition 3. *A* complete *$n-$level binary tree is the tree with 2^k nodes on level k for $k = 0, \ldots, n - 1$.*

The 2^n-leaf complete binary tree has a level hierarchy (levels $0, 1, \ldots, n$). The root is associated with level zero and its two children are on level one, and the j-th node on level i is denoted by (i, j) for $0 \leq j < 2^i$. In the complete binary tree, each node is either a leaf or has degree 2. The complete binary 3-level ($n = 3$) tree is given in Figure 1.

Definition 4. *An Ordered Binary Decision Diagram (OBDD) for Boolean function f with respect to ordering π is a directed acyclic graph consisting of non-terminal nodes labelled with the variables of f and terminal nodes labeled with 0 and 1.*

OBDD has a number of useful properties. In particular, an arbitrary Boolean function can be represented by an OBDD, and using reduction rules, the OBDD can be transformed into a canonical form of the Boolean function, i.e., ROBDD.

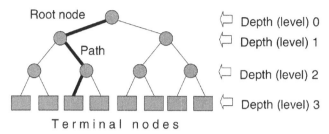

Figure 1. Complete binary tree.

Definition 5. *Hypercube is a graph data structure that consists of d dimensions and 2^d vertices, encoded by $(g_0 \ldots g_{d-1})$. Two vertices $(g_0 \ldots g_{d-1})$ and $(g'_0 \ldots g'_{d-1})$ are connected by an edge if and only if they differ in exactly one bit (Gray code). There are $d \times 2^{d-1}$ edges in a d-dimensional hypercube.*

The *fan-out* (i.e., degree) of every node is n, and the total number of communication links is $\frac{1}{2}N \log_2 N$. Each node of the n-dimensional hypercube can be specified by the binary address $(g_{n-1}, g_{n-2}, \ldots, g_0)$, where bit g_i corresponds to the i-th dimension in a Boolean space. The dimensions are specified by the set $\{0, 1, \ldots, n - 1\}$.

3. Embedding a Binary Tree into a Hypercube

The crucial phase to design a hypercube space topology is a mapping of a 2D graph into a hypercube. This procedure is called *embedding* of a guest graph into a hypercube (host graph).

3.1. *Forms of a binary tree representation*

Definition 6. *An **embedding**, $\langle \varphi, \alpha \rangle$, of a guest graph G into a host graph H is the one-to-one mapping $\varphi: V(G) \to V(H)$, along with the mapping α that maps an edge $(u, v) \in E(G)$ to a path between $\varphi(u)$ and $\varphi(v)$ in H. The embedding of a graph G into a graph H is an injection (one-to-one mapping) of the nodes in G to the nodes in H.*

For simplification but without loss of generality, we will discuss embedding a complete binary tree into a hypercube. The form of representation of the binary tree (specification) and conditions of embedding are critical to choosing the method of embedding.

Definition 7. *Node specification of an arbitrary binary tree includes a function of the node (abstract machine that computes the outputs accordingly to some algorithm) and its topology information (root, internal node,* *terminal node, coordinates). **Edge specification** includes a function of the edge (abstract data or specified value transmission) and its topology information (indexes of related nodes). Node and edge specifications have different additional **attributes** such as the level number, the path number, and the value of terminal node.*

The combination of a node and edge specifications together with attributes allows generating different forms of a binary tree representation. We distinguish three forms of the binary tree representation.

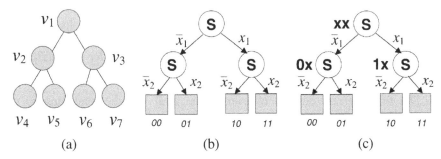

Figure 2. The forms of representation of a binary tree: Specification 1 is typical for communication problems (a), Specification 2 is used for Boolean function manipulation and decision diagram design (b), and Specification 3 is used for representation of a decision diagram in 3D space (c).

Specification-1: The binary tree is a particular network architecture, that simply carries information about dual connections of each node (Figure 2a). This formulation is typical for many communication problems.

Specification-2: The binary tree carries information about functionality of the logic circuit (Figure 2b). A node of the binary tree is associated with Shannon expansion with respect to a variable (denoted by S). Algebra of relationships of nodes (variables) is known as a cube algebra to manipulate Boolean functions. In this formulation, a number of optimization problems is solved in terms of logic circuit design. Simplification of this tree based on reduction rules gives a ROBDD.

Specification-3: The binary tree carries information about functionality of the logic circuit and, in addition to Specification-2, the topology (coordinate). The nodes of the binary tree are associated with Shannon expansion with respect to each variable and coordinate in 3D space (Figure 2c).

The algorithms of embedding a complete binary tree given in Specification-1 and Specification-2 are introduced in many publications, in particular, (Saad et al. 1988; Öhring et al. 1995). We are focused on the unsolved problem of embedding the tree given in Specification-3 into a hypercube.

3.2. *Specification-1*

Let the embedding conditions be formulated as follows: (i) each node of a tree is labeled by v_i, and (ii) the number of nodes in a tree is equal to or less than the number of nodes in the hypercube. Even in the simplest form of a binary tree description, Specification-1, the effectiveness of the embedding strongly depends on a number of parameters. The *dilation cost* of an embedding of the graph G in the hypercube H is the maximum distance in H between any two neighbor nodes in G. This cost gives a measure of the proximity in H of the neighbor nodes in G under the embedding. The *expansion* is the ratio $|V(H)|/|V(G)|$. The *load* of the embedding is the maximum, over all host nodes, of the number of guest nodes mapping into it. The *edge-load* of edge $e \in E_H$ in the embedding is the maximum number of edges of G that are routed by g over the edge e. The *edge-congestion* is the maximum edge load, over all host edges.

In our design methodology and experimental study we use these parameters to measure the designed 3D structures. In addition, we have found that 3D Voronoi diagrams are very useful tools for measuring in 3D space (Okabe et al. 1992).

In terms of Specification-1, the task is formulated as finding a balance between edge load and communication time. It requires to minimize such cost parameters as dilation, expansion, and edge-congestion. For example, there exists an optimal embedding of a complete binary tree with load and expansion 1, and dilation and edge-congestion 2 into a hypercube (Öhring et al. 1995).

Hence, in Specification-1 functions of nodes are different. Their functions are not important, and only edges between the nodes carry useful information. The task of embedding is formulated in terms of edges, i.e. distance, dilation cost, *etc.*

3.3. *Specification-2*

The idea of a hypercube in a logic circuit design is that an n-bit string can be visualized geometrically, as a vertex of a hypercube that consists of 2^n vertices, each of which is labelled with an n-bit string (Roth 1999). 0D hypercube is a point, a 1D hypercube is a line, a 2D hypercube is a square, and 3D hypercube is a cube. The truth table of an n-input Boolean function can be represented in an n-dimensional Boolean space. There is one axis for each variable which can take two values: 0 and 1. Each node is labeled with its coordinate in the n- dimensional space. Specification-2 is utilized in the minimization a Boolean functions.

The minterm of an n-variable Boolean function can be represented by an n-bit integer (the minterm number) where each bit indicates whether the corresponding bit variable is complemented or uncomplemented. To deal with product terms that are not minterms, where some variables do not appear at all, we must represent three possibilities for each variable in a general product term: 1 – uncomplemented, 0 – complemented, and **x** – does not appear. For example, given a function of six variables, $x_1, x_2, x_3, x_4, x_5, x_6$, we write the following product terms and their cube representations: $x_1, x_2, \overline{x}_3 = 110\mathbf{xxx}$, $\overline{x}_1 x_5 x_6 = 0\mathbf{xxx}11$ (the hypercube notation is conventional in logic design).

A node in the binary decision tree realizes the Shannon decomposition

$$f = \overline{x}_i f_0 \oplus x_i f_1, \tag{1}$$

where $f_0 = f(x_i = 0)$, and $f_1 = f(x_i = 1)$, for all variables in f.

Consequently, the embedding conditions can be formulated as follows:

(i)   Each node realizes Shannon expansion (1),

(ii)   The nodes are distributed over levels,

(iii)   Each level, except level 0, corresponds to a variable x_i,

(iv)   Each path from the root node to a terminal nodes corresponds to a minterm in SOP representation, determined as the product of labels at the edges, and

(v)   A terminal node is labelled with 1 if the path from the root node to the terminal node (minterm) is equal to 1.

Note, that since the variables appear in a fixed order, such tree is an *ordered* binary decision tree.

Example 2. *Five minterms of the Boolean function $f = x_1 \overline{x}_2 \vee \overline{x}_3$ given by the truth table (Figure 3a) are represented by the tree with five active terminal nodes (the value 1) (Figure 3b). The cube representation (Figure 3c) implies representation of the function in cube terms: $< \mathbf{xx}0,\ 01\mathbf{x} >$.*

It follows from the above description of input data that, in addition to the Specification-1 of a tree, levels become the carriers of information. Based on this extension, effective methods to manipulate edges have been developed with the goal to minimize the number of hypercubes to represent a given Boolean function.

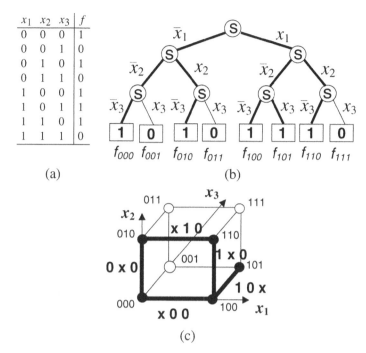

(a)   (b)

(c)

Figure 3. Truth table of the function $f = x_1\bar{x}_2 \vee \bar{x}_3$ (a), the binary decision tree (b), and hypercube representation (c).

3.4. *Specification-3*

The complete binary decision tree in Specification-3 is given in Figure 4. We observe that this tree is different from Specification-2 by labels of non-terminal nodes. The embedding conditions for this tree are the same as for Specification-2 accept that each node is labelled with a coordinate.

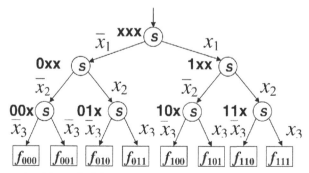

Figure 4. A complete binary decision tree to represent an arbitrary Boolean function of three variables; each node is labelled with a distinguished coordinate.

4. $\mathcal{N}$ -hypercube

As mentioned above, there is a number of reasons to develop a new data structure based on hypercube topology. The main reason is to overcome limitations of the hypercube topology while embedding a binary decision tree. Specification-3 is of practical interest because it satisfies technological requirements of nanodevice fabrication. In this section, we focus on modification of a classical hypercube with the purpose to solve the formulated problem.

4.1. *Topology*

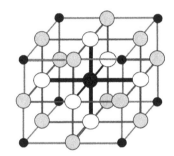

Figure 5. The complete $\mathcal{N}$ -hypercube.

The $\mathcal{N}$ -hypercube topology is different from the topology of a classical hypercube, in the number of nodes, types of nodes, and connections (Figure 5).

Number of nodes. The classical hypercube contains 2^n nodes, $n = 0, 1,$ $\ldots, M$, while the $\mathcal{N}$ -hypercube includes $\sum_{i=0}^{n} 2^{n-i} C_i^n$ extra nodes, so it is possible to manipulate $2^n + \sum_{i=0}^{n} 2^{n-i} C_i^n$ nodes while designing the 3D circuit.

Type of nodes. In the classical hypercube, the nodes are not differentiated by their type, i.e. each node can be a generator or receiver of information. In contrast, in $\mathcal{N}$ -hypercube there are three types of nodes, and their functions are different.

Connections. In traditional design methodology, hypercubes can communicate via appropriate nodes accordingly to a given strategy, i.e. there are no special requirements to the nodes. Communication in the $\mathcal{N}$ - hypercubes is scheduled by intermediate nodes.

It should be noted that the connection strategy is closely related to the well-studied problem of the ROBDD design (Bryant and Meinel 2002).

In Figure 6 various types of $\mathcal{N}$-hypercube nodes are illustrated.

Example 3. *Let* $n = 3$. *The 3D hypercube consists of 8 nodes and* $\mathcal{N}$*-hypercube involves* $\sum_{i=0}^{3} 2^{3-i} C_i^3 = 27$ *nodes: 8 terminal nodes, 12 intermediate nodes, 6 face nodes and one root node.*

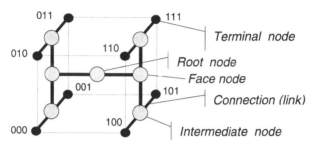

Figure 6. $\mathcal{N}$-hypercube topological structure consists of four types of nodes: terminal nodes, intermediate nodes, face node, and the root node.

In addition, we note that $\mathcal{N}$-hypercube can be considered as technology *dependent* model because of its close relationship to the SET topology.

4.2. $\mathcal{N}$-hypercube as technology dependent data structure

Consider two simplest topological structures: a node of a binary decision tree and SET (Figure 7). In the binary tree, a *messenger electron* travels along the path and reaches 1 or 0 terminal (Asahi et al. 1997). A value of the Boolean function is recognized by observing which terminal the electron reaches. The Shannon expansion of a node is implemented as two-way clocked and switching controlled variable x. Physically, it consists of four tunnel junctions. A messenger electron travel through the entry branch to the exit branch that corresponds to the binary value of the input signal: if input signal x is a positive (1), then the electron is transported to 1-branch, otherwise, 0-branch.

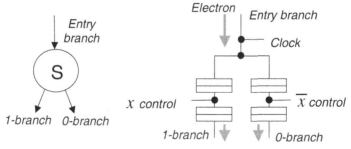

Figure 7. The binary decision tree (a) and a model of a SET (b).

The set of two SETs corresponds to a binary decision tree with four nodes that can be embedded into a 2D hypercube. If the set contains three SETs, its 2D model is a tree with eight nodes that is embedded into a 3D $\mathcal{N}$-hypercube. To design a 4D $\mathcal{N}$-hypercube from the binary tree with 16 nodes, we need to add new links to connect intermediate nodes.

4.3. $\mathcal{N}$-hypercube design

Definition 8. *The $\mathcal{N}$-hypercube is a hypercube that consists of* **terminal nodes, intermediate nodes** *and* **root.** *These nodes are connected accordingly to the following rules:*
Rule 1: *The terminal node can be connected with one intermediate node only.*
Rule 2: *The root can be connected with two intermediate nodes located symmetrically in opposite faces.*
Rule 3: *Configurations of the terminal and intermediate nodes on the opposite faces are symmetric, these two faces are connected via the root.*

Any n-dimensional structure can be represented by a 3D $\mathcal{N}$-hypercube which is a hierarchy of $\lceil n/3 \rceil$ of a $3D$ $\mathcal{N}$-hypercubes, so that each corresponds to a hypercube representation of a 3-variable Boolean function.

Definition 9. *The $\mathcal{N}$-embedding is a complete binary tree with 2^k terminal nodes embedded into k-dimensional $\mathcal{N}$-hypercube.*

For instance, the 3D $\mathcal{N}$-hypercube carries information about three-variable Boolean function.

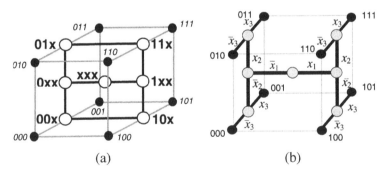

(a)   (b)

Figure 8. $\mathcal{N}$-hypercube: coordinate description (a) and embedding of a complete binary tree given in Figure 4 (b).

4.4. *Coordinate description*

To design and manipulate $\mathcal{N}$-hypercubes, we need to define a coordinate description of the topological components of $\mathcal{N}$-hypercubes. For this, we utilize principles of Specification-2 and denote links by $\{0, 1, \mathbf{x}\}$.

There are two possible configurations of the intermediate nodes. The first configuration in coordinate description is:

$$x00 \iff x01 \iff xx1 \iff x11 \iff x10 \iff xx0 \iff x00.$$

The second configuration is relevant to the symmetric face (Figure 8a):

$$00x \iff 0xx \iff 01x \iff 11x \iff 1xx \iff 10x \iff 00x.$$

$\mathcal{N}$-hypercube includes two configurations with respect to the root:

$$\text{Link 1: } xx0 \iff xxx \iff xx1,$$
$$\text{Link 2: } 0xx \iff xxx \iff 1xx.$$

The coordinate of the root node is xxx. There are two types of links in $\mathcal{N}$-hypercube: links between terminal nodes and intermediate nodes, and links between intermediate nodes, including the root node.

Example 4. *Link <00,00x> indicates the connection of the terminal node 000 and intermediate node 00x. By analogy, if two intermediate nodes x10 and xx0 are connected, this is indicated by <x10,xx0> (Figure 8a).*

4.5. *Relationship between hypercube and $\mathcal{N}$-hypercube*

There is a simple relationship between the classical hypercube utilized in Boolean function manipulation, and $\mathcal{N}$-hypercube: the coordinates of both hypercubes correspond to each other as follows:

```
Link in the hypercube ≡ Intermediate node of the 𝒩-hypercube
    Face of the hypercube ≡ Face node of the 𝒩-hypercube
```

Example 5. *The edges x00, 0x0, x10, and 10x in Figure 3c correspond to the intermediate nodes x00, 0x0, x10, and 10x in the $\mathcal{N}$-hypercube in Figure 8a. Further, the faces xx0, xx1, 0xx, 1xx, x1x, and x0x in Figure 3c correspond to the face nodes 0xx, 1xx, x1x, and x0x in Figure 8a.*

4.6. *4D $\mathcal{N}$-hypercubes*

Consider two 3D $\mathcal{N}$-hypercubes. To design 4D $\mathcal{N}$-hypercube, those two 3D $\mathcal{N}$-hypercubes must be linked. Then, the number of bits in the coordinate description of both $\mathcal{N}$-hypercubes must be increased by one bit on the right (Figure 9). Suppose that $\mathcal{N}$-hypercubes are connected via link <xxx0,xxx1> between the root nodes xxx0 and xxx1. The resulted topological structure is

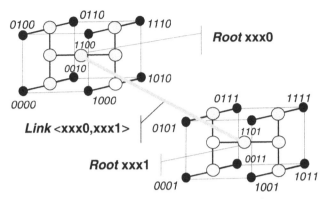

Figure 9. Coordinates of the 4D $\mathcal{N}$-hypercube.

called 4D $\mathcal{N}$-hypercube. Thus, the designs of the 4D classical hypercube and $\mathcal{N}$-hypercube differ.

5. Embedding Binary Decision Tree into $\mathcal{N}$-hypercube

In this section we will show that an arbitrary binary decision tree, or ROBDD can be embedded into a $\mathcal{N}$-hypercube. We keep in mind that an arbitrary threshold network can be represented by a binary decision tree or ROBDD, and our final task is to convert these graphical structures into 3D space.

5.1. *Basics statements*

Briefly, embedding a complete binary tree into a classical hypercube is the mapping of the leaf nodes of the tree with q levels to 2^q nodes of a classical hypercube, where two nodes are adjacent if and only if their q-bit binary codes differ in exactly one position.

Lemma 1. *The number of terminal nodes of the complete binary tree which are mapped into the intermediate nodes of a $\mathcal{N}$-hypercube is equal to 2^{q-1}, while the number of such embeddings (the number of all wires) is $q \times 2^{q-1}$.*

Proof. The proof follows from the fact that the number of intermediate nodes in the second last level of the binary tree is 2^{q-1}. The number of possible orders of variables is equal to the number of variables, q, so, the number of embeddings is $q \times 2^{q-1}$. □

Lemma 2. *The number of nodes of the binary tree embedded into the face nodes of a $\mathcal{N}$-hypercube is equal to 2^{q-2} while the number of such embeddings (the number of all faces) is $q \times 2^{q-2}$.*

The proof is adequate to the proof of Lemma 1.

Corollary 1. *The total number of intermediate nodes and face nodes in a $\mathcal{N}$-hypercube is $\sum_{i=0}^{n} 2^{n-i} C_i^n$. The total number of connections between these nodes is $\sum_{i=0}^{n} 2i \cdot 2^{n-i} C_i^n$.*

It is obvious that the total number of intermediate and face nodes in $\mathcal{N}$-hypercube is equal to the number of all nodes but leaves in the complete binary decision tree of a Boolean function of n variables, and the total number of connections is equal to the number of edges in the complete binary decision tree.

Theorem 3. *An arbitrary complete 2^q - terminal node binary tree defined by* `Specification-3` *can be embedded into a q-dimensional $\mathcal{N}$-hypercube.*

Proof. The proof follows from Definition 8, Lemma 1 and 2, and Corollary 1.   □

Theorem 4. *A ROBDD of an arbitrary Boolean function can be embedded into a $\mathcal{N}$-hypercube.*

Proof. The proof follows from one-to-one correspondence between ROBDD and the complete binary tree.   □

The process of embedding a ROBDD turns to expanding the ROBDD to the complete binary tree, then embedding this tree into an $\mathcal{N}$-hypercube.

5.2. *Algorithm for embedding a binary decision tree into a $\mathcal{N}$-hypercube*

Input data: A complete binary decision tree (`Specification-3`).

Step 1. Embed 2^q leaves of the binary tree (nodes of level $q - 1$) into a 2^q-node $\mathcal{N}$-hypercube; assign codes (q-tuple) to the nodes so that each node is adjacent with another one which code differs in exactly one position.

Step 2. Embed 2^{q-1} nodes of the binary tree (nodes of the level $q - 2$) into edges that connect the existing nodes of the $\mathcal{N}$-hypercube, taking into account the polarity of variables.

Step 3. Embed 2^{q-2} nodes of the binary tree (nodes of the level $q - 3$) into faces of the $\mathcal{N}$-hypercube, taking into account the polarity of variables.

This recurrent strategy can be used to embed any binary decision tree into a $\mathcal{N}$-hypercube. Consider $q = 1, 2$ and 3.

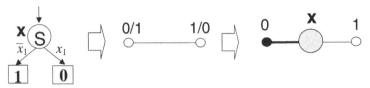

Figure 10. Embedding the 1-variable binary tree into the $\mathcal{N}$-hypercube

Case $q = 1$. Embedding the binary decision tree that represents a Boolean function of one variable ($q = 1$) is illustrated in Figure 10. The function takes value 1 while $x = 0$ and value 0 while $x = 1$. These values are assigned with two leaves of the tree; it is mapped to the $\mathcal{N}$-hypercube, which nodes are assigned with the values of variable/function (0/1 and 1/0). Finally, it is mapped to a two-node graph (1-dimensional $\mathcal{N}$-hypercube) with the node embedded into the edge.

Case $q = 2$. Let us consider a complete binary decision tree that represents a Boolean function of two variables ($q = 2$). It follows from Lemmas 1 and 2 that four leaf nodes of the tree can be embedded into the $\mathcal{N}$-hypercube with four nodes (a quadrant). The nodes of the tree can be embedded into two opposite edges at a time. The choice of the pair of edges depends solely on the order of variables in the tree. The obtained intermediate nodes are associated with two middle-level nodes of the tree. The face nodes are embedded into the faces, each is connected to both intermediate nodes and is associated with the other variable. Figure 11 illustrates this case. Note, the axes are associated with the polarity of variables (complemented, uncomplemented).

Case $q = 3$. Figure 8b illustrates embedding a binary decision tree of a three-variable Boolean function into a $\mathcal{N}$-hypercube. First of all, embed 8 leaves of the tree (Gray code ordered). Then, embed four intermediate nodes of the tree into the edges connecting 8 nodes of the $\mathcal{N}$-hypercube; this is a sub-cube to implement a sub-function with respect to x_3. Next step, embed two face nodes of the tree into faces of the $\mathcal{N}$-hypercube and connect them to the intermediate nodes; x_2 is introduced. Finally, embed the root of the tree into the center of the $\mathcal{N}$-hypercube and connect it to two face nodes; the function is completed with x_1.

5.3. *Connections in n dimensional space*

A wire is associated with the tunnel junction controlled by the voltage that is direct (value of x_i) or inverse (value of $\overline{x}_i$).

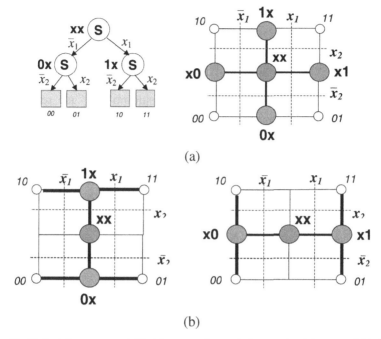

Figure 11. Embedding a two-variable binary tree into a $\mathcal{N}$-hypercube: (a) two-variable binary tree and the corresponding $\mathcal{N}$-hypercube (b) possible variants of embedding.

Definition 10. *Expanding a $\mathcal{N}$-hypercubes to more dimensionals is accomplished by:*
Rule 1: *Connecting roots, and adding one bit on the right of the code for each $\mathcal{N}$-hypercube.*
Rule 2: *Assigning coordinates to the new root.*

6. Implementation of a Neuron on a Multigate Single-electron Transistor

In this section we show that $\mathcal{N}$-hypercube is a reasonable model for 3D NN computing. For this we choose a promising nanotechnology, SET technology. The fundamental physical principle of SET is the Coulomb blockade resulting from the quantization of the elementary charge in an isolated node of a double junction structure (Likharev 1999). In a multigate SET, current I_D is a function of input voltages and capacitances:

$$I_D(V_1, V_2, \ldots, V_N) = f\left(\sum_{i=1}^{N} \frac{C_i V_i}{e}\right), \qquad (2)$$

where e is a single electron charge ($e = 1.6 \cdot 10^{-19}$ Coulomb). The function (2) indicates that the device implements a neuron function. However, the threshold understanding is different from the traditional one. In a traditional understanding, the neuron output is 1 (a device is on) if the output superceds some threshold value. In multigate SET, the current I_D takes a minimum when $\sum_{i=1}^{N} \frac{C_i V_i}{e}$ is an integer because the Coulomb blockade sets in. Here, $\frac{C_i V_i}{e}$ corresponds to the number of exceeds electrons on the i-th input (gate electrode).

When $\sum_{i=1}^{N} \frac{C_i V_i}{e}$ is a half-integer, the current flows because the Coulomb blockade is lifted. So, the multigate SET will implement the following threshold function

$$I_D(V_1, V_2, \ldots, V_N) = f\left(\sum_{i=1}^{N} \frac{C_i V_i}{e}\right) = \begin{cases} 1, & \sum_{i=1}^{N} \frac{C_i V_i}{e} = \frac{2l-1}{2}, \\ 0, & otherwise, \end{cases}$$

where l is an integer, C_i is a weight, and V_i is an input signal.

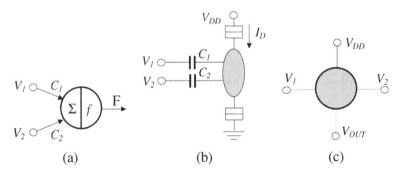

Figure 12. 2-input neuron (a), 2-input multigate SET (b) and 2D two-input neuron model (c)

Consider the 2-input threshold gate (Figure 12a) described by the function

$$F = f\left(\frac{V_1 C_1}{e} + \frac{V_2 C_2}{e}\right).$$

The corresponding multigate SET and its 2D model is given in Figure 12b and Figure 12c accordingly.

The four-input threshold element can be designed based on three multigate SETs. This set of multigate SETs is represented by a complete binary decision diagram that represent a Boolean function of two variables. Embedding this diagram into $\mathcal{N}$-hypercube results in a 2D $\mathcal{N}$-hypercube.

7. $\mathcal{N}$-hypercube Threshold Network Design

To design an arbitrary threshold networks, we have developed a library of a 3D models. This library includes threshold primitives to model logic primitives (AND, OR, NOR, NAND, EXOR), and macro threshold components (full adder, comparator, multiplexer, etc.).

The design strategy to represent a given 2D feedforward NNs in 3D space includes the steps as follows:
(i)   The initial 2D network is levelized.
(ii)   The threshold gates in each level are replaced with the corresponding $\mathcal{N}$-hypercubes
(iii)   Levels are connected accordingly to the initial netlist
Note that another strategy can be used to partition the initial 2D network, for example, cascading strategy.

8. Experiments

The goal of the experimental study is to represent large NNs in 3D space and evaluate their characteristics such as *space size* and *topological parameters* of nanoICs. Our experiment was organized as follows:
(i)   A common model of NN that is the feedforward (with no feedback) multilayer network with a threshold gate as basic processing unit.
(ii)   The linear threshold gates in this network are described by 3D models ($\mathcal{N}$-hypercubes) and, thus, by a set of connected $\mathcal{N}$-hypercubes.
(iii)   A multilevel combinational circuits over the library of NAND, NOR and EXOR gates from the ISCAS85 benchmarks set (combinational circuits)[1] were used.
Notice that we do not consider the problem of optimization of the 3D model.

Experiments were conducted on PIII 800 MHz, 256Mb RAM computer. A fragment of the results is shown in Table 1. The first three columns include the benchmark **Name**, the number of inputs and outputs **I/O**, and the number of gates **#G**. In four columns **Space size** we give the size of nanoICs in **X, Y** and **Z** coordinates. Coordinate **X** indicates the number of levels in the network. The value of the resulting nano-solid is calculated as $V = X \times Y \times Z$. The topological characteristics are evaluated by the total number of terminal nodes $\#N_T$, the total number of intermediate nodes $\#N_T$, and the total number of connections $\#C$. The last column contains the runtime in CPU seconds per cube.

Consider, for example, 3D feedforward NN that was designed based on 2D benchmark c6288. This 16 bit multiplier (32 inputs and 32 outputs) includes

2416 gates. These gates were grouped in 124 levels that were interpreted as layers. Each gate in ith level c6288 was represented by the corresponding $\mathcal{N}$-hypercube. Then, the number of incompletely specified $\mathcal{N}$-hypercubes was minimized. The $\mathcal{N}$-hypercube in the i-th layer was connected to the corresponding $\mathcal{N}$-hypercubes in $(i-1)$-th and $(i+1)$-th layers. The obtained structure was distributed in 3D space as $X \times Y \times Z = 124 \times 2 \times 2$. The value of the resulting nano-solid is equal to $V = X \times Y \times Z = 124 \times 2 \times 2 = 496$ hypercube[3]. The total number of terminal nodes and intermediate nodes are equal to $\#N_T = 9600$ and $\#N_T = 7184$ correspondingly. To combine all NN layers, 14368 connections were generated. The design of the structure was implemented in $t = 0.93$ seconds.

We observe that the nanodimensional size and the number of nodes are both near linearly proportional to the number of logic gates in the conventional netlist. Notice, that in our experiments we used various metrics to evaluate the designed 3D structure, including the mentioned above traditional parameters (diameter, dilation cost, expansion, load, *etc.*) and specific parameters, in particular, the average number of variables in the logic function described by the $\mathcal{N}$-hypercube, the average number of links, the average fan-out of the intermediate nodes, and statistical distribution of different parameters.

Table 1. Experimental results on 3D feedforward NN

Test			Space size				Nodes, connections			
Name	I/O	#G	X	Y	Z	V	$\#N_T$	$\#N_I$	$\#C$	t
c17	5/2	6	3	2	2	12	24	18	36	<0.01
c432	36/7	160	44	8	8	2816	2508	2348	4696	0.06
c499	41/32	202	14	4	4	224	1072	870	1740	0.05
c880	60/26	383	34	2	4	272	1614	1231	2462	0.06
c1355	41/32	546	27	4	4	432	2384	1838	3676	0.17
c1908	33/25	880	51	8	8	3264	5374	4494	8988	0.22
c2670	233/140	1193	50	4	4	800	4842	3649	7298	0.27
c3540	50/22	1669	81	8	8	5184	10258	8589	17178	0.44
c5315	178/123	2307	81	8	8	5184	11728	9421	18842	0.72
c6288	32/32	2416	124	2	2	496	9600	7184	14368	0.93
c7552	207/108	3512	61	4	4	976	14636	11124	22248	1.10

9. Conclusion

The presented design methodology for the design of large-scale feedforward 3D threshold network based on $\mathcal{N}$-hypercube topology implies 3D representation and spatial input/output interpretation. This is coherent with the requirements, nanotechnology poses to implementation of networks of gates. Taking this into consideration, we believe that SET devices are potential candidates for prototyping the NN in nanodimension. We have demonstrated two strategies for 3D structure design. The first strategy is based on embedding a decision tree or decision diagram into a $\mathcal{N}$-hypercube:

$$\text{Decision tree (diagram)} \Rightarrow \mathcal{N}\text{-hypercube structure}$$

The second strategy comprises the design procedure over the library of 3D primitives.

$$\text{Library of 3D primitives} \Rightarrow \mathcal{N}\text{-hypercube structure}$$

The primary features of our methodology are the following:
(i)   It offers technology independent data structure compatible with decision diagrams, an advanced structure to represent Boolean functions, and
(ii) It becomes technology dependent since the decision diagram structures can be directly mapped to the topology of a network of SETs or quantum-dot arrays.

We tested our approach on the large size networks. The reasonable results on space size evaluation, the number of nodes and interconnections have been obtained.

It should be noted, that the proposed design methodology has a generic character. It is possible to generate 3D structures based on another type of decision tree and decision diagram (Bryant and Meinel 2002). In particular, we conducted experiments on so called *linear word-level* decision trees and diagrams (Yanushkevich et al. 2002; Yanushkevich et al. 2004). This data structure provides excellent embedding conditions. However, the resulting 3D structures are technology independent. This is because manipulation of word-level (integer) values is not available in today's nanotechnological solutions.

Acknowledgements

This work was partially supported by NATO through the Collaborative Linkage Grant PST-CLG.979071, and Natural Sciences and Engineering Research Council of Canada (NSERC) through grant 239025-02. The help of

Dr. *P. Dziurzanski* in conducting the experiments is acknowledged. Discussions with Professor Ottar Johnsen, Ecole d'ingénieurs et d'architectes de Fribourg, Switzerland, were extremely useful.

Note

¹ http://www.cbl.ncsu.edu/CBL_Docs/iscas85.html

References

Asahi, N., Akazawa, M. & Amemiya, Y. (1997). Single-Electron Logic Device Based on the Binary Decision Diagram. *IEEE Transactions on Electron Devices* **44**(7): 1109–1116.

Bryant, R. E. & Meinel, C. (2002). Ordered Binary Decision Diagrams: Foundations, Applications and Innovations. In Hassoun, S. & Sasao, T. (eds.), *Logic Synthesis and Verification*, 285–307. Kluwer Academic Publishers.

Fiesler, E. & Beale, R. (eds.) (1997). *Handbook on Neural Computation*. IOP Publishing Ltd and Oxford University Press.

Frank, M. P. & Knight, T. F., Jr. (1998). Ultimate Theoretical Models of Nanocomputers. *Nanotechnology* (9): 162–176.

Likharev, K. (1999). Single – Electron Devices and Their Applications. *Proceedings of IEEE* **87**(4): 606–632.

Okabe, A., Boots, B. & Sugihara, K. (1992). *Spatial Tessellations. Concept and Applications of Voronoi Diagrams*. John Wiley & Sons.

Roth, J. P. (1999). *Mathematical Design: Building Reliable Complex Computer Systems*. IEEE Press.

Siu, K-Y., Roychowdhury, V. P. & Kailath, T. (1991). Depth-Size Tradeoffs for Neural Computation. *IEEE Transactions on Computers* **C-40**(12): 1402–1411.

Saad, Y. & Schultz, M. H. (1988). Topological Properties of Hypercubes. *IEEE Transactions on Computers*. **37**(7): 867–872.

Takahashi, Y., Fujiwara, A., Ono, Y. & Murase, K. (2000). Silicon Single-Electron Devices and Their Applications. In *Proceedings of 30th IEEE International Symposium on Multiple-Valued Logic*, 411–420.

Öhring, S. & Das, S. K. (1999). Incomplete Hypercubes: Embeddings of Tree-Related Networks. *Journal of Parallel and Distributed Computing* (26): 36–47.

Yamada, T., Kinoshita, Y., Kasai, S., Hasegawa, H. & Amemiya, Y. (2001). Quantum – Dot Logic Circuits Based on Shared Binary – Decision Diagram. *Journal Applied Phys., Japan* **40**(7), Part 1: 4485–4488.

Yanushkevich, S., Shmerko, V., Malyugin, V. & Dziurzanski, P. (2002). Linarity of Word-Level Models: New Understanding. In *Proceedings of IEEE/ACM 11th International Workshop on Logic and Synthesis*, 67–72. New Orleans.

Yanushkevich, S., Shmerko, V. & Dziurzanski, P. (2004). Logic Circuit Representation by Linear Decision Diagrams with Extention to Nanosructures. *Automation and Remote Control*. Special Issue on *Arithmetical Logic in Control Systems*. Plenum Academic Publishers, in press.

Wasshuber, C. (2001). *Computational Single-Electronics*. Springer Verlag.

Webster, J. (ed.) (1999). *Encyclopedia of Electrical and Electronics Engineering*. John Willey & Sons.